EARTHQUAKE HAZARDS AND THE DESIGN OF CONSTRUCTED FACILITIES IN THE EASTERN UNITED STATES

ANNALS OF THE NEW YORK ACADEMY OF SCIENCES
Volume 558

EARTHQUAKE HAZARDS AND THE DESIGN OF CONSTRUCTED FACILITIES IN THE EASTERN UNITED STATES

Edited by Klaus H. Jacob and Carl J. Turkstra

The New York Academy of Sciences
New York, New York
1989

Library of Congress Cataloging-in-Publication Data

Earthquake hazards and the design of constructed facilities in the eastern United States/edited by Klaus H. Jacob and Carl J. Turkstra.
    p. cm.—(Annals of the New York Academy of Sciences, ISSN 0077-8923 ; v. 558)
    Bibliography: p.
    Includes index.
    ISBN 0-89766-494-9 (alk. paper).—ISBN 0-89766-495-7 (pbk. : alk. paper)
    1. Earthquake engineering.   2. Earthquake resistant design—East (U.S.) 3. Earthquakes—East (U.S.) I. Jacob, K. (Klaus), 1936– . II. Turkstra, Carl J. III. Series.
Q11.N5 vol. 558
[TA654.6]
500 s—dc20
[624.1'762'0974]   89-9363
                                                  CIP

BiC/PCP
Printed in the United States of America
ISBN 0-89766-494-9 (cloth)
ISBN 0-89766-495-7 (paper)
ISSN 0077-8923

ANNALS OF THE NEW YORK ACADEMY OF SCIENCES

Volume 558
June 16, 1989

EARTHQUAKE HAZARDS AND THE DESIGN OF CONSTRUCTED FACILITIES IN THE EASTERN UNITED STATES[a]

Editors

KLAUS H. JACOB AND CARL J. TURKSTRA

CONTENTS

[a] This volume is the result of a conference entitled Earthquake Hazards and the Design of Constructed Facilities in the Eastern United States, which was held on February 24–26, 1988 in New York, New York and was sponsored by the New York Academy of Sciences and the National Center for Earthquake Engineering Research.

Part II. Seismic Ground Motions

Part III. Probabilistic Seismic Hazards Assessment

Part IV. Soil Liquefaction: Principles and Examples from Earthquakes in the Eastern United States

Part V. Theoretical Basis for Earthquake Engineering Design in Regions with Moderate Seismic Hazards

Major funding for this conference was provided by:
- THE NATIONAL CENTER FOR EARTHQUAKE ENGINEERING RESEARCH, SUNY at BUFFALO, N.Y. (funded by the NATIONAL SCIENCE FOUNDATION and the STATE OF NEW YORK)

Additional support was provided by:
- AMMAN & WHITNEY
- MICHAEL BAKER, JR. OF NEW YORK, INC.
- THE OFFICE OF IRWIN G. CANTOR, PC
- EDWARDS & KELCEY, INC.
- HOWARD NEEDLES TAMMEN & BERGENDOFF
- IFFLAND KAVANAGH WATERBURY, PC
- OVE ARUP & PARTNERS
- PARSONS, BRINCKERHOFF, QUADE & DOUGLAS, INC.
- THORNTON-TOMASETTI, PC
- WEIDLINGER ASSOCIATES, INC.
- WEISKOPF & PICKWORTH
- WOODWARD-CLYDE CONSULTANTS

ROBERT L. KETTER
(1929–1989)

This volume is dedicated to the memory of Dr. Robert L. Ketter, who devoted his life to making others' lives safer and more enlightened.

Preface

KLAUS H. JACOB

Lamont-Doherty Geological Observatory of
Columbia University
Palisades, New York 10964

CARL J. TURKSTRA

Department of Civil Engineering
Polytechnic University
Brooklyn, New York 11201

On August 10, 1884, almost 104 years ago, a magnitude 5 earthquake shook New York City. Fortunately it was centered some 15 miles offshore, south of Jamaica Bay on the Atlantic shelf, and thus did only modest damage on land. Nevertheless, it toppled chimneys and broke windows not just in Brooklyn, but from New Jersey to Connecticut.

Two years later, in 1886, large sections of South Carolina were shaken by a magnitude 7 earthquake, with Charleston taking the brunt of the damage. A week after the quake, an article appeared in *Harper's Weekly,* published on September 11, 1886 here in New York, which begins as follows:

THE EARTHQUAKE: Although during the year 1885 fifty-one earthquakes were recorded as having taken place in the United States, no one of the sixty millions of inhabitants of this country was prepared for the awful phenomenon which manifested itself last week. The science of seismology and the invention of delicate instruments for the detection and measurement of earth shocks had made geologists familiar with the fact that the crust of our quarter of the globe does not enjoy complete immunity from "faults," and that mild disturbances of the earth's surface are frequently occurring on this hemisphere; but that so vigorous an earthquake centre would be established on the Atlantic coast as that which last week wrought havoc and destruction in South Carolina was not anticipated or believed to be within the bounds of probability. . . .

. . . *was not believed to be within the bounds of probability.* It seems incredible that this sentence could be almost as valid today as it was one hundred and four years ago. Have scientists failed to communicate their case? Have economic forces ignored the scientists' ominous warnings?

Despite the progress that has been made in the fields of seismology and earthquake engineering during the last few decades, the public remains largely unaware of earthquake risks. For instance: Seismologists tell us that a magnitude 5 earthquake is likely to occur within 50 miles of New York City every 100 to 300 years. Similarly we know roughly the probabilities of destructive earthquakes that can affect Boston, Toronto, Ottawa, Quebec, St. Louis, Charleston, or Memphis (Tennessee).

Moreover, those working in the field of earthquake engineering have made great strides in developing an understanding of the response of buildings and other engineered structures to earthquakes. A variety of very cost-effective solutions have been developed to protect buildings from collapse, and their occupants from life-threatening effects.

However, a major problem in the U.S. is that very few of these earthquake-protective measures are in place outside of California, and many dangerous buildings remain, even in regions where building codes with earthquake provisions have been in effect for many years. We note with great concern that this very City of New York has no seismic provisions in its building code. Except for the Commonwealth of Massachusetts, and thus Boston, few if any states or municipalities east of the Rocky Mountains seem to have moved closer to adopting seismic code provisions.

This is so despite the fact that very positive cost-benefit effects can be seen when one compares case histories where earthquakes have struck cities with and without code provisions that have been in effect for some time. Compare, for instance, the relatively small damage from the October 1987 Whittier-Narrows earthquake in California, where seismic codes have been in effect for many years, to the devastating effects of the much smaller earthquake that struck El Salvador's capital, San Salvador, in 1986.

California's Whittier earthquake, on the eastern edge of Los Angeles, measured 5.9 on the Richter scale and had a 5.2 aftershock, which together caused 3 deaths, injured 1000 persons, displaced up to 10,000 from their homes, and caused about four hundred million (400,000,000) dollars worth of damage and business losses. In contrast, the considerably smaller San Salvador earthquake of 1986, measuring 5.1 to 5.4 on the Richter scale, caused a staggering 1500 deaths, at least 7000 injuries, and left up to 100,000 homeless. Damage and business losses combined are reported to have exceeded 1 billion dollars!

The lesson to be learned is that even during modest earthquakes our increasingly complex urban systems can experience devastating losses, with high fatality rates, if buildings, lifelines, and other structures have little or no earthquake resistance. On the other hand, fatalities and injuries can be greatly reduced, and damage and business losses contained to more acceptable levels, when structures with earthquake-protective design features are in place.

The continued lack of earthquake-resistant design and construction even for new buildings in the eastern United States, combined with the fact that a very large stock of unsafe older buildings exists, is of great concern to those who understand the risks to which the eastern parts of this country are exposed. It is therefore with a sense of utmost urgency that the National Center for Earthquake Engineering Research convened this conference, and we are grateful that the New York Academy of Sciences has been willing to host and organize it here in the City of New York.

The organization of this conference asks two basic thematic questions: (1) What are the earthquake hazards in the eastern United States as defined by the seismological community? (2) What should be the response of the engineering community to these risks?

As conveners of the conference we are especially grateful to the staff of the Academy who organized this meeting and will see the proceedings through the press as an *Annals* of the New York Academy of Sciences. We also wish to acknowledge the members of the Organizing Committee who prepared the program, the Earthquake Engineering Research Institute, the Metropolitan Section of the American Society of Civil Engineers for their support, the consulting engineering firms that provided financial support and the National Center for Earthquake Engineering Research and its sponsor, the National Science Foundation, who provided major funding.

Opening Remarks

ROBERT L. KETTER[a]

National Center for Earthquake Engineering Research
State University of New York at Buffalo
Buffalo, New York 14261

I would like to begin my remarks by stating a few truisms:

1. Earthquakes tend to recur where they have occurred in the past.
2. During recorded history, the eastern part of the United States has experienced major seismic events. (In fact, the most severe one to hit the continental United States occurred not in California, but in the central Mississippi valley.)
3. On the basis of current thinking, the probability of a major earthquake occurring in the eastern part of the United States is relatively low. However, moderate to severe earthquakes have occurred, and will certainly recur in the future.
4. At the same time, major devastating earthquakes have occurred in areas where no previous events have been recorded, and where current knowledge would suggest the existence of a "fundamentally quiescent area." (The Tangshan, China event of 1976, which caused in excess of a quarter-million deaths, is an example of such an area.)
5. In short, there are few if any areas of the United States that are immune from the effects of earthquakes. Earthquake hazard reduction is a problem for all of us.

The problem of earthquake hazard in the East is compounded by high population densities, many older buildings, and a high degree of industrialization. As noted earlier, the potential threat of a single significant event occurring at a particular location is relatively low. The potential threat to people and property, however, is sufficiently high that it cannot be ignored. Moreover, it must be recognized that an earthquake of a given magnitude can shake, with the same intensity, an area 100 times greater in the East than in California. This same earthquake will have higher frequencies, and cause a much larger area to experience architectural and structural damage. Soils and rock formations in a number of locations in the East appear to have characteristics that can cause amplification of ground motions, and this, too, magnifies the problem.

Since 1977, the U.S. Geological Survey, in cooperation with other Federal and State agencies and institutions, has held a number of workshops and conferences. Their primary purpose was to bring together knowledge producers and users with the general goal of improving utilization of knowledge on earthquake hazards. Many of these conferences were concerned with problems associated with non-West coast events. In 1986, for example, one was held in Albany, New York and dealt with the question of continuing actions to reduce potential losses from future earthquakes in New York and nearby states.

"Seismic Hazard Assessment in the Eastern United States" was defined as a part of the first-year program of the National Center for Earthquake Engineering Research. This conference is a direct outgrowth of that program and is jointly

---

[a] Deceased April 18, 1989.

sponsored by the Center and the New York Academy of Sciences. It is an effort to bring together scientists, engineers, architects, educators, builders, developers, city and state officials, planners, insurers and managers of our highways, ports and utilities, to review what is known, and to define what remains to be learned. That is why we are all here!

The Center and its sponsors, the National Science Foundation and the State of New York, are pleased that we could play a role in bringing you this material. Much has been accomplished, but much remains to be done.

Earth-Science Issues of Seismic Hazards Assessment in Eastern North America[a]

P. W. BASHAM

Geophysics Division
Geological Survey of Canada
Ottawa K1A 0Y3, Canada

INTRODUCTION

FIGURE 1 shows the fundamental data base that is the starting point for any discussion of seismic hazards in eastern North America. One does not need to be an earth scientist to see the gross geographical trends in these data. The earthquakes tend to surround what appears to be a stable region in the central continent. One does, however, have to be an excellent, perhaps clairvoyant, earth scientist to tell our decision-makers where the next damaging and likely fatal earthquake will occur in the next few decades.

The halo of larger earthquakes around the stable Precambrian craton of North America has a reasonably well-established basis in the large-scale geological structures that remain from the last two episodes of North Atlantic plate tectonics. These episodes tore and battered the craton from the east while the proto-Atlantic Ocean was opening and closing. At the present stage of opening, the Mid-Atlantic Ridge is producing compressional stresses that can reactivate these older geological structures.

The current stage of opening has produced reasonably uniform compression for tens of millions of years. Eastern North American society has observed and recorded the effects of the larger resultant earthquakes during only the last 300 years. Because so many of the larger historical earthquakes have been single events at their particular locations, we must assume that we have seen only a brief snapshot of the long-term earthquake potential. In the few locations where we have observed repetitive large earthquakes, the lack of surface geological evidence suggests that the causative faults have not been similarly active over thousands or tens of thousands of years; they may be currently active zones that have been turned on only in the very recent geological past.

Around and between the large earthquakes represented in FIGURE 1, almost at random, are numerous earthquakes of magnitude 5. The observational history for magnitude 5 earthquakes is not uniform throughout the map area. Some of the events in the eastern U.S. and southeastern Canada are known from written records in the eighteenth and nineteenth centuries; all of the events in uninhabited northern Canada date only from the time of reasonable global seismographic coverage in the 1930s. This probably accounts for the greater number of magnitude 5 earthquakes in the southeast than in northern Canada.

---

[a] Contribution from the Geological Survey of Canada No. 49087.

This is the observational framework within which we must estimate seismic hazards in eastern North America. In the following sections I elaborate with examples on some of the knowns and unknowns about the earthquake potential; I describe a conceptual model based on plate-tectonic rifting that can be used to address this potential; and I end with a brief discussion of how this affects our immediate need of providing decision-makers with defendable estimates of seismic hazards. Other contributions in this volume contain additional details on most of the earthquake zones discussed below.

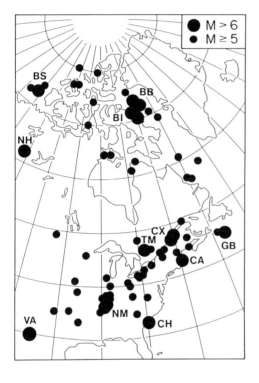

FIGURE 1. Historical earthquakes having magnitude greater than 5 (*small dots*) and greater than 6 (*large dots*) in eastern North America. Earthquakes discussed in the text are identified as follows: NM, New Madrid; CH, Charleston; CX, Charlevoix; TM, Timiskaming; CA, Cape Ann; GB, Grand Banks; BB, Baffin Bay; BI, Baffin Island; BS, Beaufort Sea; NH, Nahanni; VA, Valentine, Texas.

THE LARGE EARTHQUAKES OF EASTERN NORTH AMERICA

New Madrid

Although magnitude estimates for pre-instrumentally recorded earthquakes are uncertain, there is no question that the Mississippi Embayment in the Missouri-Arkansas-Kentucky-Tennessee border region experienced great (magnitude 8) earthquakes in the winter of 1811–1812. These events are unique for eastern North America and rare elsewhere; only two other similar-sized earthquakes are known to have occurred in similar tectonic environments, in 1604 on the southeast China coast and in 1819 in northwest India.[1]

Through mainly geophysical studies and accurate locations of recent lower-magnitude seismicity, the 1811–1812 earthquakes have been associated with the

New Madrid rift complex, which formed as a failed arm of a triple junction during a late Precambrian continental break-up[2] (this can be thought of as a tensional crack extending into the craton from the margin where the craton separated). There are at least two other similar Precambrian rift structures with similar origins in the southern U.S.—the Delaware aulacogen (failed rift) of west Texas and the Oklahoma aulacogen—but these have not experienced large earthquakes during historical times.

Charlevoix

The first earthquake (1534) in the historical earthquake catalogs of eastern North America occurred in the St. Lawrence Valley, probably in the Charlevoix zone. Six large earthquakes have subsequently occurred in this zone: in 1638 (location somewhat uncertain), 1663, 1791, 1860, 1870 and 1925, all with estimated magnitudes of 6 or greater; the 1663 and 1925 events were about magnitude 7. Present seismicity at Charlevoix has been shown[3] to align on northeast-striking planes that dip to the southeast at depths between 5 and 25 km. When the hypocenters are projected to the surface, they are confined between Paleozoic rift faults that have been mapped on the north shore and a bathymetric feature near the river's south shore, which is assumed to be a river-bottom expression of a parallel rift fault. These faults are part of the system remaining from the last-but-one breakup of the North American continent (of similar age to the New Madrid rift complex noted above). The faults have been mapped along the St. Lawrence Valley, with "failed arms" extending northwestward up the Ottawa River valley and southward through Lake Champlain.[4]

If earthquakes at Charlevoix had been occurring at the historical rates during recent geological time, we would expect to see surface geological deformation on the rift faults that would amount to kilometers over a million years. That we do not see such deformation implies that the earthquake activity must be recent at Charlevoix, and perhaps intermittent along the remaining rift system, possibly with a time constant of thousands to tens of thousands of years.

Other earthquakes (magnitude about 6 near Montreal in 1732, magnitude 6.2 near Temiskaming in 1935, magnitude 5.6 at Cornwall-Massena in 1944, and close to magnitude 5 in the lower St. Lawrence) are indicators that this rift system is capable of producing significant earthquakes elsewhere along its extent.

All other historical earthquakes in eastern North America with magnitudes greater than 6 can be associated with the geological features that remain from the last stage of rifting of the continental mass. This rifting produced the present continental margin (the zone of transition between continental and oceanic crust) and various fault systems that extend toward the continent as remnants of an "attenuated" (thinned) continental crust resulting from the "pull apart" rifting of the continent. I proceed from south to north in describing these earthquakes.

Charleston

The 1886 Charleston, South Carolina, earthquake, with estimated magnitude of 6.5 to 7, is the largest known along the eastern U.S. seaboard. Extensive studies of this earthquake have included geological and geophysical investigations,[5] intensity distribution,[6] and paleo-earthquakes.[7,8] In contrast to the Grand Banks earthquake discussed below, the Charleston earthquake occurred not at the edge of the continental margin but in a region of attenuated continental crust that underlies the southeastern coastal plain and continental shelf.

Many studies have been undertaken to test the hypothesis that the tectonic environment of the Charleston earthquake is unique. Although much has been learned, its significance in terms of earthquake hazards remains controversial; the characteristics of the source region remain largely unknown and the earthquake has not yet been associated with any clearly identified geological feature.[9] Paleoseismic evidence suggests that at least three prehistoric earthquakes have shaken the Charleston region within the past 7200 years, but earthquakes of different ages have also occurred elsewhere along the South Carolina coast.[10]

Cape Ann

Because of its age, both the magnitude and the location of the 1755 Cape Ann earthquake are poorly known. Interpretation of contemporary accounts, including observations from ships at sea, suggest an epicenter about 100 km east of Cape Ann.[11] Macroseismic information would suggest a magnitude of about 6. If the location is reasonably accurate, this places the earthquake in the attenuated crust on the continental shelf.

Grand Banks

As we proceed northward, the large historical earthquakes appear to move further offshore. The 1929 Grand Banks earthquake (magnitude 7.2) occurred right at the edge of the continental shelf, and triggered a large submarine slump that slid down the continental slope. The slump in turn became a turbidity current that flowed a thousand kilometers onto the deep ocean floor. The sudden displacement of the water column by the slump produced a tsunami which washed ashore on the south coast of Newfoundland, 12 m above high tide, devastating the fishing villages and causing the loss of 27 lives. This is the only locally generated tsunami known to have been produced on the east coast of North America.

Although the available instrumental seismic data are too poor to establish a definitive mechanism, the epicentral trends suggest that the earthquake may have ruptured westward along a fault about 70 km long,[12] although it has also been suggested that much of the seismic energy release was produced by the slumping itself.[13] I prefer the former mechanism, as it appears that subsequent seismicity, including four magnitude-5 earthquakes since 1951, may be occurring in the 1929 rupture zone.

The seismic hazard model for the current National Building Code of Canada[14] assumes that 1929-sized earthquakes occur at this location approximately every 300 years. However, recent geophysical surveys and sediment sampling of the 1929 slump suggest that large slumps are very infrequent at this location and the triggering earthquakes may have return periods of the order of 10,000 years. That other such earthquakes may have occurred in the past is suggested by prehistoric submarine slumps that have been found elsewhere along the margin.[15,16]

Baffin Bay

The magnitude-7.3 1933 Baffin Bay earthquake is the largest earthquake known to have occurred in northeastern Canada. Magnitude-6 events have since occurred in the Bay in 1945, 1947, and 1957. There is evidence for an extinct seafloor spreading center in the deep central region of Baffin Bay, a remnant of the Tertiary separation of Greenland and Baffin Island, but the large earthquakes are not associated with it. The 1933 earthquake occurred on the landward side of the 2000-m bathymetric contour which delimits a thick sequence of sediments. It occurred at a depth of about 10 km and had a thrust mechanism.

A model has been developed[17] that suggests that the recent removal of surface loads by deglaciation is a major source of local stress, comparable in magnitude to and superimposed on the compressional stress in the North American plate due to Mid-Atlantic Ridge spreading. These stresses are likely reactivating the faults that remain from the rifting episode.

The northeastern Baffin continental margin is a more recent (about 60 million years) equivalent to the late Precambrian/Paleozoic (approximately 500 million years) rifting of the Precambrian craton discussed above (New Madrid, St. Lawrence), and might yield more information than the buried rifts in the south. Unfortunately, because of its remoteness it has been very little studied.

Beaufort Sea

The Arctic Ocean margin was produced by rifting in Cretaceous time when Alaska rotated anticlockwise away from Arctic Canada. The earthquakes are clustered in regions where the transitional crust (continent-ocean) is loaded by thick sequences of sediments. The most active cluster is in the Beaufort Sea beneath a thick sequence of sediments deposited by the Mackenzie River; it includes a magnitude-6.5 earthquake in 1920.

Although the 1920 Beaufort Sea earthquake was identified as "transitional,"[1] the geological history of the Arctic Ocean margin is similar (in style but not in age) to that of the other margins in eastern North America. The Nahanni and Valentine, Texas earthquakes shown in FIGURE 1 are also in this category.

A CONCEPTUAL FRAMEWORK FOR THE LARGE EARTHQUAKES

Each of the large historical earthquakes in eastern North America has been spatially associated with some geological feature that remains from the Paleozoic (approximately 500-million-year-old) rifting of the Precambrian proto-continent, or the Mesozoic (less than 200-million-year-old) last stage of rifting that formed the present eastern and northern North American continental margin.

The original work that has made these correlations so apparent was the excellent study by Coppersmith, Johnston and Arabasz[1] of the seismicity of "stable continental interiors" (SCI) worldwide. This study was undertaken as part of the Electric Power Research Institute (EPRI) seismic hazard program for the eastern United States, to provide a scientifically supportable base for maximum earthquake assessments. They developed a working definition of SCI with the restriction that the defined regions be geologically and tectonically similar to North America east of the Cordilleran thrust and fold belt. For nine SCI regions worldwide they compiled information on historical earthquakes greater than 5.0 (or epicentral intensities greater than VIII for older events with no assigned magnitude). They found that 71 percent of the SCI seismicity was associated with imbedded continental rifts and continental passive margins (one-sided rifts). Further, all of the 17 SCI earthquakes of magnitude 7 or greater are strongly associated with the imbedded rifts or passive margins.

We have adopted this framework for purposes of discussion of the large earthquake potential in eastern Canada.[18,19] A further generalization of the concept is shown in FIGURE 2.

The schematic representation of the rifting of the passive margins (FIG. 2) is easy to draw; it can simply follow the continent-ocean transition, but it is intended to encompass adjacent regions of attenuated continental crust such as occur under the southeastern U.S. coastal plain. To refine this scheme, we would need to know the inland boundary of the significant rift faults in the attenuated crust. With this information the band showing potential locations for large earthquakes would no doubt have variable width over its 10,000-km extent.

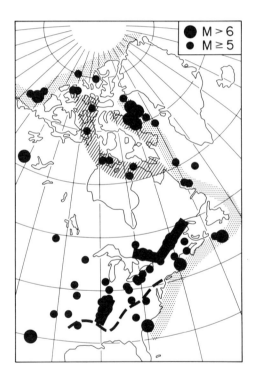

FIGURE 2. Schematic outline of Paleozoic rifting of the Precambrian craton (*black*); the buried edge of the craton in the southeastern U.S. (*dashed line*); Mesozoic rifting of the Arctic and Atlantic margins (*stippled*); and postglacial rebound reactivation of Paleozoic structures in the eastern Arctic (*diagonal hatch marks*) superimposed on the seismicity of FIGURE 1.

A complete schematic representation of the through-going Paleozoic rifts of the Precambrian craton is more difficult. The St. Lawrence system of rift faults is exposed on the Precambrian Canadian Shield. The more distinct faults can be spatially associated with the earthquakes only along the St. Lawrence and Ottawa river valleys (the sections shown in FIGURE 2), but there is geological evidence that associated rifts may extend northeast toward the Labrador coast, south through Lake Champlain and northwest into northern Ontario.

In the southeastern U.S., equivalent features are buried beneath many kilometers of Phanerozoic sedimentary and overthrust Appalachian rocks, and are known only from remote sensing geophysical techniques. Only the New Madrid rift complex is shown in FIGURE 2. Perhaps it is best known because it has been most studied to find a source structure for the New Madrid earthquakes.

THE UBIQUITOUS MAGNITUDE-5 EARTHQUAKES

A number of magnitude-5–6 earthquakes have occurred at or near the locations of the large earthquakes discussed above: for example, southeastern Illinois, 1968, and northeastern Arkansas, 1976; Charlevoix, 1939, 1952, and 1979; Beaufort Sea, 1975 and 1987; some[15] have argued that these could be considered long-delayed aftershocks of the larger historical earthquakes which caused a major perturbation to the ambient stress regime in a volume of the crust much larger than their rupture zones. Other earthquakes of magnitude 5 are clearly displaced from the large earthquakes but covered by the rift framework of FIGURE 2 (recall the 71 percent of SCI seismicity noted above); for example, in the Lower St. Lawrence and at Cornwall-Massena on either end of the well-defined portion of the St. Lawrence rift; on the continental shelves of Labrador and the Arctic Ocean.

But there are numerous magnitude-5–6 earthquakes that cannot be accounted for by the rift framework of FIGURE 2. They occur in a wide variety of geological environments that can be only briefly summarized here. The following summary moves north to south to describe many of these earthquakes. They occur:

(a) beneath 10 km of sedimentary rocks in the Sverdrup Basin of the Arctic archipelago, perhaps on Cretaceous normal faults;

(b) on shallow faults in Precambrian terrain on northeastern Baffin Island as a result of postglacial uplift and tilting of Baffin Island;

(c) as a reactivation of Paleozoic structures near the Boothia Penninsula and northern Hudson Bay, suggestive of differential crustal block motion in response to postglacial uplift;

(d) on extinct spreading ridges and fractures zones in the Labrador Sea;

(e) as thrust faults in the Appalachian overthrust sheet in New Brunswick, and perhaps on the detachment surface between the overthrust sheet and the underlying Precambrian basement;

(f) in an apparent zone of weakness in the Precambrian of western Quebec which may have been thermally stressed by a mantle hot spot 130 million years ago;

(g) on an unknown structure beneath the prairies near the Saskatchewan-North Dakota border;

(h) in New Hampshire in a region where the earthquake potential may result from an intersection of faults and intrusive bodies;

(i) at relatively shallow depth in the Precambrian Adirondacks on what may be old, brittle fracture zones with some surface expression as lineaments, but little accumulated displacements;

(j) in western New York state, possibly on the Clarendon-Linden fault, which offsets Paleozoic rocks at depth but whose surface expression is a gently dipping monocline;

(k) beneath 2 km of flat-lying Paleozoic sediments in Ohio on what may be a Precambrian tectonic boundary known only from its aeromagnetic anomaly signature;

(l) in the Precambrian basement of Virginia beneath the Appalachian detachment on what may be ancient rift faults near the edge of the craton.

When viewed as a whole, these magnitude-5 earthquakes are associated with a wide variety of geological features. It is not clear that we could estimate future locations of magnitude-5–6 earthquakes even if we knew the geological structure in some detail.

ISSUES CONFRONTING SEISMIC HAZARD ESTIMATES

It is clear that the description of eastern North American earthquakes in the previous two sections presents significant difficulty to those with responsibility for seismic hazard assessment. The larger earthquakes appear to be accommodated by the rifting framework model; but there is no conclusive evidence that these rifts should be considered uniformly active over their entire extent, certainly not in the next few decades. The smaller (magnitude-5) earthquakes occur in a large variety of geological environments; it appears that if we wait long enough we can expect to see one almost anywhere in eastern North America (although we may exclude much of the apparently stable central region).

A key problem in the hazard assessments is the relatively small number of large historical earthquakes for such a large area. If we claim to know nothing about the geological controls on such earthquakes, for hazard estimates we could assume that in future they will occur at random. The hazard at moderate probabilities, such as used for building codes, would then be negligible at all locations and not protect against even a moderate nearby earthquake.

On the other hand, earth scientists are very uncomfortable with the suggestion that our (short) earthquake history will simply repeat itself during the next three centuries. Each of the four large earthquakes along, or adjacent to the eastern margin was an isolated event: Cape Ann, 1755; Charleston, 1886; Grand Banks, 1929; Baffin Bay, 1933. What grounds do we have for saying that the next one will occur in one of these locations?

While the New Madrid rift complex may have released its once-in-a-millenium seismic energy, the Charlevoix zone in the St. Lawrence valley should be expected to continue its repetitive shocks. For the Charlevoix hazard model it would be prudent to expect at least one more; but the lack of geological deformation suggests that the activity has not persisted in the geological past and may turn off and start somewhere else along the rift system in the future.

The hazard models developed by earth scientists must reflect these uncertainties. Recent examples of how uncertainties can be accommodated were the projects undertaken by the U.S. Nuclear Regulatory Commission and EPRI for seismic hazard estimates at sites of nuclear power stations in the eastern United States. These studies used "expert opinion" to incorporate a broad range of options into multiple earthquake-source-zone models of future earthquake potential. Each of the possible models is assigned a (judgmental) probability that it is the "correct" interpretation of the known earthquakes. The hazard calculation then appropriately weights each of these models in deriving the final "hazard curve" (seismic ground motion as a function of probability) for a particular site. The genuine uncertainty on a computed ground-motion parameter can be a factor of 3 or more, and this must be recognized by the end-user.

Existing hazard maps of eastern North America, computed at moderate probabilities for application to common structures, such as buildings, provide reasonable protection against large earthquakes in those few locations (New Madrid, Charleston, Charlevoix) where large earthquakes have occurred historically.

They do not provide such protection at other locations, that is, a large earthquake in a new location will likely seriously exceed the design loads of existing nearby structures. Until earth scientists can learn much more about the earthquake potential in eastern North America, and change the hazard maps accordingly, this fact must also be recognized by the end-user.

SUMMARY

Earthquake hazard estimates in eastern North America are accompanied by a large degree of uncertainty. The larger earthquakes (M > 6), which dominate the hazard estimates, appear to be confined to major rifts that surround or break the Precambrian craton (continental margin, St. Lawrence Valley, Mississippi embayment). The principal earth-science issue is to establish why these features do not appear to be uniformly active throughout their extent. Is it simply because our observation history is much too short compared to repeat times that may be thousands of years? Or is it that in only some regions are the rift features favorably oriented with respect to the current stress regime? Elsewhere in eastern North America smaller earthquakes (M < 6), are occurring in the Appalachian overthrust sheet (New Brunswick, Virginia), in lesser zones of weakness in the craton (Ohio, New York, western Quebec), or as crustal block motion showing differential movement in response to glacial unloading (eastern Arctic). For these, the earth-science issue is that each new event is somewhat of a surprise: the earth scientists had not previously clearly identified the potentially active feature.

REFERENCES

1. COPPERSMITH, K. J., A. C. JOHNSTON & W. J. ARABASZ. 1987. Methods for Assessing Maximum Earthquakes in the Central and Eastern United States. EPRI Research Project 2556-2, Working Report. Electric Power Research Institute. Palo Alto, CA.
2. BRAILE, L. W., W. J. HINZE, G. R. KELLER, E. G. LIDIAK & J. L. SEXTON. 1986. Tectonic development of the New Madrid Complex, Mississippi Embayment, North America. Tectonophysics **131:** 1–21.
3. ANGLIN, F. M. 1984. Seismicity and faulting in the Charlevoix zone of the St. Lawrence Valley. Bull. Seismol. Soc. Am. **74:** 595–603.
4. KUMARAPELI, P. S. 1985. Vestiges of Iapetan rifting in the west of the northern Appalachians. Geosci. Canada **12:** 55–59.
5. GOHN, G. S., Ed. 1983. Studies related to the Charleston, South Carolina, earthquake of 1886: Tectonics and seismicity. U.S. Geol. Surv. Prof. Pap. **1313.**
6. BOLLINGER, G. A. 1977. Reinterpretation of the intensity data for the 1886 Charleston, South Carolina, earthquake. U.S. Geol. Surv. Prof. Paper **1028:** 27–32.
7. OBERMEIER, S. F., G. S. GOHN, R. E. WEEMS, R. L. GALINAS & M. RUBIN. 1985. Geologic evidence for recurrent moderate to large earthquakes near Charleston. Science **227:** 408–411.
8. TALWANI, P. & J. COX. 1985. Paleoseismic evidence for recurrence of earthquakes near Charleston, South Carolina. Science **229:** 379–381.
9. SEEBER, L. & J. G. ARMBRUSTER. 1987. Seismicity along the Atlantic seaboard of the U.S.: Intraplate neotectonics and earthquake hazard. In The Atlantic Continental Margin, U.S., R. E. Sheridan & J. A. Grow, Eds. Geological Society of America, The Geology of North America, Vol. I-2: 565–582.
10. OBERMEIER, S. F., R. E. WEEMS & R. B. JACKSON. 1987. Earthquake-induced liquefaction features in the coastal South Carolina region. In Proceedings of NCEER Symposium: Seismic Hazards, Ground Motions, Soil-Liquefaction and Engineering

Practice in Eastern North America. National Center for Earthquake Engineering Research. Buffalo, NY. Technical Report NCEER-87-0025, 480–493.

11. WESTON GEOPHYSICAL CORPORATION. 1976. Historical Seismicity of New England. Report BE-SG7601 prepared for Boston Edison Company.

12. ADAMS, J. 1986. Changing assessment of seismic hazard along the southeastern Canadian margin. Proceedings of the Third Canadian Conference on Marine Geotechnical Engineering. St. John's, Newfoundland: 41–53.

13. HASEGAWA, H. S. & H. KANAMORI. 1987. Source mechanism of the magnitude 7.2 Grand Banks earthquake of November 1929: Double couple or submarine landslide? Bull. Seismol. Soc. Am. **77:** 1984–2004.

14. BASHAM, P. W., D. H. WEICHERT, F. M. ANGLIN & M. J. BERRY. 1982. New probabilistic strong seismic ground motion maps of Canada: A compilation of earthquake source zones, methods and results. Earth Physics Branch (Ottawa) Open File Rep. **82–33.**

15. BASHAM, P. W. & J. ADAMS, 1983. Earthquakes on the continental margin of eastern Canada: Need future large events be confined to the locations of large historical events? U.S. Geol. Surv. Open File Rep. **83–843:** 456–467.

16. PIPER, D. J. W., J. A. FARRE & A. SHOR. 1985. Late Quaternary slumps and debris flows on the Scotian Shelf. Bull. Geol. Soc. Am. **96:** 1508–1517.

17. SLEEP, N. H., G. KROEGER & S. STEIN. 1988. Canadian passive margin stress field inferred from seismicity. J. Geophys. Res. In press.

18. ADAMS, J. & P. W. BASHAM. 1988. The seismicity and seismotectonics of eastern Canada. *In* Decade of North American Geology, Vol. CSMV-1, Neotectonics of North America, E. R. Engdahl, D. Blackwell, D. Schwartz & M. Zoback, Eds., Geological Soc. Am. In press.

19. ADAMS, J. 1988. Seismicity and seismotectonics of southeastern Canada. This volume.

Engineering Issues of Earthquake Hazard Mitigation in Eastern North America

ROBERT V. WHITMAN

Department of Civil Engineering
Massachusetts Institute of Technology
Cambridge, Massachusetts 02139

INTRODUCTION

In the broadest sense, there are three engineering issues as regards earthquake hazard mitigation:

1. What, if any, earthquake-resisting requirements should be placed upon new construction and what is their cost?
2. What, if anythinq, should be done about the earthquake-resistant deficiencies in existing structures?
3. How can the functioning of essential equipment within structures best be assured?

These aspects of the earthquake problem concern architects as well as engineers.

The challenge of providing adequate seismic resistance for new construction is quite different from that of existing, potentially hazardous buildings and facilities. The principles governing design for earthquake resistance are generally agreed upon. Moreover, the costs of incorporating such seismic resistance in new construction are generally small or modest. However, the cost can be very substantial when making an existing structure as resistant as new construction, and for this reason there is still little agreement as to what measures should be required, even in highly seismic regions. Protecting critical equipment within structures is a concern that has received major attention only recently.

SCOPE OF THE PROBLEM

Discussion about engineering for earthquake hazard mitigation usually focuses first upon buildings. This is because the regulation of building safety is a clearly established function of government, and the procedures and institutions for carrying out this function are well established.

However, of equal concern are "lifelines"—those support systems whose functioning is essential to a city or region during the minutes, hours, and days following a sudden disaster. Included under this heading are the electrical generating and distribution system; water and sanitary systems; fire, rescue, and medical facilities; and transportation networks (including local roadways) and communication systems. Responsibility for these diverse lifelines is diffused among various private- and public-sector organizations; procedures and rules for regulation often are lacking; and standards for performance vis-à-vis earthquakes are generally still not well developed.

I too will emphasize buildings, because discussion is clearest when one concentrates upon this part of the problem, but much of what I say applies also to lifelines.

DESIGN EARTHQUAKES

The engineering approach to design generally is quantitative. Forces caused by prescribed loadings are evaluated, and the members of a structure are proportioned to have resistances greater than these forces. With such an approach, a design earthquake must be specified quantitatively.

The preceding keynote address, by Peter Basham, discusses the state of knowledge concerning the occurrence and distribution of earthquakes in eastern North America (ENAM). The possibility that a major earthquake will strike any location in ENAM (except for perhaps Florida and a few other areas) cannot be ruled out. However, for most places in ENAM, the likelihood of such a large event during any one person's lifetime is very small, although there is a significant probability of experiencing a smaller earthquake that might cause some damage. This situation leads to major questions about the selection of earthquakes to be used for design of buildings and other facilities.

Designing within ENAM for the largest credibly imagined earthquake seems unreasonable and is politically unacceptable, except for the most critical and controversial facilities. Thus, design requirements, if adopted at all, are set at some level smaller than those associated with the largest possible earthquake. However, there is no totally objective and generally agreed-upon basis for selecting a suitable design earthquake. The most up-to-date of the seismic zoning maps appearing in model building codes and standards give good guidance based on careful thought, but the basis for them involves considerable subjective judgment.

All of this means that the choice of earthquakes for design is an important earth science *and* engineering issue. And, to adapt a common saying: This choice is too important to be left to either earth scientists or to engineers alone.

Collateral Geological Hazards

Another problem area where the interests of engineers and earth scientists overlap concerns earthquake-caused instabilities of the earth. Landslides and rockfalls are common occurrences in steep terrain, but more damaging in heavily built areas are phenomena known as lateral spreading or lurching. Such permanent horizontal movements can damage buildings and are especially damaging to pipelines (buried or on the surface), roadways, railroads, and canals. Liquefaction—a phenomenon involving loss of strength in saturated sandy soils during shaking—contributes to lateral spreading and may also cause settlement and tilting of buildings.

Identification and location of earthen masses that might fail is a typical task for geologists, whereas engineers have been involved in developing methods for deciding whether a specified level of ground-shaking might cause failure. Seismologists use evidence of liquefaction to help date pre-historical earthquakes.

There are few regulations or standards for control of these collateral geological hazards. Geotechnical engineers in the United States are, as a group, philosophically opposed to such standards, having been influenced by the teachings of Terzaghi, Casagrande, and Peck that almost every geotechnical problem is different and that such diversity defies standardization of practice. However, it is possible to develop maps for a city or region that serve as useful guides to land-use planning. The State Building Code in Massachusetts does contain a provision for evaluation of sites for liquefaction, but it is more a screening test to identify potentially troublesome sites than a regulation limiting construction.

NEW CONSTRUCTION

Architects, engineers, and building owners, in order to obtain a license for new construction, must comply with the requirements of a building code. The constitution of the United States reserves to the states the regulation of public safety, and states often pass this authority on to individual counties, cities, and towns. Thus there potentially (and actually) are a myriad of different building codes in this country, and there is no such thing as a "national building code."

These individual, local codes are often based, at least in part, upon one or the other of several "model codes." The most important of these are the Uniform Building Code (UBC),[7] the BOCA Basic/National Code,[4] the Standard Building Code,[11] and that of the American National Standards Institute.[1]

All of these model codes have provisions governing seismic design. Generally speaking, such provisions (or changes to them) have appeared first in the UBC, and then (with some time lag) were picked up by the other model codes—usually with some modifications. Recently, a consensus standard, intended to be the basis for model and actual codes, has been recommended by the Building Seismic Safety Council (BSSC)[6] on the basis of a landmark study—known as ATC-3—by the Applied Technology Council.[2] Many of these model codes and standards refer to other standards developed by trade associations, such as the American Concrete Institute (ACI) and the American Iron and Steel Institute (AISC).

Several important points concern these model codes and standards. First, they have no legal force until adopted into a state or local building code. It is quite possible to adopt portions of a model code while excluding other parts, and a number of states and communities have intentionally excluded the earthquake design provisions.

Second, insofar as earthquake resistance is concerned, the intent is to protect life safety. Indeed, it is specifically recognized that a building designed to the seismic provisions of a code might experience expensive damage should the design earthquake actually occur. However, such provisions do reduce damage compared to what it would be otherwise.

Third, the model codes incorporate very simplified approaches for evaluating the forces in the structural members and walls of a building. These methods work reasonably well for "regular" buildings, but may not be adequate for structures with irregular layout or unusual structural framing. Dynamic structural analysis seldom, if ever, is required.

Fourth, in addition to prescribing forces for use in design, model codes require certain types of detailing, such as concern connections or minimum amounts of steel reinforcement in concrete and masonry. These so-called "ductility provisions" are intended to ensure that a structure that is overstressed during strong shaking will be able to support the normal dead and live load when shaking ceases.

Fifth, when provisions of the model codes are applied to cities in ENAM, it frequently is found that wind forces exceed earthquake forces. It is then often concluded that "wind prevails" and that the seismic provisions may be ignored. This is not necessarily correct, since the relatively low lateral forces assigned for earthquakes apply *only if the "ductility requirements" are met;* if these requirements and limitations are not respected, larger seismic forces should (and, according to some codes, *must*) be used in computation. This point is often overlooked.

Sixth, as previously mentioned, the model codes imply for most of ENAM earthquake ground motions less than may actually some day occur. Indeed, the ratio of the largest credibly imagined earthquake shaking to the recommended design earthquake shaking is considerably greater in ENAM than in an area such as California. This, coupled with the fact that the recommended ductility provisions are less stringent for the less seismic than the more seismic zones, actually means that the margins of safety are smaller for the less seismic zones.

Seventh, model codes are constantly changing, and, it is hoped, improving, as knowledge about earthquakes and structural response to earthquakes increases, and as new structural arrangements are developed. For example, the UBC has just undergone a major restructuring, adopting many of the new concepts suggested in ATC-3 and recommended by BSSC. Other model codes can be expected to pick up these changes in the near future. Thus, where a model code is adopted locally by reference, local requirements automatically change.

What Can States/Cities in ENAM Do?

The easiest step is simply to adopt the seismic provisions of a model code by reference. This still leaves the question: Which model code? In ENAM, the BOCA code is most widely used as regards nonseismic provisions, and it would be most natural to use all of it. In the not-too-distant past, the BOCA code was generally regarded as being weak in its seismic requirements, but currently it is reasonably strong in this regard and likely will be kept well up-to-date in the near future. I believe that no state/city would go badly wrong simply to adopt BOCA's seismic requirements.

There are, however, several other questions that might be asked:

Will you stay with the seismic zone(s) recommended in the code for your state/ city, or do you want to declare for either a lower or a higher zone?

Are you satisfied that the strategy embodied in the code is really the proper one for your area?

Obviously, considerable thinking and work is implied if a state/city attempts to deal with these questions.

I would certainly caution against arbitrarily lowering a zone from that recommended in the code. There can be many pressures to do so, inspired by fears over the costs implied by the adoption of seismic requirements. However, as discussed subsequently, these costs may be more imagined than real.

The concern about the strategy implied by the existing model codes arises primarily because existing recommended provisions have been heavily influenced by California-based engineers. There have been arguments that these provisions are too conservative for less-seismic ENAM. On the other hand, it may also be argued, in view of the discussions above, that they are actually non-conservative for these areas. It has also been suggested that there is too much emphasis upon evaluation of earthquake-induced forces (which means that a design earthquake must be established), and that the proper approach is to ensure that structures are "well tied together." The trouble, of course, is that it is difficult to regulate "tying together" without telling engineers the forces for which the "ties" must be designed.

I have urged that these questions of the appropriateness of existing model codes for ENAM be a major focus for the National Center for Earthquake Engineering Research at SUNY Buffalo.

Do Code Provisions Actually Do Any Good?

There is clear evidence that the answer to this question is *yes*. Generally, in every major earthquake it is the older buildings, not meeting modern requirements, that experience extreme damage and collapse partially or completely, while buildings designed according to modern codes usually perform well. A statistical study was made of buildings greater than 4 stories after the San Fernando Earthquake of 1971; modern buildings had much less damage than did comparable buildings without seismic design features. In Viña del Mar, Chile in 1985, large buildings designed to California standards performed very well during the extremely strong shaking of an earthquake; there was damage, but the essential functioning of these buildings was not impaired. While there was a tragic loss of life in Mexico City as a result of the earthquake there in 1985, well-designed modern buildings performed remarkably well during the stronger-than-thought-possible ground shaking. The great majority of such buildings, even those in the most affected area, had little or no structural damage. There were even examples of buildings with all-glass exterior where few, if any, panes were broken.

The Whittier earthquake of last October has provided further proof of the resistance of new construction designed to current codes.

What Costs Are Implied by Seismic Requirements?

An owner bears the burden of increased construction cost. The only way to evaluate this potential cost is to design the same structure with and without seismic design requirements (or with two different levels of requirements). I was involved in one such set of studies[12] at MIT during the early 1970s. Both steel and reinforced concrete structures were designed with and without the seismic resistance required in Zone 2. The increases in structural cost were 2% to 4%. More recently, BSSC had trial designs made using the provisions suggested in ATC-3. For designs carried out for New York, Chicago and St. Louis, the estimated increases in structural cost ranged from 2.5% to 7.3%.[5] The cost of the structural system is usually only about one-quarter to one-third of the total cost of a building. Thus a 7% increase in the cost of the structural frame is only a 2% cost increase overall. For most structures located in seismic zones 1 and 2, the cost increase will not be more than about 1%. Larger increases may be necessary if the goal is to ensure no damage rather than to protect life safety, or where the framing of a structure is very irregular.

Another cost of adopting seismic provisions is that to the designing architect and engineers. Many engineering firms in ENAM fear this cost. Certainly there is a transient cost of learning how to deal with a new set of requirements. However, since seldom is a dynamic analysis mandatory, the long-range increase in design cost is minimal.

A final potential cost is that to segments of industry who may, as a result of new requirements, be placed in a less competitive position. Requiring that all brick/masonry construction be reinforced, with grouting in the case of hollow tile construction, obviously has a potential impact upon the suppliers of masonry units. Corresponding statements may apply concerning prestressed concrete. Because of the "ductility requirements," reinforced concrete construction may appear to be more affected than steel construction. However, the concrete industry has long been cooperating in the development of sound but reasonable seismic design requirements, carrying out and supporting considerable experimental research, and recently the masonry block industry has adopted this same approach. While concerns will continue to be expressed when a new state or community considers the adoption of new requirements, I believe the days of blind resistance are now past.

None of these costs is large enough to be a basis for reasonable objections when life safety is of serious concern.

EXISTING CONSTRUCTION

What to do about existing hazardous construction is really the most important challenge facing the earthquake hazard mitigation effort. It is made difficult because many of these buildings house and provide support services for the less affluent portion of our population. Despite pioneering efforts, such as that in Long Beach, little serious effort has been given to this problem until the past decade.

On one hand, it can be enormously expensive to make an existing structure comply with the seismic requirements suitable for new construction. At the same time, although the approach of having existing structures strengthened to requirements less stringent than those for new construction has often been suggested, there are no generally-agreed-upon model code provisions for this purpose. This latter fact is, in my judgement, a major impediment to action. In Massachusetts, for example, at present no alternative is open to an owner renovating a building but to bring the structure all the way up to code—except to apply for a waiver from all seismic requirements, which is routinely granted.

Identifying Existing Hazardous Construction

Strictly speaking, all existing construction that does not comply with the currently applicable code is hazardous. However, many structures that cannot be shown to so comply have in fact performed quite well during strong earthquakes. In addition, we usually do not expect all existing structures to be altered each time a small change is made in a code. Hence, the problem is one of identifying the most hazardous structures.

In a general way we know those classes of construction that may be most hazardous. Unreinforced masonry, precast concrete without adequate continuity and poorly reinforced concrete frames are all "seismically suspicious." However, structures made of these materials are not all equally hazardous; the quality of construction, the degree of deterioration, the lengths of spans, and the degree of redundancy all are important. Thus we really only have identified the types of construction most in need of attention.

There have been a number of attempts to develop procedures whereby the seismic resistance of buildings can be assessed individually. The most recent and comprehensive approach is set forth in ATC-14.[3] Typical deficiencies are described and illustrated in detail, and methods for rapid screening as well as detailed evaluation are described. ATC now has under way two efforts (ATC-21 and 22) to field-test and validate these methods and to produce handbooks to facilitate their application. These studies, which are intended for nationwide use, will be completed in 1988 and 1990 respectively.

What Can be Done to Strengthen Existing Construction?

At the simple end are steps to tie a building together better. For example, quite often the floors and roofs of brick buildings are not anchored into the interior and exterior walls. Anchoring these horizontal elements to the walls can significantly increase seismic resistance, preventing damage during modest-sized earthquakes and reducing the probability of catastrophic damage during major earthquakes. Quite often it also is desirable to stiffen floors, by addition of another layer of flooring, so that they can better distribute loads to the walls.

Procedures that can increase the seismic resistance of masonry structures even further include adding reinforced concrete columns on the outside of exterior brick walls, placing gunite over existing walls, and constructing reinforced masonry shear walls. Techniques for accomplishing this have been worked out in connection with the law in the City of Los Angeles requiring upgrading of all such buildings. Some old buildings treated in this way were strongly shaken during the Whittier Earthquake of October 1987, and they performed very well.

One of the most difficult problems is the strengthening of poorly reinforced concrete frame buildings. Perhaps this is best handled by inserting concrete shear walls into selected bays.

At the exotic end is the use of base isolation devices, as has been done to preserve the historic City Hall in Salt Lake City.

The matter of parapets deserves special mention. Unreinforced masonry parapets have long been recognized as a threat, during earthquakes, to people in the narrow streets of cities. While strengthening (or partial or total removal) of such parapets does nothing to increase the resistance of the basic building, such action can increase life safety significantly.

What are the Costs of Strengthening Existing Construction?

The cost of simple steps to tie together brick masonry buildings is so small as to be lost within the total cost of a major renovation. The cost of major upgrading to brick masonry buildings in Los Angeles, to requirements roughly one-half those for new construction, has come down to about $5/ft². Costs for strengthening masonry buildings elsewhere have ranged from $5/ft² to $10/ft². Where existing buildings have been strengthened, often to meet code requirements for new construction, costs have typically ranged from $10/ft² to $20/ft², and in extreme cases to more than 100% of the replacement cost.[10]

What Can be Done Politically about Existing Construction?

Putting in place new regulations or laws requiring the seismic upgrading of a group of existing buildings is a task fraught with political difficulties. Any such action can mean significant costs to the affected building owners. Even in Los Angeles, where the danger posed by older masonry buildings is clearly evident, there was strong political resistance before (and since) passage of the law requiring upgrading. Most other cities in California with similar problems have generally been unwilling to face up to political action to reduce the hazard. Torrance, a city within Los Angeles County, has very recently adopted the unique and promising plan of selling city bonds the proceeds of which will be made available as low interest loans to owners of buildings required to upgrade.

I am convinced that it is hopeless to attempt to require privately owned structures in Eastern cities to be strengthened for seismic resistance, except as part of major renovations being taken to increase the economic value of a structure. Possibly some governmentally owned structures, including some buildings of great historical value, will be upgraded as a result of specific political pressures to do so.

I also am satisfied that it makes sense to require less seismic resistance in upgraded existing structures than in new construction. Since, at least in ENAM, the design earthquake for new buildings generally will be smaller than the largest earthquake that might occur, we are not providing absolute safety in new construction. Thus, there is at least an implicit cost/benefit consideration that, when applied to existing construction, argues for a reduced level of requirements.

What is needed, then, is a scheme that requires a politically palatable degree of strengthening consistent with the extent of renovation under consideration. The Seismic Design Advisory Committee to the Massachusetts State Building Code Commission is now developing one such possible scheme. It uses a matrix, with one axis being occupancy class and the other being the cost of renovation in relation to the value of the building. Occupancy class is determined from the function of the structure, the maximum number of occupants and the occupancy averaged over a week. Intersections within the matrix give the minimum required seismic resistance as a fraction of that for new construction.

This is a piecemeal approach to dealing with existing hazardous construction, but I believe it represents a step that is politically acceptable and will, however slowly, help to abate the hazard.

PREVENTING DAMAGE AND PROTECTING CONTENTS

As noted previously, requirements in building codes are aimed primarily at protecting life safety. If the owner of a structure wishes to avoid the costs and interruptions associated with damage caused by a major earthquake, it is necessary to do more than meet the minimum requirements of codes. By using stronger and stiffer designs, it is possible to reduce greatly—and perhaps even eliminate—structural damage during a very severe earthquake. It is also possible to take steps to ensure the functioning of most critical equipment, primarily by tying it down and bracing it to prevent overturning. Essential supplies stored on shelves can be restrained so as to prevent their loss. However, all such steps can be expensive and are not always practically feasible, and will not necessarily protect very sensitive equipment from internal damage.

There has long been interest in "base isolation" as a method for protecting buildings from earthquakes. This approach involves placing soft mountings at or near the base of a structure, so as to reduce (but not eliminate) the accelerations and forces induced in the structure as a result of base motion. In the past such methods have not proved cost-effective for new construction as compared with simply fulfilling the requirements of codes. However, with increased concern over protecting sensitive equipment and in retrofitting existing hazardous structures, there is now major interest in base isolation. A number of patented schemes have been developed and tested in the laboratory. About 100 structures worldwide have been base-isolated. In the United States, ten projects have either been completed or are underway, including bridges, a hospital, a plant manufacturing "high-tech" products and the City Hall at Salt Lake City. One of these buildings—in San Bernardino, California—has been shaken by at least two moderate earthquakes, and strong motion recordings within the building have shown it to be behaving as expected.

Base isolation is still in its infancy as a method, and it is not at this time clear just what role it will play in reducing the earthquake hazard in ENAM. Engineers in this part of the country must be aware of this technique, but should not place undue emphasis upon it at the expense of neglecting more traditional approaches.

GETTING STARTED

The Federal Emergency Management Agency (FEMA) and BSSC have embarked upon a number of efforts intended to guide and assist states and cities in moving forward to adoption of earthquake hazard mitigation practices. One has been the publication of a handbook for communities considering steps to improve seismic safety of buildings.[5] This publication provides an excellent starting point.

If earthquake design requirements are to come into law in states and cities, and other places where they do not now exist, I believe it is essential that the local professional structural engineering and architectural organizations be strongly behind them. In the terminology of mathematics, this is not a sufficient condition, but it is a necessary one. Unless the initiative comes from this sector and with strong backing from the majority of engineers and architects, government will find too many reasons not to act.

Hence, the first step is to convince these professional groups that (1) there is a problem, and (2) that something can be done at reasonable cost. This conference is a beginning in this direction, especially for New York City and its metropolitan region. I believe, though, that several follow-up steps are required.

One step is to complete a set of earthquake damage scenarios for this area (and for any other cities/regions that wish to begin action). Such scenarios set forth the state of affairs in the area following earthquakes of specified size and location, tabulating expected casualties and damage and describing the status of lifelines and facilities critical to rescue, recovery and living in the time immediately after the earthquake. It is desirable to have several such scenarios prepared, for different earthquakes of increasing magnitude (and, of course, decreasing probability of occurrence). Preparation of such scenarios requires close cooperation between earth scientists and engineers, and it is essential that local governmental officials (who must ultimately react to the results) be intimately involved throughout the study.

A second step is for local professional societies to consider in detail the various possible code provisions that might be adopted, so as to develop a first-hand impression of their implications for design and for cost. This is a situation where familiarity will breed comfort and not contempt. Work aimed at developing code language applicable to existing construction should begin, plus the collection of details that might be used to reinforce weak existing structures.

At the same time, it is essential that research into the behavior of structures during earthquakes continue—especially experimental research aimed at evaluating the resistance of the types of construction common in ENAM. Experiments on shaking tables will be very valuable, and testing of methods for tying together otherwise unreinforced masonry construction should be a part of the effort. A recent workshop sponsored by the National Science Foundation has mapped out a research program oriented specifically to existing structures.[8]

CONCLUSION

I believe that the time is ripe for steps to reduce the earthquake hazard in eastern North America. There is public interest in the earthquake problem in this region. Many of the necessary tools, such as model building code provisions, are available. Information needed to persuade others of the importance of action is available, such as that concerning the risk of earthquake losses and the costs of providing increased seismic resistance, and, while such information is inevitably incomplete, it can be put to use in decision-making. Research aimed at providing the engineering know-how for implementing new requirements is under way. I hope that local engineers and architects will seize this opportunity and ensure that action begins.

REFERENCES

1. AMERICAN NATIONAL STANDARDS INSTITUTE. 1982. Minimum Design Loads for Buildings and Other Structures, ANSI A58.1-1982. New York, NY. (Responsibility for this standard has now been assumed by the American Society of Civil Engineers, and a revision is due in 1988.)
2. APPLIED TECHNOLOGY COUNCIL. 1978. Tentative Provisions for the Development of Seismic Regulations for Buildings. ATC-3 Report to National Science Foundation and National Bureau of Standards. Redwood City, CA.
3. APPLIED TECHNOLOGY COUNCIL. 1987. Evaluating the Seismic Resistance of Existing Buildings. ATC-14 Report to FEMA. Redwood City, CA.
4. BUILDING OFFICIALS AND CODE ADMINISTRATORS INTERNATIONAL. 1987. BOCA Basic/National Building Code. Country Club Hills, IL.
5. BUILDING SEISMIC SAFETY COUNCIL. 1986. Improving Seismic Safety of New Buildings: A Community Handbook of Societal Implications. FEMA 83, Federal Emergency Management Agency, Washington, DC. (Design and construction costs are summarized in Chapter 5.)
6. BUILDING SEISMIC SAFETY COUNCIL. 1986. NEHRP Recommended Provisions for the Development of Seismic Regulations for New Buildings. Earthquake Hazards Reduction Series 19, FEMA 97, Federal Emergency Management Agency, Washington DC. (Appendix 3 deals with existing buildings.)
7. INTERNATIONAL CONFERENCE OF BUILDING OFFICIALS. 1988. Uniform Building Code. Whittier, CA.
8. JIRSA, J. O. 1987. Repair and Retrofit of Existing Structures. Report to NSF by Department of Civil Engineering, University of Texas. PMFSEL Report 87-6.
9. RABIN, J. L. 1987. Torrance plan to renovate quake-prone buildings believed to be first of its kind. Los Angeles Times December 13: 3.
10. SABOL, T. A., G. T. ZORAPAPEL & G. C. HART. 1988. Typical Costs for Seismic Rehabilitation of Existing Buildings. ATC-14 Report to FEMA. Englekirk and Hart, Inc., Los Angeles, CA.
11. SOUTHERN BUILDING CODE CONGRESS INTERNATIONAL. 1987. Standard Building Code. Birmingham, AL.
12. WHITMAN, R. V., J. M. BIGGS, J. E. BREMAN III, C. A. CORNELL, R. L. DE NEUFVILLE & E. H. VANMARCKE. 1975. Seismic design decision analysis. J. Structural Div., Proc. Am. Soc. Civ. Eng. **101**: 1067–1084.

Low-Displacement Seismogenic Faults and Nonstationary Seismicity in the Eastern United States[a]

LEONARDO SEEBER AND JOHN G. ARMBRUSTER

Lamont-Doherty Geological Observatory of
Columbia University
Palisades, New York 10964

INTRODUCTION

Engineers designing structures must rely on proven principles; in contrast, seismologists investigating geologic processes have the privilege of proposing, testing, and discarding erroneous hypotheses. This difference in approach may be in part responsible for difficulties in communication. Propositions that are only working hypotheses for seismologists may be given undue credit by engineers. This is particularly likely in the eastern United States where the state of knowledge on seismogenesis is rather primitive. This paper discusses recent results that raise questions regarding some long-standing assumptions in procedures to estimate earthquake hazard. These results point out basic differences between the eastern and western U.S. in terms of the role of preexisting faults in neotectonics. Although the geologic basis for intraplate seismogenesis is still poorly understood, the results suggest that geologic information will play an increasingly important role in the estimate of earthquake hazard in the eastern U.S.

Spatial Clustering of Earthquakes and Faults

Seismicity in the eastern U.S. is spatially clustered over a wide range of scales. Some of these clusters are related to preexisting structural features and probably relfect long-term characteristics of the seismicity. Some of the results from detailed correlations between structural and lithologic features and recent aftershock data in the eastern U.S. include the following: (1) Ruptures can be delineated by the aftershocks and they tend to be smaller than expected when scaled to moments according to criteria based on data from plate boundaries. Thus, stress-drops tend to be high and small faults may be capable of damaging earthquakes. (2) Earthquakes can be generated by faults clearly manifested at the surface by zones of concentrated brittle fractures, which, however, show little or no accumulated displacement of pre-Cenozoic markers. Thus, recency-of-faulting criteria to evaluate fault capability may not be applicable in intraplate regions. (3) Spatial correlation between seismicity and faults is not always the result of reacti-

[a] This work was supported by Grants 86-1014 and 87-1303 from the National Center for Earthquake Engineering Research and Grant 04-85-111-02 from the Nuclear Regulatory Commission.

vation and, by itself, should not be considered evidence for fault capability. (4) The correlation of seismicity with structural features is not systematic since similar geologic environments can be associated with drastically different seismic regimes.

Temporal Clustering of Earthquakes and Stationarity

Seismicity is also clustered in time. Changes in seismicity patterns with characteristic periods from hundreds to thousands of years are close to the duration of the historic period and cause seismicity to be nonstationary. An example of such nonstationarity is found in the South Carolina–Georgia seismic zone. This seismic zone is recognized in the recent instrumental data and still reflects the pattern of aftershocks immediately after the 1886 Charleston, S.C. event. In contrast, for almost a century prior to 1886, seismicity in the Coastal Plain of South Carolina is not higher than elsewhere along the southeastern Coastal Plain. If this drastic change in seismicity coinciding with the 1886 main shock is representative of the relation between large earthquakes and background seismicity, then (1) spatial clustering of historic seismicity is not stationary; (2) the relation between historic seismicity patterns and structural features is not systematic as a result of temporal clustering; and (3) the location of future large earthquakes is generally not marked by a high level of historic seismicity. Hazard analyses based on the assumption of stationarity may be appropriate for small- and intermediate-magnitude events, but is probably misleading for the large events.

SEISMICITY VERSUS GEOLOGIC SLIP RATES

The relation between seismicity and displacement on specific faults has been the main guiding principle for the study of seismogenesis in the western United States. Earthquakes in the eastern U.S. are also generated by slip-on faults, but the specifics of the seismogenic process and of the behavior of active faults in this area are still largely unresolved. Available data suggest fundamental differences in the relation between seismicity and faults in the East and in the West. Lack of correlation between seismicity and known faults has often been claimed in the East. This paper presents recent results illustrating clear cases of correlation between seismicity and previously known structural features in the eastern U.S. and discusses the nature of this correlation and its implication for earthquake hazard.

The Meers Fault

No historic earthquake has been associated with a surface fault rupture in the East, but Holocene fault scarps indicating surface ruptures have been recognized. The most prominent of these scarps was recently documented along the Meers fault in Oklahoma. This scarp is the result of one or more prehistoric surface ruptures associated with large ($M \geq 7$) earthquakes.[1] This evidence for a high overall rate of seismic moment release along the Meers fault during the Holocene is contrasted, on the one hand, by low historic seismicity, and, on the other hand, by evidence for little accumulated neotectonic displacement. Thus, the Meers fault is clearly active, but this activity appears to be concentrated in short bursts separated by long periods of dormancy. Similar behavior is inferred for other intraplate faults worldwide, particularly in Australia (e.g., the 1988 Tennant Creek earthquake with a 30-km-long surface rupture).

1811–12 and 1886 Source Zones

The two major earthquake sequences in the eastern United States—in New Madrid, Missouri (1811–1812) and in Charleston, South Carolina (1986)—were centered in areas capped with flat-laying Neogene sediments. Subsurface investigations in these areas have revealed a remarkable absence of significant structural features in these sediments, and only minor deformation in older Cenozoic sediments.[2,3] The lack of accumulated vertical deformation in the 1811–12 and 1886 mesoseismal areas is in contrast with the abundance of secondary near-surface deformation associated with mobilization of unconsolidated sediments resulting from liquefaction.

In the lower Coastal Plain of South Carolina, secondary deformation has been associated with the widespread liquefaction event triggered by the 1886 earthquake, and to other distinct previous events, inferred to be related to large prehistoric earthquakes.[4] In both New Madrid and Charleston, paleoseismic data suggest repeat times of one to two millennia for events similar to the historic ones. The locations and attitudes of the faults that generated these historic and prehistoric events are weakly constrained. Thus, it is possible to account for the low rates of geologic deformation on any single steeply-dipping fault by either distributing the seismicity and related moment-release rate on many faults[5,6] or by generating it on a shallow-dipping fault or detachment.[7,8] Neither of these explanations is satisfactory in terms of better known seismogenic faults in the northeastern United States.

The 1983 Adirondacks Earthquake and the Catlin Lake Lineament

Recent results from the northeastern U.S. also raise the issue of low geologic rates of deformation. In these cases, low geologic rates are found on steeply dipping faults, which are known to be seismogenic. The 1983 Goodnow earthquake in the central Adirondacks (Ms = 5.3) ruptured a NNW striking fault dipping steeply to the west, according to first motions and to the distribution of aftershocks (FIGS. 1 and 2). The 1.5-km-wide circular rupture is centered at a depth of about 8 km. The surface extrapolation of this inferred rupture nearly coincides with the 15-km-long Catlin Lake Lineament.[9] Detailed geologic investigation has revealed a concentration of fractures along this lineament, including slickensided surfaces parallel to the lineament (FIG. 3). The data suggest that the topographic lineament and the 1983 rupture are both related to the same brittle fracture zone or fault.

The detailed structural data also indicate, however, that Grenville age markers cannot be laterally displaced across the fracture zone along the Catlin Lake lineament by more than a few tens of meters (FIG. 3). In this and other respects, then, the Catlin Lake fracture zone resembles the "zero-displacement faults" described by Isachsen et al.[10] The age of these brittle features is poorly constrained, but they are probably old, possibly Precambrian, on the basis of their relation to ductile structures of Grenville age.[11,12] Thus, relatively old faults with an insignificant amount of accumulated displacement may generate significant earthquakes.

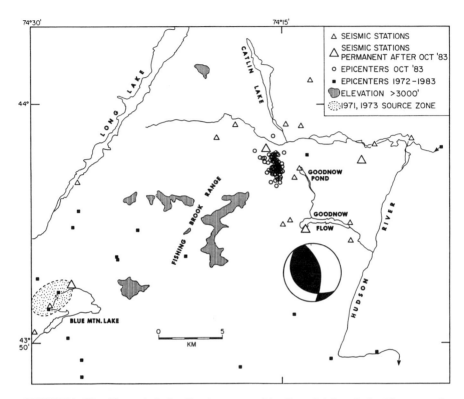

FIGURE 1. Blue Mountain Lake–Goodnow area of the Central Adirondacks. Three sets of epicenters are indicated: *open circles* = aftershocks of the 1983 Goodnow earthquake (October 7–29, 1983; data from the temporary seismic stations shown); *stippled area* = epicenter zone of the 1972–73 Blue Mountain Lake swarms; *solid squares* = 1972–83 epicenters located by the regional seismic network. Long Lake, Catlin Lake and the Hudson River mark geomorphic features that reflect zones of fractures, possibly brittle faults. The north-northeast trend is dominant in the morphology of the Central Adirondacks, but the north-northwest trend seems to reflect the seismogenic faults.

1985 Westchester Earthquake and the Dobbs Ferry Fault

A close spatial correlation was also found between a mapped fault, the Dobbs Ferry fault, and the inferred rupture of the 1985 Westchester earthquake (Mblg = 4.0). Aftershocks and first motions suggest a small subvertical and left-lateral rupture, centered about 5 km deep (FIG. 4). A map compiled by Leo Hall[13] shows early Paleozoic features cut by this fault with no detectable displacement (FIG. 5). New detailed surface data for the Dobbs Ferry fault confirm the small accumulated displacement (< 30 m).

The Dobbs Ferry fault is part of a set of northwest-striking brittle faults of Mesozoic age in the Manhattan Prong. This fault set includes most of the known brittle features in the New York City area and in Westchester County and is believed to account for most of the seismicity. The 125th Street fault in Manhattan is one of these faults. Subsurface studies of this fault reveal a thick breccia zone,

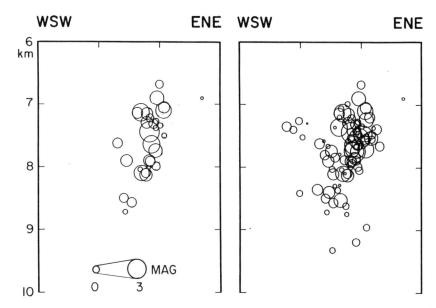

FIGURE 2A. Aftershocks of the 1983 Goodnow earthquake in a vertical section perpendicular to strike (see FIGURE 1). *Left:* aftershocks during the first week only; *right:* aftershocks during the first month. Note growth of seismic zone on antithetic fault.

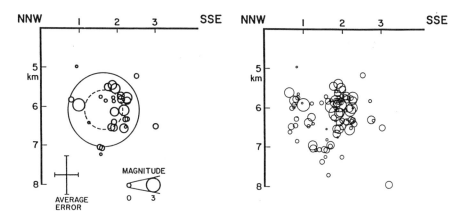

FIGURE 2B. Aftershocks of the 1983 Goodnow earthquake. Obliquely dipping section, parallel to the inferred rupture surface. *Left:* aftershocks during the first week only; *right:* aftershocks during the first month. *Circles* cover the range in values for the dimension and depth of the rupture inferred from a moment-tensor inversion.[20] Note annular pattern defined by aftershocks. This annulus probably defines the outer edge of the rupture. Compare with FIGURE 8.

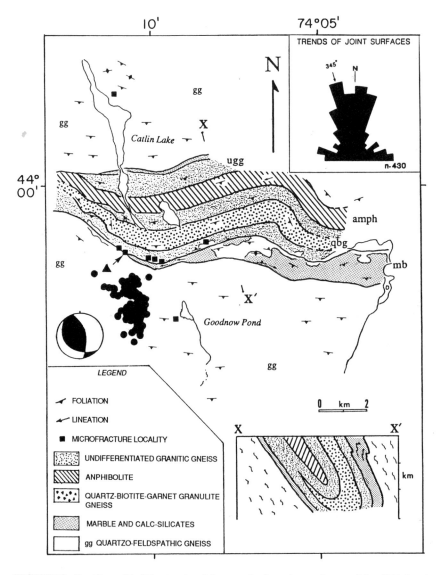

FIGURE 3. Ductile and brittle structural features in the epicentral area of the 1983 Good-now earthquake (*circles* delineate aftershock epicenters). Surface extrapolation of the rupture falls approximately along the Catlin Lake lineament, which is marked by a narrow 10-km-long valley and a pronounced fracture set (**inset**). Slickensided surfaces parallel to the inferred rupture and microfractures are concentrated at the intersection of the lineament with the marble layer. Although this lineament has many of the characteristics of a brittle fault, no displacement of the Grenville structures can be detected. The resolution is better than 50 meters. (From Dawers.[12] Reproduced by permission.)

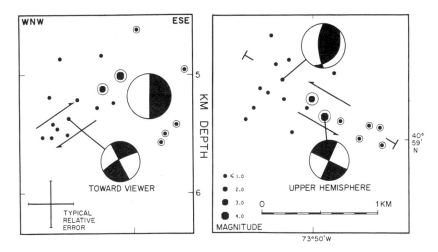

FIGURE 4. Aftershocks of the 19 October 1985 Westchester Co. earthquake. *Right:* map view; *left:* vertical section parallel to the inferred rupture. Circled hypocenters occurred during the first week; the other hypocenters occurred later. Aftershock and first-motion data constrain the rupture to be vertical, strike northwest, and to be left-lateral. The distribution of early aftershocks suggests that the rupture is about ¾ of a kilometer across. Comparison with FIGURES 2, 6 and 8 reinforces this conclusion.

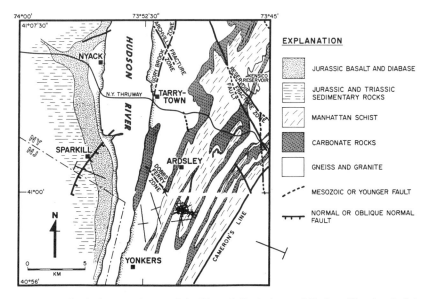

FIGURE 5. Geologic map of parts of the Newark Basin (west of Hudson River) and of the Manhattan Prong in the epicentral area of the October 19, 1985 Westchester Co. earthquake. The northwesterly striking main rupture is marked by the tight cluster of aftershocks (*crosses* indicate possible location errors). The Dobbs Ferry fault is aligned with the rupture. Most other brittle faults in the Manhattan Prong have the same trend. (After Ratcliffe [1980 and 1981] and Hall [1981].)

suggesting large accumulated slip. The northwestern portion of this fault intersects the Palisades Sill, but shows relatively little displacement (< 100 meters) of this Jurassic Age feature, suggesting that much of the displacement on these faults predates the intrusion of the sill. Finally, no vertical displacement (i.e., < 30 meters) is recognized on the Cretaceous erosional surface above the sill.

In contrast to the little displacement accumulated on brittle faults in the Manhattan Prong during the Cenozoic, earthquake data suggest a significant rate of moment release during the historic period. If a set of NW faults with a total seismogenic surface of 500 km^2 is considered to be the source of an event similar to the one in the Adirondacks in 1983 (average slip, 30 cm; rupture area, 2.5 km^2) every one and half century (i.e., the New York City area events in 1884 and 1737), these events alone would account for a displacement rate of about 10 meters per million years, or ½ km since the Cretaceous. This back-of-the-envelope calculation for the inferred slip rate does not consider the contribution to the moment release from smaller events, like the one in 1985, from events larger than any of the historic events and from aseismic slip. Thus, if the current regime is extrapolated backward through the Cenozoic, accumulated displacements on the seismogenic faults should be significantly larger than the displacements observed.

The 1986 Ohio Earthquake and the Akron Magnetic Lineament

The data for the 1986 Ohio earthquake (Ms = 5.0) and related seismicity present another case of a seismogenic fault with evidence of little accumulated displacement. The aftershocks and first motions for this event suggest a small (1 × 1 km) vertical rupture with right slip centered at a depth of about 5 km (FIG. 6). The inferred rupture is aligned with the Akron Magnetic Lineament, a major linear feature which strikes NNE and can be followed for at least 50 km. Many of the other epicenters in northeastern Ohio fall within error estimates of this feature (FIG. 7). Aeromagnetic data reflect primarily rock type in the Precambrian basement. The feature in question is expressed as a boundary between domains with different magnetic signature and is probably a major Precambrian fault marking a lithologic boundary. The earthquake data suggest that this fault is reactivated, at least locally, in the current tectonic regime.

About 2 km of flat-lying Paleozoic sediments cap the basement in northeastern Ohio. These sediments can serve as a marker for vertical displacement since the early Paleozoic. An effort to detect any such displacement using the abundant subsurface data from wells in the 1986 epicentral area has been unsuccessful so far.[14] Once again there is evidence for a seismogenic preexisting fault with little accumulated neotectonic displacement.

Detailed data on recent Mb = 4 to 5 earthquakes in the northeast U.S. suggest that even for these relatively small events, ruptures are often located on faults that can be recognized from other data. Some of these faults, however, appear to have little or no accumulated slip. An important practical implication of these results is that lack of recent movement along a fault is not necessarily an indication of low earthquake potential. Recency of movement, then, is a proven criterion for fault capability in interplate zones, such as the Western U.S., but it is not a reliable criterion in intraplate areas. Models where active faults are coupled with an elastic rather than a ductile substratum may account for the combination of a significant rate of seismic moment release and no accumulated displacement.[15]

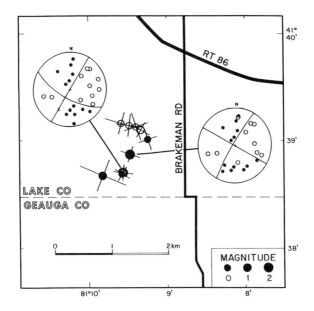

FIGURE 6A. Epicenters for the nine aftershocks of the January 31, 1986 Chardon, Ohio earthquake for which reliable data are available. These data are from temporary seismic stations operated by many institutions. *Solid circles* indicate aftershocks during the first week of the sequence. Hypoinverse error bars are shown.

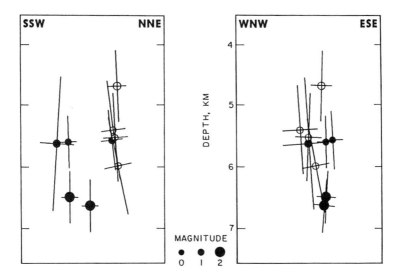

FIGURE 6B. Perpendicular vertical sections with hypocenters for the nine aftershocks shown in FIGURE 6A (same symbols). These sections are parallel (*left*) and perpendicular (*right*) to the inferred rupture.

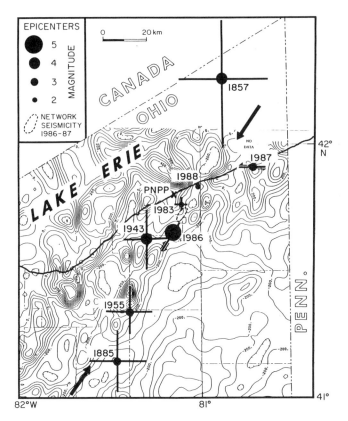

FIGURE 7. Epicenters of the most prominent known earthquakes and residual total magnetic map[21] for the area in northeastern Ohio centered near the Perry nuclear power plant (PNPP). All known historic earthquakes with magnitudes M ≥ 3 and all instrumental earthquakes with M ≥ 2 in that area are represented. *Lines* and *arrows* associated with the 1986 Chardon and 1987 Ashtabula epicenters indicate strike and sense of movement of the corresponding subvertical ruptures. The *dashed outline* includes most of the recent epicenters for small earthquakes recorded by a seismic network operated by Cleveland Electric between the Perry Plant and the 1986 epicenter. The seismicity is concentrated along a northeast trending zone that traverses northeastern Ohio and appears to be correlated with a major linear boundary in the magnetic field (the Akron Magnetic Lineament; indicated by *large arrows*). The 1986 rupture, within this zone, is also parallel with it. The 1987 rupture, which is off the active zone, was probably induced by fluid injection.

RUPTURES DELINEATED FROM DISTRIBUTION OF AFTERSHOCKS

A close spatial correlation between ruptures and aftershock zones has been demonstrated for ever smaller ruptures, as location uncertainties have continued to decline. By considering both hypocentral locations and first-motions, it has been possible to differentiate aftershocks that occur on the main rupture, the aftershocks proper, from aftershocks occurring beyond the limits of the rupture and on other faults. Moreover, particularly well constrained aftershock zones of intermediate-magnitude events suggest a tendency for the aftershocks to populate the perimeter of the rupture, leaving the central part of the rupture relatively unpopulated. This characteristic is also instrumental for identifying the rupture from the aftershock distribution. These characteristics of aftershock zones have been illustrated by recent intermediate-magnitude events in California (FIG. 8).

Results from detailed studies of recent aftershock sequences of relatively small events in the northeastern U.S. suggest relationships between ruptures and aftershock zones that resemble the one observed for larger events in California. FIGURES 2B, 4, and 6B show the aftershock zones of the 1983, 1985, and 1986 earthquakes (discussed above) viewed in sections parallel to the inferred ruptures (i.e., the ruptures are viewed face on). An annular distribution of the seismicity can be clearly recognized for the 1983 sequence. A similar pattern may then be seen in the other two events. In each case, the annular pattern is the only active feature in the early part of the sequence. Subsequent aftershocks show a tendency for activity spreading to other faults as the sequence progresses. These results are significant in terms of the mechanics of the aftershock process, but also in terms of earthquake hazard.

The rupture areas inferred from the aftershock distributions appear to be roughly circular with diameters of 1.5 km, 0.8 km, 1.0 km for the events in 1983, 1985 and 1986, respectively. These ruptures are surprisingly small, significantly smaller than predicted by the Nuttli[16] relation between moment and rupture size for the eastern United States. One direct implication for earthquake hazard estimates is that the fault-dimension vs. maximum-earthquake relation may need to be altered toward larger earthquakes for a given fault dimension. Another one is that stress-drops may tend to be higher for earthquakes in the East than for earthquakes in the West.[17] Finally, a detailed spatial correlation between rupture and fault patterns may illustrate how fault morphology controls rupture dimensions. Structural data may then offer more stringent constraints to the determination of maximum earthquakes.

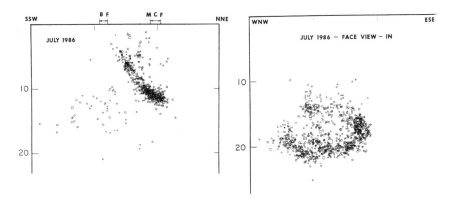

FIGURE 8. Aftershocks of the July 7, 1986 North Palm Spring earthquake (Ms = 5.9) on the Banning fault in southern California. *Left:* vertical section is perpendicular to the strike of the rupture; the section on the *right* is tilted and parallel to the inferred rupture plane. Note the concentration of aftershocks around the perimeter of the rupture. Note also that some of the "aftershocks" are not located on the rupture. (Data from the Caltech/USGS network; relocation by Doug Given.)

SEISMICITY ALONG PREEXISTING FAULTS: ZONES OF WEAKNESS OR STRESS CONCENTRATION?

Geologic data are relatively abundant and easily acquired. An understanding of the relation between seismicity and structural features inherited from previous orogenies may greatly expand the use of geologic data to infer characteristics of the seismicity. Earthquake data show that preexisting structural features often control the spatial distribution of seismicity and are generally consistent with the concept of reactivation of preexisting faults that persist as zones of weakness.[18] For example, the data for northeastern Ohio in FIGURE 7 illustrate a case of possible reactivation of a Precambrian fault in the current stress regime. A similar correlation between preexisting faults and seismicity has been proposed for the New Madrid seismic zone. When both large- and small-scale elements are considered, however, the mechanical basis for the correlation between seismicity and structure often appears to be more subtle than simple reactivation.

FIGURES 9 and 10 show two broad seismic zones related spatially to ancient structural features in the central Adirondacks and in the northern Piedmont,

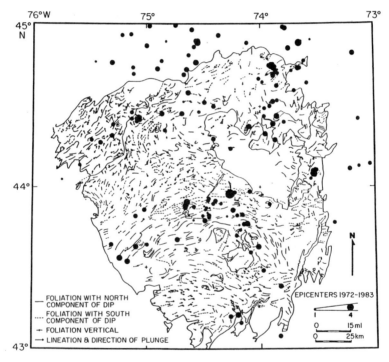

FIGURE 9. Epicenters from the New York State Seismic Network, 1972–1983 (*black dots*) superimposed on foliation and lineation data extracted from the 1:250,000 N.Y. State geologic map.[22] Two arcuate belts of seismicity in the central and northwestern Adirondacks are related to Precambrian (Grenville) structural trends. Large domains where dips of foliation have either a north or south component can be recognized; boundaries between these domains may be tectonic. The seismic zone in the central Adirondacks, which contain the Goodnow and Blue Mountain Lake epicentral areas (FIG. 1), is associated with a foliation-dip boundary.

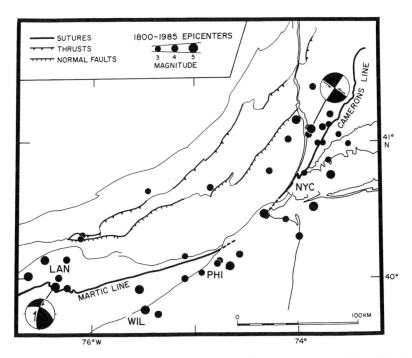

FIGURE 10. Epicenters of significant earthquakes (felt-area magnitudes Mfa ≥ 3) in the area of the Newark Basin during almost two centuries. Most of the epicenters fall in a broad belt on the southeast side of the basin centered along the Martic-Camerons line, an early Paleozoic suture. The ruptures in the two most recent events on this map, the Lancaster (LAN) event in 1984 and the Westchester Co. event north of New York City (NYC) in 1985, were resolved from aftershock data and found to be at large angles to the trend of the belt. *Balloons* represent fault-plane solutions for these events (upper hemispheres); *arrows* indicate the fault plane and the sense of slip. Faulting in few smaller events have trends similar to the ones in 1984 and 1985. Reactivation of the Martic-Camerons line cannot be the cause of the spatial association between this feature and the seismic belt.

respectively. The central Adirondacks seismic zone follows Precambrian structural trends and probably the locus of a major structural boundary. Exhumation has now brought to the surface the ductile deformation level of Precambrian (Grenville) deformation. Similarly, the seismic zone along the southeast side of the Newark Basin is centered along the Martic–Camerons line, a lower Paleozoic suture, also exhumed at the formerly ductile deformation level. The geometry of faulting along each of these zones has been resolved for several events, including the 1983 and 1985 events discussed above (FIGS. 2 and 4). In all these cases, the inferred seismogenic faults strike at large angles to the trend of the seismic zones and to the preexisting features along these zones. A similar relation between the overall trend of the seismicity, Precambrian structure, and seismogenic faults is found in eastern Tennessee.[19] The spatial correlation between seismicity and some of the major ductile structural features, then, does not appear to be caused by renewed motion on these features in the current stress regime (i.e., not by reactivation).

If seismic zones can be associated with preexisting faults by a process other than reactivation, the length of these zones and the related faults lose their connotation in terms of the maximum possible length of a rupture in that zone. It is possible, moreover, that these apparently inactive preexisting faults are associated with seismicity because they are zones of stress concentration, rather than zones of weakness. This working hypothesis is inspired by the probable association of some of these structural features with deeply buried calcitic marble, which is expected to yield ductilely and transfer the load to the surrounding brittle rock.[15]

DISTRIBUTION OF SEISMICITY IN TIME AND SPACE

The coverage of seismicity during the historic period in the eastern U.S. has greatly improved in recent years by both detailed instrumental hypocenter coverage from regional and local seismic networks, and by reexamination of archival

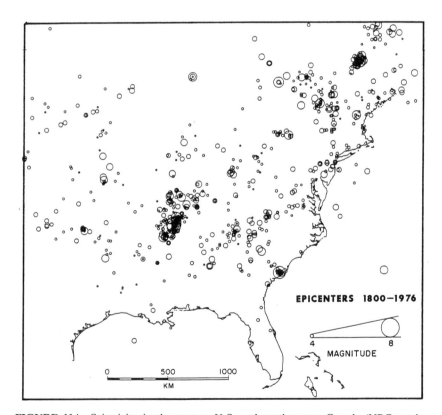

FIGURE 11A. Seismicity in the eastern U.S. and southeastern Canada (NRC catalog; Barstow *et al.*, 1981). Seismicity is not randomly distributed, but tends to be clustered, leaving broad regions with low seismicity. Do structural features control the pattern of seismicity? What elements in this pattern are stationary and what elements are time-dependent?

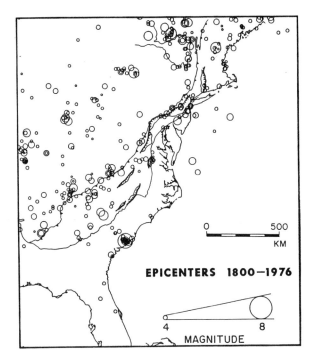

FIGURE 11B. Seismicity along the Appalachians and the Atlantic Seaboard of the United States (same data as in FIG. 11A) and some of the main geologic features (from southeast to northwest: Fall Line, Mesozoic rift basins, western limit of the crystalline sheet in Paleozoic thrusts, Precambrian shield in the Adirondacks). Seismicity is spatially related to structural features in some areas, but it is not associated with similar features in other areas.

data for felt reports.[8] The improved control on the distribution of epicenters confirms that seismicity has been unevenly distributed during the historic period. Well-defined seismic zones are separated by areas where seismicity is low or absent (FIG. 11).

Along the Appalachians and Atlantic Seaboard of the United States, epicenters are concentrated in northeastern and in southern New England (including the offshore areas), in the Adirondacks, around the Newark Basin, in central Virgina, in western Virginia (Giles Co.), in eastern Tennessee, and in southwestern South Carolina–northern Georgia. In the central United States, seismicity is concentrated in the Reelfoot Rift area of the lower Mississippi Valley. Several of these seismic zones are found to correlate spatially with geologic provinces and prominent structural features. These correlations, however, tend to be unsystematic because seismicity is correlated with portions of these features, but is not correlated with other portions which appear to be geologically equivalent.

The Newark Basin, for example, is associated with a prominent zone of seismicity, but the Gettysburg Basin and other Mesozoic basins are not. The Piedmont is active in South Carolina and Virginia, but is not active in North Carolina. The feather edge of the crystalline overthrust sheet is active in the Hudson Highlands and in the southern Blue Ridge, but not elsewhere along the Appalachians. The autochthonous basement is active in the Adirondacks and in eastern Tennessee, but is not active along the central Appalachians. The Eucambrian rift system is active in the Reelfoot Rift area, but is not active in other areas, such as in the Rome Trough. The lack of systematics in the spatial relation between seismicity and structure may, in part, be resolved by more detailed knowledge of the structure. An equally likely factor in this apparent lack of systematics is a nonstationary temporal behavior of the seismicity.

Generally, the pattern of seismicity derived from recent short-term instrumental data resembles the pattern of seismicity derived from long-term samples of historic data. This similarity has been considered evidence that seismicity is time-stationary so that the future distribution of seismicity can be inferred directly from the historic distribution. Some changes in the pattern of seismicity, however, have been detected. The most prominent is the change associated with the 1886 Charleston, S.C. earthquake. The reexamination of intensity data from the southeastern U.S. has demonstrated that the burst of aftershocks that followed immediately the 1886 main shock established a pattern of seismicity that has subsided in level, but has persisted in shape to the present, and is known as the South Carolina–Georgia seismic zone. But the distribution of seismicity in the same area was markedly different for at least 80 years before 1886, with a pronounced lack of seismicity in most of the area of the forthcoming aftershock zone, including the 1886 epicentral area (FIG. 12). Thus, the large earthquake in 1886 coincided with a pronounced and long-lasting change in the pattern of seismicity.

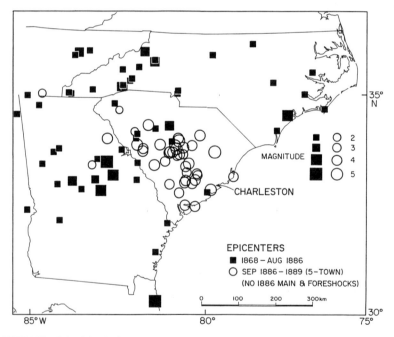

FIGURE 12. Seismicity before (*solid squares*) and after (*open circles*) the August 31, 1886 Charleston, S.C. earthquake. Epicenters and magnitudes of these preinstrumental earthquakes were obtained by an algorithm (MACRO) from felt reports in contemporary newspapers. The seismicity after the main shock is represented by the larger aftershocks (reported by at least five towns). They indicate a widespread aftershock zone measured in hundreds of kilometers. Except for immediate foreshocks, the level of seismicity prior to the main shock in the epicentral zone near Charleston was low and similar to that of other areas of the southeastern Coastal Plain.

The scenario suggested by the historic data from the Atlantic Seaboard is a relatively stationary seismicity between major events and large changes set off by these events. If the low level of seismicity in the 1886 epicentral area before that event is also representative, the assumption of stationarity would be particularly misleading since it would tend to yield low estimates of hazard in epicentral areas of impending large events. The assumption of stationarity would only lead to realistic hazard assessments for small and intermediate-magnitude earthquakes.

A more complex model where seismicity patterns can vary with characteristic periods measured in thousands of years (i.e., somewhat longer than the historic period and similar to the inferred repeat times of large intraplate earthquakes) is also consistent with the spatial pattern of seismicity. Seismicity is correlated with portions of structural features, while other portions of the same features appear to be aseismic. Major earthquakes with repeat times in the thousands of years may set off long-lasting "aftershock" sequences. In the historic snapshot view of seismicity, these sequences may appear as stationary active areas. In a hypothetical view of the seismicity covering a period substantially longer than the repeat times of the large events, however, seismicity would appear continuous along the structural belts. In such a model, the probability of a large event may be high in areas of low historic seismicity.

FINDINGS

1. Improved geologic and earthquake data reveal that the spatial distribution of seismicity in the eastern U.S. is often controlled by preexisting structure, but the association of seismicity with structural features is nonsystematic;
2. Old ductile faults may be the locus of concentrated seismicity without being reactivated;
3. Brittle faults or zones of fracture concentration giving rise to topographic anomalies are found at the surface extrapolation of ruptures from small- to intermediate-magnitude earthquakes. These features, however, show little accumulated displacement; and
4. The pattern of seismicity is mostly invariant through the historic period, except in the source zone of the 1886 Charleston event, where seismicity was low for at least a century prior to the event, but has been relatively high since then.

CONCLUSIONS

1. Seismogenic faults or fault sets may be identified by combined detailed investigations of geology and seismicity;
2. Recency-of-faulting to assess the earthquake potential of a fault are not applicable to intraplate regions;
3. Seismic zones associated with large preexisting faults may not necessarily imply large maximum earthquakes. The dimension of individual brittle faults in the source zone need to be resolved in order to constrain the size of maximum earthquakes; and

4. Major changes in the pattern of seismicity can be triggered by large events. In particular, the epicentral areas prior to these events may be characterized by low seismicity, rendering hazard analyses based on stationary seismicity particularly unrealistic with regard to large earthquakes.

ACKNOWLEDGMENTS

This paper greatly benefited from a review by Klaus Jacob and from discussions with Tish Tuttle, Nancy Dawers, and Klaus Jacob.

REFERENCES

1. Luza, K. V., R. F. Madole & A. J. Crone. 1987. Investigation of the Meers fault in southwestern Oklahoma, NUREG/CR-4937: 55. U.S. Nuclear Regulatory Commission, Washington, DC.
2. Stewart, R. M. 1986. Review of geological, geophysical, and seismological data relevant to seismotectonics of the Charleston, South Carolina area (abstract). Earthquake Notes **57:** 16.
3. Hamilton, R. M., J. C. Behrendt & H. D. Ackerman. 1983. Land multichannel seismic-reflection evidence for tectonic features near Charleston, South Carolina. U.S. Geol. Surv. Prof. Pap. 1313I: 18.
4. Obermeier, S. F., G. S. Gohn, R. E. Weems, R. L. Gelinas & M. Rubin. 1985. Geologic evidence for recurrent moderate to large earthquakes near Charleston, South Carolina. Science **227:** 408–411.
5. Wentworth, C. M. & M. Mergner-Keefer. 1983. Regenerate faults of small Cenozoic offset: Probable earthquake sources in the southeastern U.S. U.S. Geol. Surv. Prof. Pap., 1313S, 18.
6. Anderson, J. G. 1986. Seismic strain rates in the central and eastern United States. Bull. Seismol. Soc. Am. **76:** 273–290.
7. Behrendt, J. C., R. M. Hamilton, H. D. Akerman & V. J. Henry. 1981. Cenozoic faulting in the vicinity of the Charleston, South Carolina 1886 earthquake. Geology **9:** 117–122.
8. Seeber, L. & J. G. Armbruster. 1987. The 1886–1889 aftershocks of the Charleston, S.C. earthquake: A widespread burst of seismicity. J. Geophys. Res. **92:** 2663–2696.
9. Isachsen, Y. W. & W. G. McKendree. 1977. Preliminary brittle structure map of New York, 1:250,000 N.Y.S. Museum Map and Chart Series 31A.
10. Isachsen, Y. W., E. P. Geraghty, & R. W. Weiner. 1986. Fracture domains associated with a neotectonic basement-cored dome—the Adirondack Mountains, New York. *In* Proceedings of the Proc. Fourth International Conference on Basement Tectonics, Oslo, Norway: Gabrielsen *et al.*: Eds. 287–305.
11. McLelland, J. & Y. W. Isachsen. 1980. Structural synthesis of the southern & central Adirondacks: A model for the Adirondacks as a whole and plate tectonics interpretation. G.S.A. Bull. 91, Part II: 208–292.
12. Dawers, N. H. 1987. The Basement Geology of the Catlin Lake-Goodnow Pond Area, Adirondack Mountains, New York, and Its Relationship to the October 7, 1983 Goodnow Earthquake. M. S. thesis, University of Illinois, Urbana, IL.
13. Ratcliffe, N. 1986. Personal communication.
14. Hansen, M. 1986. Personal communication.
15. Seeber, L. & J. G. Armbruster. 1986. A study of earthquake hazard in New York State and adjacent areas. NUREG Publication CR-4750.
16. Nuttli, O. W. 1983. Average seismic source—parameter relations for midplate earthquakes. Bull. Seismol. Soc. Am. **73:** 519–535.
17. Scholz, C. H., C. A. Aviles & S. G. Wesnousky. 1986. Scaling differences between large intraplate and interplate earthquakes. Bull. Seismol. Soc. Am. **76:** 65–70.

18. RATCLIFFE, N. M., W. C. BURTON, R. M. D'ANGELO & J. K. COSTAIN. 1986. Low-angle extensional faulting, reactivated milonites, and seismic reflection geometry of the Newark basin margin in eastern Pennsylvania. Geology **14:** 766–770.
19. JOHNSTON, A. C., D. J. REINGOLD & S. I. BREWER. 1985. Seismotectonics of the southern Appalachians. Bull. Seismol. Soc. Am. **75**(N1): 291–312.
20. NABELEK, J. & H. SUAREZ. The 1983 Goodnow earthquake in the central Adirondacks, N.Y.: A broadband teleseismic analysis. *In* Proceedings from the Symposium on Seismic Hazards, Ground Motions, Soil Liquefaction and Engineering Practice in Eastern North America. K. H. Jacob, Ed.: 300–317.
21. HILDENBRAND, T. G. & R. P. KUCKS. 1984. Residual total intensity map of Ohio, Map GP-961, U.S. Geol. Survey, Reston, VA.
22. ISACHSEN, Y. W. & D. W. FISHER. 1970. Geologic map of New York (scale 1 : 250,000), New York State Museum Map and Chart Series 15.

Seismicity and Seismotectonics of Southeastern Canada[a]

JOHN ADAMS

Geophysics Division
Geological Survey of Canada
Ottawa K1A 0Y3, Canada

INTRODUCTION

Within the southern part of the continental region of eastern Canada, seismicity is clustered in four zones. In three of these zones—western Quebec, Charlevoix, and the lower St. Lawrence—most of the earthquakes are occurring at depths of 5 to 25 km within the Grenville basement, apparently chiefly through reactivation of a Paleozoic rift fault system along the St. Lawrence and Ottawa rivers. The fourth zone, the northern Appalachians, includes the Miramichi earthquakes of 1982, which represent shallow thrust faulting in a sheet of rocks that have been thrust over the older basement.

Along the eastern margin of the continent, the seismicity includes the 1929 magnitude (M) 7.2 Grand Banks and 1933 M7.3 Baffin Bay earthquakes. These and smaller earthquakes appear to be concentrated at the ocean-continent transition, perhaps by reactivation of the Mesozoic rift faults created when the North Atlantic was formed. In the Labrador Sea, earthquakes are associated with the extinct spreading ridge and its associated transform faults.

In the present paper I briefly discuss the seismicity and seismotectonics of southeastern Canada as it is currently understood,[1,2] proceeding from west to east, before summarizing inferences about the causes of these earthquakes. All the earthquakes in eastern Canada appear to be occurring within a regional stress field dominated by northeast-to-east compression, and most large earthquakes have occurred near Paleozoic or younger structures, that is, in places where the continent has been most recently weakened.

REGIONAL SEISMICITY AND SEISMOTECTONICS

Western Quebec

A significant cluster of earthquakes occurs in the Grenville Province of the Canadian Shield, predominantly in western Quebec, but extending across the Ottawa River into eastern Ontario. An earthquake with magnitude about 6 occurred at or near Montreal in 1732 and during this century earthquakes of M6.2 occurred near Lake Timiskaming in 1935 and M5.6 near Cornwall, Ontario, in 1944.

---

[a] This paper is Geological Survey of Canada Contribution Number 49587.

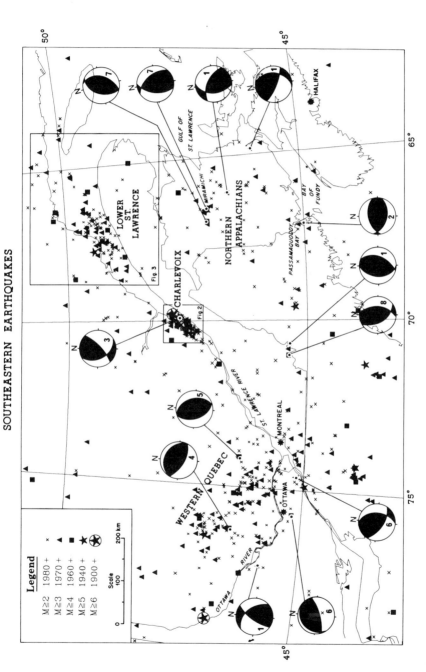

FIGURE 1. Seismicity of southeastern Canada (legend shows the magnitude/time period criteria for the earthquakes shown), with representative focal mechanisms.[1,2]

For the last four decades all earthquakes have been M4.6 or less and most have been located north of the Ottawa River. In detail (FIG. 1), the seismicity occurs in two bands. The first band, trending slightly west of northwest, lies along the Ottawa River and includes the Timiskaming, North Gower (1983), Cornwall, and Montreal earthquakes. The second band trends slightly north of northwest and extends from Montreal to the Baskatong Reservoir, about 200 km north of Ottawa.

Field monitoring of aftershock sequences has provided good estimates for the focal depth of some earthquakes, and these together with some approximate depths computed from the Eastern Canada Telemetered Network suggest that most earthquakes occur between 5 and 20 km. Focal mechanisms have been determined for about 40 earthquakes in the zone, and a selection is shown in FIGURE 1. Almost all mechanisms have near-horizontal compression axes, and represent mainly thrust earthquakes. This evidence for high horizontal compression is confirmed by other evidence for regional stresses in eastern Canada.[3,4]

Forsyth[5] has shown that the earthquakes in the first band, which includes the larger historical earthquakes, may be associated with rift faults along the Ottawa River that were last active in the Paleozoic. The rift faults are part of a large structure that extends from the Ottawa Valley and from rifts along Lake Champlain, down the St. Lawrence River[6] and, in related structures, through southern Labrador to the Labrador Sea. Other related structures may extend northwest from Lake Timiskaming to Kapuskasing, Ontario. The second, more northerly-trending, band of seismicity is believed to be related to the passage of a hotspot 130 million years ago.[2]

Charlevoix

The Charlevoix zone is historically the most active in eastern Canada with at least five earthquakes of magnitude 6 or greater (1663, 1791, 1860, 1870, and 1925). Most recent earthquakes are confined to a zone that is about 80 km long by 35 km wide (FIG. 2), mainly under the St. Lawrence River,[7] and most are at depths between 5 and 25 km (FIG. 2, *inset*) in the Precambrian basement and beneath a Paleozoic sedimentary wedge (unshaded area on FIG. 2 *inset*). Stereo plots[7] demonstrate that most of the microearthquakes are occurring on northeast-striking planes that dip to the southeast. A projection of the hypocenters to the surface along the postulated faults (FIG. 2) suggests that the activity is confined between mapped Paleozoic rift faults on the north shore and a bathymetric feature near the river's south shore. Further, the earthquakes do not extend downriver beyond the cross-cutting Saguenay graben faults.

Charlevoix, the Ottawa River, and the lower St. Lawrence seismic zones all lie along the same rift system. However, at Charlevoix the rift structures have been complicated by a late Devonian meteorite impact that caused ring faults and distributed fracturing (seen in FIGURE 2). Considerable attention is being paid as to whether the earthquakes at Charlevoix occur because of the associated impact structure (in which case the seismicity could be considered to be localized and unlikely to occur elsewhere) or whether the earthquakes just happened to be coincident with the impact (in which case other parts of the rift fault system could become similarly active). Current opinion is that the impact structure is not the controlling factor in the seismicity.

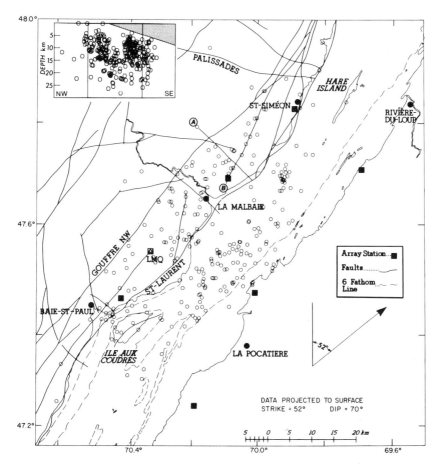

FIGURE 2. Microearthquakes in the Charlevoix area. In order to show their relation to the mapped surface faults, the epicenters have been migrated to the surface up the dip of the regional faults. The inset is a NW–SE cross section of the hypocenters to show their depth distribution relative to the Paleozoic wedge (*shaded*). (After Anglin.[7])

Lower St. Lawrence

Similar to those in the Charlevoix zone, the Lower St. Lawrence earthquakes also occur mainly under the St. Lawrence River and may involve the old rift faults. The record of felt earthquakes extends back less than a century in this sparsely populated area, and none is likely to have been much larger than M5. Despite the lack of known larger earthquakes, magnitude 3 to 4 earthquakes occur at a rate similar to that of the more confined Charlevoix zone. Early epicenter maps[8,9] show a scattering of earthquakes extending onto the north and south shores. Relocation of some of these early epicenters has affirmed that most actually occurred under the river (FIG. 3). Reliable estimates of earthquake depth have been obtained from depth phases for two earthquakes recorded by the Yellowknife array, 30 degrees to the northwest. Phases interpreted as pP and sP gave depths of 19 km and 21 km. Like the Charlevoix earthquakes, this places them within the Grenville basement and well beneath the onlapping St. Lawrence platform and the overthrust Appalachian sedimentary wedge.

Focal mechanisms have been derived for seven earthquakes.[10] They indicate mostly thrust faulting in response to compression from the eastern quadrant. Significantly, five of the mechanisms have a common plane that strikes parallel to the river and dips to the southeast. Taken together, the northeast-striking focal planes, the position of the larger earthquakes relative to the coastline, and the distribution of the smaller earthquake epicenters strongly suggest that the Paleozoic rift faults are the chief active structure.

Northern Appalachians

The northern Appalachian region, which includes most of New Brunswick and which extends into New England, is a zone of relatively uniform seismicity. Significant earthquakes include those near Passamaquoddy Bay on the New

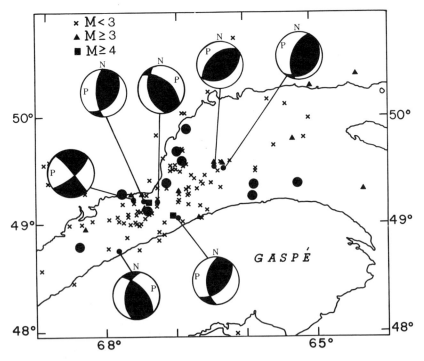

FIGURE 3. Lower St. Lawrence seismicity, 1981–1987. *Large circles* represent relocated epicenters of M3.5 to M4.8 earthquakes from the period 1944–1968 (Sharp and Adams, unpublished work, 1987). Focal mechanisms shown are those derived by Adams *et al.*[10]

Brunswick/Maine border in 1817 (M4.8), 1869 (M5.2), and 1904 (M5.7), near Moncton in 1855 (M5.2),[11] and at Miramichi and Trousers Lake in 1982.[12] Within New Brunswick there is an apparent concentration of earthquakes in the Miramichi Highlands and near Passamaquoddy Bay. Available focal mechanisms (some are shown in FIGURE 1) represent dominantly thrust faulting, most in response to northeast- to east-directed compression.

The Miramichi earthquakes, because of their size and numerous aftershocks, have been unusually well studied. The 9 January 1982 mainshock, M5.7, occurred as a thrust with rupture up-dip on a west-dipping plane. A magnitude 5.1 aftershock occurred 3.5 hours later, probably on the lower northern portion of this plane. On 11 January, an M5.4 earthquake ruptured (probably up-dip) a conjugate east-dipping plane and was followed by an intense aftershock sequence. Finally the 31 March M5.0 earthquake occurred as a repeat rupture on the upper northern portion of the west-dipping plane.[12,13] Although trenches were dug to locate a surface rupture, none was found, suggesting that the near-surface vertical offset was distributed over a zone several hundred meters wide and not confined to a single plane.

Although geological and geophysical investigations have confirmed that all the earthquakes occurred within a single granodiorite pluton, they provided no strong reason as to why the earthquakes occurred at the Miramichi site. If Miramichi-type earthquakes had occurred regularly at the site, more surface evidence for thrusting, such as a degraded fault scarp, would be expected. Therefore, despite considerable effort, it is not understood why the earthquakes occurred where they did, and there is no evidence that they occur often at the Miramichi site. Thus, Miramichi-sized earthquakes must be considered possible anywhere in the northern Appalachian zone.

While aftershocks of the Miramichi and Trousers Lake earthquakes were shallow, all less than 9 km, earthquakes near Passamaquoddy Bay are deeper, perhaps 10–16 km. The northern Appalachian seismicity is thus significantly less deep than the earthquakes at Charlevoix, perhaps reflecting a thinner brittle upper crust in the young Appalachian belt (FIGURE 9 of Hasegawa[14]). The shallow depths may indicate that the earthquakes lie above a shallow, sub-Appalachian detachment zone, such as has been found beneath the Appalachians of the United States. If this were the case, it might mean that the Miramichi earthquakes nucleated on the detachment and ruptured upwards to the surface, while the Trousers Lake earthquake may have occurred on the detachment itself.

Southeastern Continental Margin

Although poorly monitored and little studied, the seismicity of the southeastern margin of Canada is clearly higher than that at many comparable passive margins (FIG. 4). The description that follows is condensed from Adams.[15]

About half of the earthquakes off the southeastern continental margin occur in the Laurentian Slope seismic zone, site of the M7.2 "Grand Banks" earthquake of 1929 (FIG. 4). This earthquake caused a large submarine slump and consequent tsunami that killed 27 people on the south coast of Newfoundland. The larger historical earthquakes, and many recent smaller earthquakes, have been systematically relocated and found to lie within a 100 km E-W by 35 km N-S box at the mouth of the Laurentian Channel. In addition to the 1929 earthquake and its immediate aftershocks, there have been four M5 earthquakes since 1951, the most recent in 1975. The elongation of the seismic zone and the location of the 1929

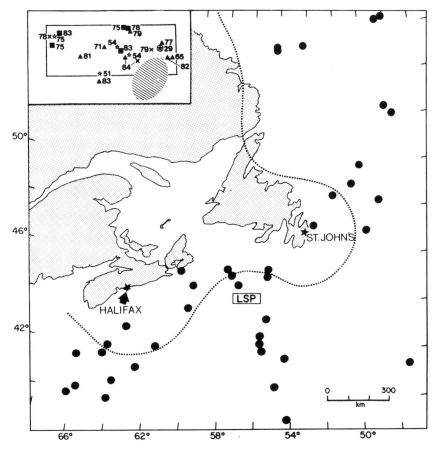

FIGURE 4. Seismicity of the southeastern margin. Magnitudes of the earthquakes (*dots*) are not distinguished. Inland of the *dashed line,* M3.5 and larger earthquakes are thought to have been completely detected and located since 1983. Box marked LSP encloses the many earthquakes near the epicenter of the 1929 "Grand Banks" earthquake. *Inset* shows in detail the seismicity of the box with the number identifying the year of the earthquake and the symbol the magnitude as in Figure 1. The *hatched ellipse* indicates the approximate location of two microearthquakes recorded by ocean-bottom seismometers in 1983. (After Adams.[15])

epicenter at the eastern end are consistent with the hypothesis that current earthquakes could represent belated aftershocks of the 1929 earthquake. If so, the earthquake appears to have ruptured westwards along a fault about 70 km long.[15] A dissenting view is given by Hasegawa and Kanamori.[16]

Outside of the Laurentian Slope zone, scattered earthquakes occur northeast of Newfoundland and seaward of Nova Scotia. A trend of seismicity follows the Laurentian Fan south from the Laurentian Slope zone. Off Nova Scotia the earthquakes represented in FIGURE 4 suggest that the transition from oceanic to continental crust is active, although not currently at a very high level. Elsewhere, the transition is too poorly monitored to show whether it is active at the low level detectable off Nova Scotia.

The cause of earthquakes along the eastern margin, and thus a rationale for their distribution, has not yet been established. Studies of stress directions from oil-well breakout data[17] confirm that the margin is subject to the same northeast-directed compression as is the rest of eastern North America. While some studies have indicated local areas of near-surface faulting (e.g., Orpheus Graben off Nova Scotia), many of the earthquakes along the margin probably occur along the ocean-continent transition on the deep-crustal rift faults formed during the opening of the Atlantic Ocean. Under the current regime these normal faults would be reactivated as thrust or strike-slip faults.

Because the numbers, location, and nature of the offshore earthquakes are poorly understood, Basham and Adams[18] produced a speculative model that suggests that rare large earthquakes can occur along the whole margin, using the pervasive Mesozoic rift faults as the causative structure, with a rate of about one M7 earthquake per thousand years per thousand kilometers of margin. This model implies that inactive parts of the margin are merely quiescent, and that recent earthquakes in active regions like the Laurentian Slope represent belated aftershocks that will diminish over the next century without producing another M7 earthquake. That other such large earthquakes may have occurred in the past is suggested by prehistoric submarine slumps elsewhere along the margin.

Southeastern Background Seismicity

Some scattered seismicity in southeastern Canada lies outside of the seismic zones discussed above, and so is briefly mentioned, once again moving from west to east (FIG. 6). The shield areas of Ontario and Quebec show very low seismicity except for earthquakes near Cochrane and Iroquois Falls in northern Ontario, which may lie on an extension of the Ottawa-Timiskaming rift structure towards Kapuskasing. Further north, the earthquakes near southern James Bay have an unknown origin. A cluster of small earthquakes in the Burlington/Niagara Falls area of Ontario is poorly understood, but in part is believed to represent very shallow stress release. Earthquakes near Quebec City may lie on the St. Lawrence rift system. Earthquakes extending from Sept-Iles across easternmost Quebec and southern Labrador may also lie on structures related to the St. Lawrence rift system.

Northeastern Canada

Although the seismicity of the northeastern part of Canada (FIG. 5) has no direct impact on United States territory, it is highly relevant to the understanding of the seismotectonics of eastern North America as a whole. The following summary is greatly condensed from Adams and Basham.[1,2] In the *Labrador Sea,* earthquakes can be associated with the extinct, central spreading ridge and its associated fracture zones, and there is a separate trend of seismicity following the continental margin off Labrador that is likely associated with pre-existing faults beneath the rifted continental margin.

There seems to be a clear separation between the activity in *Baffin Bay* and that on *Baffin Island.* Although there is evidence for seafloor spreading and an extinct spreading center in the deep central region of Baffin Bay, there is little or

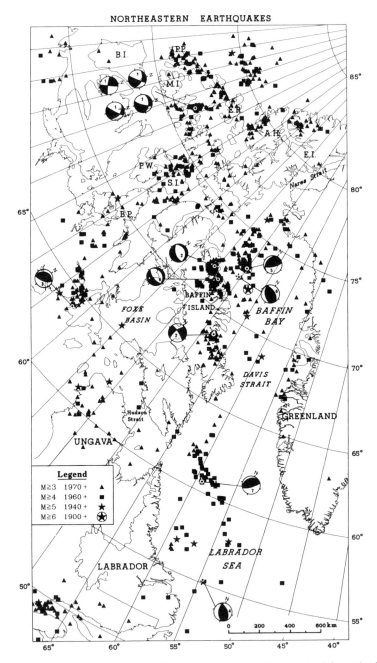

FIGURE 5. Seismicity of northeastern Canada (legend shows the magnitude/age criteria for the earthquakes shown), with all available focal mechanisms.[1,2] Geographic features are identified by initials as follows: AH: Axel Heiberg Island; BI: Banks Island; BP: Boothia Peninsula; EI: Ellesmere Island; ER: Ellef Ringes Island; LS: Lancaster Sound; MI: Melville Island; PP: Prince Patrick Island; PW: Prince of Wales Island; SI: Somerset Island.

no seismic activity in this region and the seismicity is confined along the continental margin. The highly active seismicity on Baffin Island appears to be confined to the coastal region. All available evidence on focal depths suggests that the earthquakes are shallow, and in contrast to the thrust faulting in Baffin Bay, the island earthquakes show normal faulting, which is rare for eastern Canada. In *Boothia-Ungava*, there is an arcuate band of seismicity that extends southward from the Boothia Peninsula, across northern Hudson Bay, the Ungava Peninsula, and eastward through Hudson Strait. Postglacial unloading may be causing reactivation of old faults. Most of the remaining seismicity in the Canadian Arctic archipelago can be spatially associated with the *Sverdrup Basin*, a 1000-km-long northeast-southwest regional depression in which the sedimentary rocks reach thicknesses of 10 km. Along the *Arctic Ocean Margin,* the seismicity is concentrated in distinct clusters in the Beaufort Sea and northwest of Ellef Ringnes Island, with only very scattered activity elsewhere. The rifted margin was formed in early Cretaceous time when northern Alaska rotated anticlockwise away from Arctic Canada. The earthquakes appear to occur where gravity anomalies suggest that thick sediment accumulations have loaded the ocean-continent transition.

DISCUSSION

A useful summary of the current state of knowledge regarding the causes of seismicity in eastern Canada is given by FIGURE 6. Almost all the significant earthquakes in the continental part of southeastern Canada can be spatially and probably causally associated with the Paleozoic rift system along the St. Lawrence, the chief exceptions being the northern band of seismicity of western Quebec (probably related to the hotspot), and the seismicity in the Appalachians. To a first approximation, the continent-wide stress field is uniform, represents compression from the east-northeast quadrant, and causes thrust or thrust/strike-slip earthquakes.

The earthquakes along the eastern continental margin are mainly related to the Mesozoic rift faults formed during the opening of the Atlantic, while those in the mid Labrador Sea are related to the extinct spreading ridge and its transform faults. The same structures extend into northern Baffin Bay where the earthquakes may also be partially related to stresses induced by deglaciation. The passive Arctic Ocean margin has a comparable ocean-continent transition to the Atlantic margin but it appears to be seismically active mainly where it has been recently loaded by thick sediments.

It is therefore clear from the above discussion and FIGURE 6 that most of the larger earthquakes can be associated with the rift systems that surround or break the integrity of the North American craton. By contrast, the largest historical earthquakes in the unbroken craton (in Canada, but outside the seismic zones discussed above) probably have not been much larger than M5.

Coppersmith *et al.*[19] have come to similar conclusions from a study of worldwide earthquakes in "stable continental interiors." They found that 71 percent of the seismicity of stable continental interiors was associated with imbedded continental rifts and continental passive margins (one-sided rifts). Further, all of their 17 M7 or larger earthquakes are strongly associated with the imbedded rifts or passive margins.

Both the Paleozoic and the Mesozoic rift systems in eastern Canada are continuous features, many thousands of kilometers long. Despite their continuity and

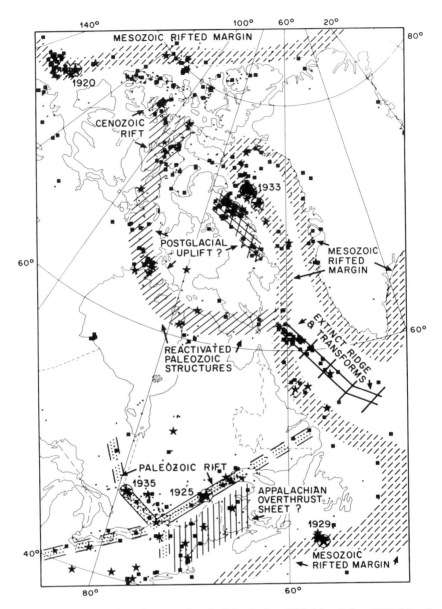

FIGURE 6. Earthquakes of eastern Canada (M > 3 since 1970; M > 4 since 1960; M > 5 since 1940; M > 6 since 1900) together with an interpretative framework for the cause of the seismicity.[2]

the uniform stress field, the rift systems are only sporadically active, showing seismicity in clusters—as at Charlevoix and the lower St. Lawrence—or single large earthquakes, such as the "Grand Banks" earthquake.

In each case there is a lack of geological evidence for a long period of prehistorical activity at present rates. I, like many others, have been confounded by the high levels of seismic activity at Charlevoix relative to other places in eastern Canada. At such high rates (M7 every few hundred years), the implied rates of geological deformation would amount to kilometers over a million years. Clearly kilometers of uplift (if due to thrusting), or even kilometers of strike-slip motion, would have been recognized at Charlevoix, had they occurred. That they have not, is probably due to intermittent activity at Charlevoix and along the remaining rift system, perhaps with a time constant of thousands to hundreds of thousands of years.

For single large earthquakes such as the "Grand Banks," a similar argument can be made for intermittent activity on comparable time scales.[18] This leads to the classic dilemma for hazard estimates that need to be made for low levels of probability: will the next large earthquake occur in a recognized seismic zone or not?

SEISMIC HAZARD AND SEISMIC RISK IMPLICATIONS

It is now 40 years since the last large, damaging earthquake in southeastern Canada (Cornwall, 1944, M5.6, damage $15,000,000 in 1987 dollars). In recent times, the 1982 Miramichi earthquakes (largest shock M5.7) have provided many seismological "surprises" (notably the swarm-like nature of the four M5 mainshocks, the large number of aftershocks, and the $1/2g$ accelerations recorded at very high frequencies), but were not really large enough to produce substantial surface faulting or to allow a definitive relationship to geological observations. Neither, fortunately, did they occur near enough to any large towns to cause any significant damage. It is clear, however, that had they occurred close to a metropolitan area there would have been considerable superficial damage (such as cracked plaster) and the possibility of some structural failure or collapse of older masonry buildings in poor condition, with attendant fatalities.

The Miramichi earthquakes were too small and too remote to have had much effect on the popular perception of earthquake risk in Canada. There is still a feeling among the general public that earthquakes are confined to western Canada (or, indeed, to California). In fact, eastern Canadian earthquakes exceed M7, shake a much larger area because of the lower attenuation in the east, and affect many more people because of the higher population density. FIGURE 7 makes a crude attempt to place earthquake *risk* in Canada into perspective. The model used multiplies the annual probability of potentially damaging ground motion (taken from the current National Building Code of Canada model for seismic hazard estimation[20] and so including the effects of weaker eastern attenuation) by the population affected (as an alias for value of structures and potential for injury) and expresses the product at each place as a percentage of the Canadian total. The pie chart shows graphically what the historical record of earthquakes has demonstrated: life in the east has an associated earthquake risk. It would be naive to suppose that the 40-year absence of damaging earthquakes in southeastern Canada should influence our expectations for the next few decades.

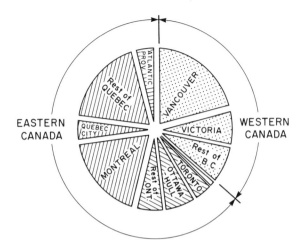

FIGURE 7. Pie chart showing relative levels of seismic risk in eastern and western Canada (Pittman and Adams, unpublished work, 1987).

SUMMARY

Eastern Canada—Canada east of the Rockies, and extending north from the United States border to the Arctic Ocean—comprises about two-thirds of the stable craton of the North American plate. Much of this large area appears to be substantially aseismic, although it contains several zones of intense seismicity. Earthquakes in eastern Canada appear to be occurring within a stress field dominated by northeast-to-east compression, and most large earthquakes have occurred near Paleozoic or younger rifts or along rifted continental margins, that is, in places where the continent has been most recently weakened.

Large or damaging earthquakes in eastern Canada during the first half of the century (1925 M7.0 Charlevoix, 1929 M7.2 Grand Banks, 1933 M7.3 Baffin Bay, 1935 M6.2 Timiskaming, 1944 M5.6 Cornwall) provide a recent history that is a useful complement to the absence of such experience in the eastern United States. Much is now being learned about the seismic zones and the nature of the seismicity through the intensive study of recent smaller earthquakes. However, seismological studies, earthquake engineering practice, and emergency-preparedness planning are all hampered by the lack of large earthquakes during the memory of most people in the eastern United States and Canada.

ACKNOWLEDGMENTS

Concepts discussed in this paper were developed with Peter Basham during the writing of our joint contribution to the "Decade of North American Geology" series.[2] Garry Rogers and George McMechan, in 1978, performed a similar calculation to that producing FIGURE 7 and obtained similar results. I thank my colleagues in the Geophysics Division for their comments on this paper.

REFERENCES

1. ADAMS, J. & P. W. BASHAM. 1987. Seismicity, crustal stresses and seismotectonics of eastern Canada. *In* Proceedings from the NCEER Symposium on Seismic Hazards, Ground Motions, Soil-Liquefaction and Engineering Practice in Eastern North America. K. H. Jacob, Ed.: 127–142. Technical Report NCEER-87-0025. National Center for Earthquake Engineering Research. Buffalo, NY.
2. ADAMS, J. & P. W. BASHAM. The seismicity and seismotectonics of eastern Canada. *In* Neotectonics of North America. D. B. Slemmons, E. R. Engdahl, D. Blackwell, D. Schwartz & M. Zoback, Eds. Geological Society of America "Decade of North American Geology" series, Vol. CSMV-1.
3. HASEGAWA, H. S., J. ADAMS & K. YAMAZAKI. 1985. Upper crustal stresses and vertical stress migration in eastern Canada. J. Geophys. Res. **90:** 3637–3648.
4. ADAMS, J. 1987. Canadian crustal stress data—a compilation to 1987. Geol. Surv. Canada Open File **1622:** 1–130.
5. FORSYTH, D. A. 1981. Characteristics of the western Quebec seismic zone. Can. J. Earth Sci. **18:** 103–119.
6. KUMARAPELI, P. S. 1985. Vestiges of Iapetan rifting in the west of the northern Appalachains. Geosci. Can. **12:** 55–59.
7. ANGLIN, F. M. 1984. Seismicity and faulting in the Charlevoix zone of the St. Lawrence Valley. Bull. Seismol. Soc. Am. **74:** 595–603.
8. SMITH, W. E. T. 1962. Earthquakes of eastern Canada and adjacent areas 1534–1927. Publ. Dom. Obs. Ottawa **26:** 271–301.
9. SMITH, W. E. T. 1966. Earthquakes of eastern Canada and adjacent areas 1928–1959. Publ. Dom. Obs. Ottawa **32:** 87–121.
10. ADAMS, J., J. SHARP & M. C. STAGG. 1988. New earthquake focal mechanisms for southeastern Canada. Geol. Surv. Canada Open File **1892:** 1–109.
11. LEBLANC, G. & K. B. S. BURKE. 1985. Re-evaluation of the 1817, 1855, 1869, and 1904 Maine-New Brunswick area earthquakes. Earthquake Notes **56:** 107–123.
12. WETMILLER, R. J., J. ADAMS, F. M. ANGLIN, H. S. HASEGAWA, & A. E. STEVENS. 1984. Aftershock sequences of the 1982 Miramichi, New Brunswick, earthquakes. Bull. Seismol. Soc. Am. **74:** 621–653.
13. BASHAM, P. W. & J. ADAMS. 1984. The Miramichi, New Brunswick earthquakes: Near-surface thrust faulting in the northern Appalachians. Geosci. Can. **11:** 115–121.
14. HASEGAWA, H. S. 1986. Seismotectonics in eastern Canada, an overview with emphasis on the Charlevoix and Miramichi regions, Earthquake Notes **57:** 83–94.
15. ADAMS, J. 1986. Changing assessment of seismic hazard along the southeastern Canadian margin. *In* Proceedings of the Third Canadian Conference on Marine Geotechnical Engineering, St. John's Newfoundland, 11–13 June, 1986: 41–53.
16. HASEGAWA, H. S. & H. KANAMORI. 1987. Source mechanisms of the magnitude 7.2 Grand Banks Earthquake of November 1929: Double couple or submarine landslide? Bull. Seismol. Soc. Am. **77:** 1984–2004.
17. PODROUZEK A. J. & J. S. BELL. 1985. Stress orientations from wellbore breakouts on the Scotian Shelf, eastern Canada. Geol. Surv. Canada Paper **85-1B:** 59–62.
18. BASHAM, P. W. & J. ADAMS. 1983. Earthquakes on the continental margin of eastern Canada: Need future large events be confined to the locations of large historical events? U. S. Geol. Surv. Open File Rep. **83-843:** 456–467.
19. COPPERSMITH, K. J., A. C. JOHNSTON & W. J. ARABASZ. 1987. Methods for assessing maximum earthquakes in the central and eastern United States. E.P.R.I. Research Project **2556-12,** Working Report, Electric Power Research Institute, Palo Alto, CA.
20. BASHAM, P. W., D. H. WEICHERT, F. M. ANGLIN & M. J. BERRY. 1982. New probabilistic strong seismic ground motion maps of Canada: A compilation of earthquake source zones, methods and results: Earth Physics Branch Open File **82–33:** 1–202.

In Situ Stress, Crustal Strain, and Seismic Hazard Assessment in Eastern North America

MARK D. ZOBACK

Department of Geophysics
Stanford University
Stanford, California 94305

MARY LOU ZOBACK

U.S. Geological Survey
Menlo Park, California 94025

INTRODUCTION

During historic time earthquakes have occurred in essentially every state of the central and eastern U.S. and throughout much of central and eastern Canada. Yet little is known about the stresses that cause these earthquakes, the faults the earthquake occur upon, or the rates of stress or strain accumulation. Recurrence rates for major earthquakes in central and eastern North America are several times as long as the historic record[1,2] and it is an unresolved question whether it is possible to evaluate seismic hazard from seismicity rates in the absence of a physical understanding of the earthquake generation process. This paper summarizes several efforts aimed at improving the physical understanding of earthquakes that occur in this region.

STATE OF STRESS IN THE CENTRAL AND EASTERN UNITED STATES

FIGURE 1 shows a generalized stress map of the U.S. and adjacent Canada. FIGURE 2 shows the quality-ranked data from which FIGURE 1 was derived (both are from Zoback and Zoback[3]). The data come from in situ stress measurements at depth, well-resolved earthquake focal mechanisms, stress-induced wellbore breakouts, and observations of fault slip. The latter are sites where evidence of young faulting can be used to estimate stress directions because both the dip-slip and strike-slip components of motion can be constrained. With the exception of the Texas-Louisiana Gulf Coast and continental margin, the state of stress in the central and eastern U.S. is compressive with strike-slip to reverse faulting observed essentially everywhere. Thus,

$$S_{H\max} > S_{H\min} \cong S_v$$

The direction of maximum horizontal compression is generally ENE. This broad-scale and uniform stress field appears to be of plate-wide tectonic origin and is consistent with the theoretical directions of *push* from the mid-Atlantic ridge[4] and absolute plate motion (these directions are essentially the same). On a plate-

54

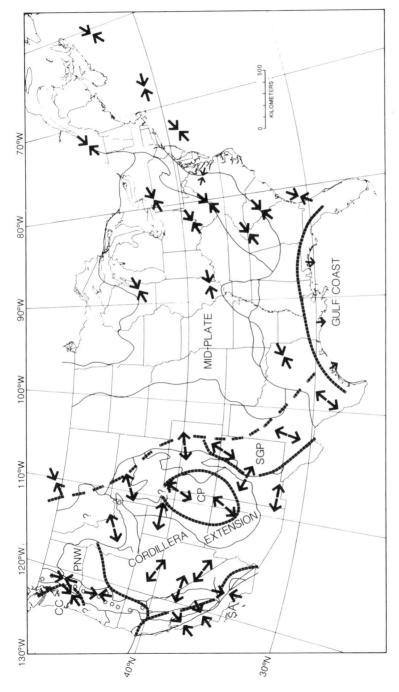

FIGURE 1. Generalized stress map of the United States. *Inward-pointed arrows* represent direction of maximum horizontal principal stress in areas of crustal compression (reverse and strike-slip faulting areas) and *outward-pointed arrows* indicate direction of least horizontal principal stress in areas of crustal extension (normal and strike-slip faulting areas). (From Zoback and Zoback.[3] Reprinted by permission.)

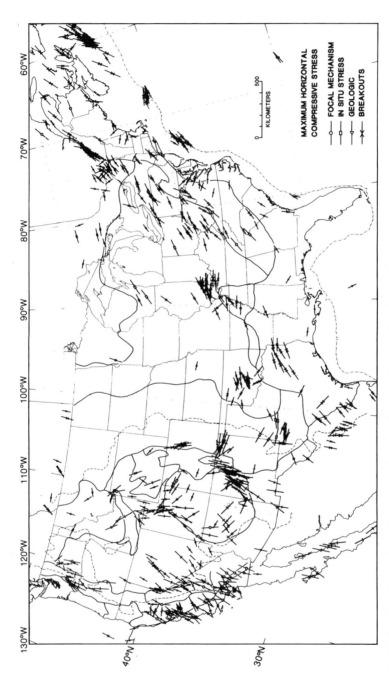

FIGURE 2. Map of directions of maximum horizontal principal stress direction gathered from a variety of sources. The different *symbols* represent the type of data and the lengths of the *arrows* indicate the data quality. (From Zoback and Zoback.[3] Reprinted by permission.)

wide scale this stress field also seems to characterize most of central and eastern Canada[5] and the western Atlantic ocean.[6]

Areas where the orientation of the stress field varies in the central and eastern U.S. are also observed. Fault slip data in the mid-Atlantic states indicates a WNW direction of maximum horizontal stress. The reason for this is not understood and there are no contemporary stress indicators in the region to confirm whether or not this stress field is still active. Earthquake focal mechanisms in New England (FIGURE 3, as summarized by Gephart and Forsythe[7]) show considerable variability in the orientations of the nodal planes and the P axes (see also Pulli and Toksoz[8]). This can be explained either in terms of true variations in stress orientation from area to area or that slip is occurring on pre-existing planes of weakness in a relatively uniform stress field.

Gephart and Forsythe preferred the latter possibility and suggested that the earthquakes represent slip on variably oriented pre-existing zones of weakness and the observed slip could form a generally uniform ENE compressive stress field. It is important to note that a "pre-existing planes of weakness" is not simply a pre-existing fault. In terms of classical Andersonian faulting theory and the Mohr-Coulomb failure criterion,[9,10] the shear stresses required to move pre-existing faults using laboratory-derived frictional coefficients[11,12] are quite high at mid-crustal depth (on the order of a kilobar) and much greater than the stress drops of crustal earthquakes (which are on the order of 100 bars). Thus, in a classical sense, faulting is likely to occur only on those faults that are optimally oriented to

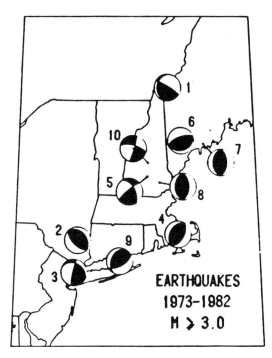

FIGURE 3. Relatively high quality earthquake focal mechanisms in New England. (From Gephart and Forsyth.[7] Reprinted by permission.)

the stress field. The "plane of weakness" argument is that some faults in nature are significantly weaker than laboratory friction experiments would suggest and that faulting can occur on nearly any "weak" plane, even those striking nearly 90° to the direction of the maximum principal stress. Zoback and others[13] and Mount and Suppe[14] present evidence that this is essentially the case along much of the San Andreas fault and that the San Andreas is a true "plane of weakness." Nevertheless, for a number of reasons discussed below, we believe that the most logical explanation of the variation of earthquake focal plane mechanisms is that it is due to local variations of the *in situ* stress field and not due to the presence of anomalously weak faults. For example, an approximately 40° change in the direction of maximum principal stress is observed between sites of borehole stress measurements in southern New York and central Connecticut.

INTRAPLATE SEISMIC ZONES AND CRUSTAL WEAKNESS

As mentioned above, there are a number of reasons why intraplate seismic zones do not appear to represent zones of anomalously low strength in the crust. First, unlike plate boundaries, stresses must be high in intraplate areas to support topography and flexure.[15] Also, about twenty cases of *in situ* stress measurements in intraplate areas around the world show high stress differences consistent with Mohr-Coulomb failure and "Byerlee" friction.[11,12,16,17] While these measurements are relatively shallow (primarily in the upper 2 km), why have none of the measurements ever detected any "weak" crust?

In seismically active areas of the eastern U.S., *in situ* stress measurements have been conducted at a number of sites using the hydraulic fracturing technique. Measurements have been made in two ~1-km-deep wells at Monticello South Carolina,[18] three 300–400-m-deep holes in northeastern South Carolina,[19] in a 1.5-km-deep hole at Auburn in west central New York,[20] in a 1-km-deep hole in southeastern New York,[21] and in a 1.5-km-deep borehole in central Connecticut (M. D. Zoback and others, unpublished data). In each of these cases (except Auburn), the measurements were made in crystalline rock and in each of these cases (except Auburn), the shallow stress magnitudes are sufficiently high to cause reverse faulting on optimally oriented planes (see FIGURE 4 from the southeastern New York site at Kent Cliffs), consistent with "classical" faulting theory as outlined above.

The most important question about these data is: how deep do the critical stress magnitudes persist? In many cases, and at many other sites around the world,[12] the critical stress magnitudes are generally restricted to shallow depths, shear stress does not increase markedly with depth below several hundred meters, and the deeper stress measurements do not imply incipient faulting. At Monticello, for example, both the stress measurements and the focal depths of the earthquakes indicate that the majority of events are in the upper 1 km. Since the stress orientations in these cases are consistent with regional orientations (Monticello is the data point in the middle of South Carolina), it seems as if the near-surface horizontal stresses are either locally amplified, possibly by a process such as denudation,[22] or that the horizontal stresses at shallow depth only seem to be high because the vertical stress is close to zero. In other words, a difference between S_{Hmax} and S_v of 20–30 MPa might be relatively insignificant at mid-crustal depth, but sufficient to cause reverse faulting in the upper few hundred meters. This means that it is possible that some of the widely distributed small-magnitude seismicity might be of shallow origin and not particularly significant with respect

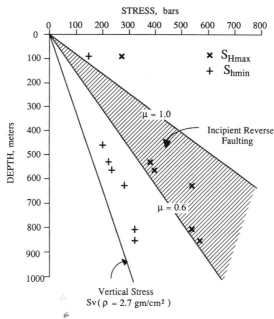

FIGURE 4. *In situ* stress measurements in the Kent Cliffs borehole. Different *symbols* represent least and greatest horizontal principal stresses and the *hachured area* indicates values for which the maximum principal stress is sufficiently greater than the vertical stress that well-oriented reverse faults would be potentially active. From Zoback *et al.*[21] Reprinted by permission.)

to estimation of earthquake hazard associated with much larger magnitude events at greater depths.

The second reason why intraplate seismic zones do not appear to be zones of anomalously low strength is that there appears to be no viable mechanism for generating weak zones within intraplate areas. Soft-inclusion (alkalic or serpentinized intrusive) models[23–25] do not hold up to careful correlations between seismicity and geology. Unlike plate boundaries, which seem to be bounded by very low strength fault zones,[13,14] total displacements along intraplate seismic zones are quite small (which seems to indicate that no pronounced zones of fault gouge would develop) and deformation rates are too low to call on pore fluid overpressure to lower the shear strength of faults as occurs along plate boundaries.[26]

Third, if a relatively small number of major weak zones did exist, it would seem that the magnitude of the regionally applied tectonic stresses would be relatively low and the intraplate deformation would be concentrated in the weak zones. Yet the regional tectonic stresses appear to be of sufficient magnitude over the entire plate that seismicity seems to occurs nearly everywhere within the North American plate. If one argues that all the earthquakes are occurring on faults that are weak, the first point above is violated.

Fourth, at New Madrid, Charleston and the Meers fault in Oklahoma there seems to be evidence for recurrent seismic activity on the order of 1000 years.[1,2] Simply allowing failure to occur at a lower stress level does not account for the fact that stress must reaccumulate for another earthquake in only about 1000 years. In other words, the rate of stress (strain) accumulation must be anomalously high in these areas to explain the recurrence data. The stress level at which the faults fail (degree of weakness) is basically irrelevant to the issue of recurrence.

Finally, in the recent paper related to the lack of frictional strength of the San Andreas fault,[13] we present evidence that the direction of maximum horizontal compression in western California is essentially perpendicular to the San Andreas fault (FIG. 5). We argue that because the San Andreas is so much weaker than the

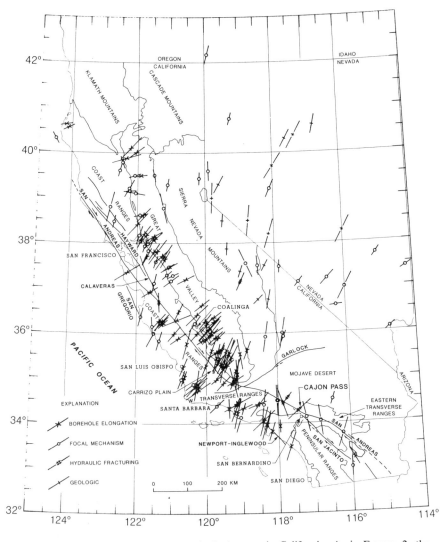

FIGURE 5. Directions of maximum principal stress in California. As in FIGURE 2, the length of the *arrows* indicates the data quality and the *symbol* indicates the type of data. (From Zoback and Zoback.[13] Reprinted by permission.)

surrounding crust, the horizontal principal stresses must be reoriented near the fault so as to minimize the shear stress on the fault. In other words, the direction of maximum compression near the fault must be either nearly perpendicular to the fault (as is currently observed in central California) or parallel to the fault (as probably existed in central California prior to a plate motion change 4 million years ago).

This argument relates directly to the issue of whether intraplate seismicity occurs along weak faults. If the fault zones along which the Charleston and New Madrid earthquakes occurred were extremely weak, one would expect the stress orientations in the epicentral zone to differ from the regional stress directions by as much as 40°. In other words, if the shear stresses in the generally "strong" crust are sufficiently high as to cause regional seismicity, the principal stresses must rotate so as to result in relatively little shear stress on the "weak" fault planes, just as along the San Andreas. The stress data shown in FIGURE 2 do not show any significant stress rotation in these areas. In fact, the stress orientation in the Charleston area is remarkably similar to that in the rest of South Carolina (as is the comparison between the stress orientation in the New Madrid seismic zone with that in the Illinois basin just to the north). In many other areas the data are regrettably sparse and it is difficult to ascertain whether significant stress orientation changes occur, and if so, why. There are many possible explanations for variations in stress orientations, some of which may be tectonic but many of which are data-related (for example, a P-axis is only an approximation of an $S_{H\max}$ direction and breakouts can be misinterpreted on four-arm caliper surveys). Nevertheless, while there are numerous minor variations in stress orientation seen in the central and eastern U.S., there does not seem to be evidence of major changes in stress orientation that would indicate the presence of major zones of crustal weakness. The consistent regional stress patterns in areas of known major earthquakes also argues strongly against major localized sources of stress being responsible for those earthquakes.

CRUSTAL STRAIN LOCALIZATION

Zoback and others[27] proposed the hypothesis that intraplate seismic zones capable of producing major crustal earthquakes on a time scale on the order of 1000 years must be areas where localized ductile strain occurs in the mid- to lower crust and high rates of elastic stress accumulate in the mid- to upper crust. They further argued that if such strain was occurring, it could conceivably produce detectable levels of geodetic strain, just as along the San Andreas fault. In a test of this hypothesis, Zoback and others[27] presented an analysis of repeated triangulation measurements in southern New York and western Connecticut that showed anomalously high strain rates, whereas the surrounding region did not (FIG. 6). Two important aspects of this analysis are that the two areas of anomalously high strain rate (New York–Connecticut and western Long Island) show similar directions of shear strain (although the rates were different) and when each of the two areas of anomalous strain rate were broken up into separate subsets of completely independent data, each subset of data reproduced the observed strain rate of the initial analysis. In other words, similar results were obtained in the analysis of four separate subsets of the data.

Nevertheless, the data used in the study by Zoback et al.[27] were poorly distributed in both space and time and the statistical significance of the finding of anoma-

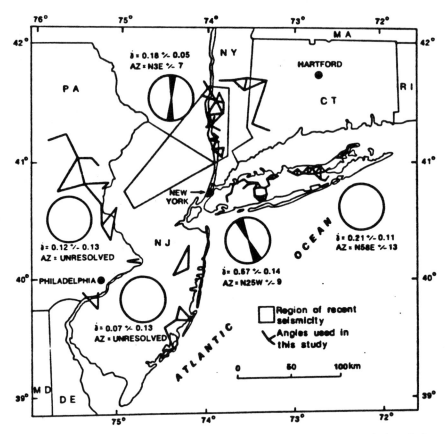

FIGURE 6. Five areas where shear strain has been computed from repeated triangulation measurements (angles are shown by *solid lines*). Only in the southern New York/western Connecticut and western Long Island areas were significant amounts of strain detected. In both areas the planes on which maximum shear occurs strike approximately N-S. (From Zoback *et al.*[21] Reprinted by permission.).

lous crustal deformation in southern New York and Connecticut was challenged by Snay.[28] In response to Snay's arguments, Prescott and others[29] pointed out that Snay's interpretation of the data was overly conservative, especially since Snay's analysis largely just replicated results of Zoback *et al.*[27] Even though Snay deleted a couple of the data points used in the first study and redefined the standard error upward, Prescott and others[29] show that a strain rate of 0.1 to 0.2 μrad/yr on north-south trending planes is still observed and the size of the signal exceeds two standard errors. Snay argued that such strain rates are not statistically significant, whereas Prescott and others[29] disagreed. Clearly, the best way to resolve this controversy is to obtain new measurements in this region aimed at understanding whether the surprisingly large strain rates that were measured are valid.

SEISMIC HAZARD ASSESSMENT IN INTRAPLATE AREA

The fact that the Charleston, South Carolina area is neither geologically nor geophysically unique and is currently characterized by very low-level seismicity is sufficient demonstration of the difficulty of assessing seismic hazard in intraplate areas. Had the 1886 Charleston earthquake not occurred during historic time, there would really be no reason to suspect it as a site of a potential major earthquake. Because of this, and equally complicated cases such as the Meers fault (an apparent site of major earthquakes in recent prehistoric time but which is currently aseismic), it is clearly uncertain whether seismic hazard can be adequately assessed simply utilizing seismicity data and first-order geologic and geophysical discriminants.

In our opinion, available evidence suggests that it is not reasonable to assume that current seismicity rates can be used to characterize long-term seismic hazard in intraplate areas. The two examples cited above would seem to be sufficient evidence to make this case. And while it would seem that use of seismicity rates for hazard assessment might tend to always lead to an underestimate of seismic hazard areas where unknown or inactive faults are potentially active, when appreciable microseismicity occurs at shallow depth, the opposite could happen. At several sites in eastern North America the shallow crust can be the source of numerous small-magnitude earthquakes completely restricted to shallow depth and potentially unrelated to the potential for larger earthquakes.

We suggest that determination of intraplate strain rates in areas of known large historic earthquakes may indicate sites of potential major earthquakes. In many areas there is excellent historic triangulation data that often go back into the nineteenth century and represent a valuable reference network for determination of crustal strain. With the aid of modern instrumentation such as Global Positioning System satellite receivers, reoccupation of the old triangulation networks may yield reliable estimates of the rates of crustal deformation. Otherwise, simply resurveying intraplate areas with modern and accurate geodetic instruments may eventually lead to detectable levels of crustal strain. In either case, the available evidence of relatively short periods of earthquake recurrence indicates that mapping the rate of strain energy accumulation in central and eastern North America may be one of the best ways to represent an accurate picture of long-term seismic hazard.

SUMMARY

Compilation of tectonic stress orientation data indicate a generally uniform ENE maximum horizontal stress direction over most of central and eastern North America and uniform relative stress magnitudes. Because the observed directions of maximum principal stress are consistent with predicted stress directions resulting from plate motion, the data suggest that the dominant stress field responsible for large earthquakes in eastern North America is of regional tectonic origin and not due to local sources. Thus, the localization of major earthquakes is primarily controlled by the occurrence of major faults and not local sources of stress. Detailed *in situ* stress measurements at shallow depth indicate that very high horizontal stresses exist in the upper ~1 km in many parts of eastern North America. These large, shallow stresses may be responsible for much widespread, low-magnitude seismicity. The occurrence of such seismicity may have nothing to do with the potential for larger earthquakes to occur and in many cases should be excluded from estimates of seismic-hazard-based earthquake magnitude-frequency relations. Finally, localized zones with high rates of elastic strain accumulation seem to be required to account for the occurrence of repeated intraplate earthquakes over relatively short time periods. Analysis of historic triangulation data shows evidence for high rates of strain accumulation in southern New York. However, as the triangulation data are sparsely distributed in space and time, further measurements of crustal deformation in this area are extremely important to either confirm or modify the implications of the triangulation data.

REFERENCES

1. Russ, D. 1979. Late Holocene faulting and earthquake recurrence in the Reelfoot Lake area, northwestern Tennessee. Geol. Am. Bull. **90:** 1013–1018.
2. Obermeier, S., G. S. Gohn et al. 1985. Geologic evidence for recurrent moderate to large earthquakes near Charleston, South Carolina. Science **227:** 408–410.
3. Zoback, M. L. & M. D. Zoback. 1988. Tectonic stress field of the continental U.S. In Geophysical Framework of the Continental United States. L. Pakiser & W. Mooney, Eds. Geol. Soc. Am. Memoir. In press.
4. Richardson, R. M., S. C. Solomon & N. H. Sleep. 1979. Tectonic stress in the plates. Rev. Geophys. Space Phys. **17:** 981–1019.
5. Adams, J. 1987. Canadian crustal stress data—a compilation to 1987. Geol. Surv. Canada Open File Rep. In press.
6. Zoback, M. L., S. Nishenko, R. Richardson, H. Hasegawa & M. D. Zoback. 1986. Mid-plate stress deformation, and seismicity. In The Geology of North America: Vol. M., The Western North Atlantic Region. Vogt, P. R. & B. Tucholke, Eds.: 297–312. Geology Society of America.
7. Gephart, J. W. & D. D. Forsyth. 1985. On the state of stress in New England as determined from earthquake focal mechanisms. Geology **13:** 70–76.
8. Pulli, J. J. & M. N. Toksoz. 1981. Fault plane solutions for northeastern United States earthquakes. Bull. Seismol. Soc. Am. **71:** 1875–1882.
9. Anderson, E. M. 1951. The Dynamics of Faulting and Dyke Formation with Applications to Britain, 2nd ed. Oliver and Boyd. Edinburgh.
10. Jaeger, J. C. & N. G. W. Cook. 1969. Fundamentals of Rock Mechanics, 2nd ed. Methuen. London.
11. Brace, W. F. & D. L. Kohlstedt. 1980. Limits on lithospheric stress imposed by laboratory experiments. J. Geophys. Res. **85:** 6248–6252.
12. Zoback, M. D. & J. H. Healy. 1984. Friction, faulting, and in-situ stress. Ann. Geophys. **2:** 689–698.
13. Zoback, M. D., M. L. Zoback, V. Mount, J. Suppe, J. Eaton, J. Healy, D. Oppenheimer, P. Reasenberg, L. Jones, C. B. Raleigh, I. Wong, O. Scotti, C. Wentworth. 1987. New evidence on the state of stress of the San Andreas fault system, Science **238:** 237–239.
14. Mount, V. S. & J. Suppe. State of stress near the San Andreas fault: Implications for wrench tectonics. Geology **15:** 1143–1146.
15. Jeffreys, H. 1962. The Earth. Cambridge University Press. Cambridge, England.
16. Batchelor, A. S. & R. J. Pine. 1986. The results of in situ stress determinations by seven methods to depths of 2500 m in the Carnmenellis granite. In Proceedings of the International Symposium on Rock Stress and Rock Stress Measurements, Stockholm, 1–3 September, 1986. CENTEK. Lulea, Sweden.
17. Stephansson, O., P. Sarkka & A. Myrvang. 1986. State of stres in Fennoscandia. In Proceedings of the International Symposium on Rock Stress and Rock Stress Measurements, Stockholm, 1–3 September, 1986. CENTEK. Lulea, Sweden.
18. Zoback, M. D. & S. Hickman. 1982. In situ study of the physical mechanisms controlling induced seismicity at Monticello Reservoir, South Carolina. J. Geophys. Res. **87:** 6959–6974.

19. ZOBACK, M. D., D. MOOS, B. COYLE & R. N. ANDERSON. 1986. In-situ and physical property measurements in Appalachian site survey boreholes. Am. Assoc. Petrol. Geol. Bull. **70**: 666.

20. HICKMAN, S. H., J. H. HEALY & M. D. ZOBACK. 1985. *In situ* stress, natural fracture distribution, and borehole elongation in the Auburn geothermal well, Auburn, New York. J. Geophys. Res. **90**: 5497–5512.

21. ZOBACK, M. D., R. N. ANDERSON & D. MOOS. 1985. In-situ stress measurements in a 1 km-deep well near the Ramapo fault zone. EOS **66**: 363.

22. VOIGHT, B. & B. H. P. ST. PIERRE. 1974. Stress history and rock stress. *In* Proceedings of the Third Congress of the International Society for Rock Mechanics, Denver. **2**: 580–582.

23. SYKES, L. R. 1978. Intraplate seismicity, reactivation of preexisting zones of weakness, alkaline magmatism, and other tectonism postdating continental fragmentation. Rev. Geophys. Space Phys. **16**: 621–688.

24. KANE, M. F. 1977. Correlation of major eastern earthquake centers with mafic/ultramafic basement masses. *In* Studies Related to the Charleston, South Carolina Earthquake of 1886—A preliminary report. U.S. Geol. Surv. Prof. Pap. 1028.

25. McKEOWN, F. A. 1982. Overview and discussion. *In* Investigations of the New Madrid Missouri, Earthquake Region. U.S. Geol. Surv. Prof. Pap. 1236.

26. DAHLEN, F. A., J. SUPPE & D. DAVIS. 1984. Mechanics of fold-and-thrust belts and accretionary wedges: Cohesive coulomb theory. J. Geophys. Res. **89**: 10087–10101.

27. ZOBACK, M. D., W. H. PRESCOTT & S. W. KRUEGER. 1985. Evidence for lower crustal strain localization in southern New York. Nature **317**: 705–707.

28. SNAY, R. A. 1986. Horizontal deformation in New York and Connecticut: Examining contradictory results from geodetic evidence. J. Geophys. Res. **91**: 12695–12702.

29. PRESCOTT, W. H., M. D. ZOBACK & S. W. KRUEGER. 1987. Comment on paper by Snay.[28] J. Geophys. Res. **92**: 2805.

General Discussion: Part I

ROBERT M. HAMILTON, CHAIRMAN (*United States Geological Survey, Reston, Virginia*): Before opening the floor to discussion, I would like to make a few summary remarks.

The ultimate cause of earthquakes in the eastern United States is plate tectonics. We've heard that argued about, but I think it's fairly generally agreed upon. And the plate tectonics, of course, derives from the movements within the earth, primarily from convection of hot materials deep within the earth. Most of the deformation of the earth's outer shell occurs where the plates collide or scrape past each other, but that's a problem for Alaska and California; it's not our problem here.

The eastern United States lies within the North American plate, which is a relatively stable area where much of the tectonic action took place well over a hundred million years ago. We've heard that such action is still going on, however, and when it does happen, because the crust is older here and because seismic waves travel further, the damage occurs over a larger area. Moreover, as we have also heard, the population density is great in the East in areas where earthquakes could occur, so the vulnerability is disproportionately greater.

The challenge we face here in the East is to translate what we know about the seismicity and tectonics into information that we can use for mitigating earthquake effects. Many of these mitigatory actions will be dealt with in the papers that follow this session, but right now there is an opportunity for us to further consider and argue some of the points that have been raised in this first session.

Anticipating that somebody might not ask a question, I decided to ask one myself. The truism was mentioned in an introductory comment that future earthquakes generally occur where ones have occurred in the past. And although there is some truth in this, I would like to argue that it's more false than true.

If one rattles off a list of past earthquakes, at least ones that have happened either in my career or earlier ones that I studied, you get a list like this: Bakersfield 1952, Long Beach 1933, Santa Barbara 1925, San Fernando 1971, Borrego Mountain 1968, Oroville, Coalinga, Superstition Hills, and looking further back, Charleston, New Madrid, San Francisco, and Fort Tejon. Now, except for the St. Lawrence Valley area, where there seems to have been a fairly persistent sequence of earthquakes, and except for a few areas where smaller earthquakes recur at a fairly good rate, I would argue that the truism isn't true.

To initiate discussion I will ask: can one base seismic risk assessments on past earthquake activity?

· ROBERT KETTER (*NCEER, SUNY at Buffalo, Buffalo, N.Y.*): I guess I have to explain why I put that statement in my comments. First, it is what I have been taught by seismologists. It would be rather difficult for anyone to say that you do not, in fact, have many earthquakes along the San Andreas fault. You do have them! They do not occur in a nice sequential geographic fashion or occur one right after the other. There may be gaps, or quiet periods, but the quakes do occur there. Your own illustration, shown earlier, shows a large concentration of events in the Mississippi Valley. These events may not have occurred sequentially or chronologically in any recognized order, but there they do occur. While I listed this as the first truism based on what geologists and seismologists had taught me— I ended with a statement saying: but don't count on it. Earthquakes do occur

66

where you least expect them. You'd better not be complacent! If you haven't had one, you should not presume that you will not have one.

ED KAVAZANJIAN (*Parsons, Brinckerhoff, Quade & Douglas, New York, N.Y.*): At one time there was some speculation that a triggering mechanism for the earthquakes around New Madrid might be the sediment load from the Mississippi River. Is this still a viable mechanism? If so, does it make sense to consider New Madrid in an East Coast hazard analysis if it can be attributed to a special geologic circumstance?

HAMILTON: I don't tend to think that the sediment load is the cause of the triggering. For one thing there are some very thick sections of sediment elsewhere in that vicinity, so why aren't the earthquakes there? In the area northwest of Memphis, Tennessee, seismic reflection data show perhaps 20,000 feet or maybe even more—perhaps 30,000 feet—of late Proterozoic and Cambrian sediment. So, in terms of sediment load, although they are old sediments, other parts of the area are more greatly loaded.

A recent paper published by Brian Mitchell and one of his students at St. Louis University argued that high fluid pressures within the rocks in the New Madrid area might be the cause of the earthquakes. About ten years ago, the U.S. Geological Survey drilled a well in the Reelfoot of Missouri, and when we went through the Cretaceous sediments and into the Paleozoics we encountered an artesian well with a hydraulic head of 110 feet on it. So, it definitely was an overpressured area. This is a common problem of wells drilled in that area. So, Brian Mitchell's and his students' hypothesis that high fluid pressures have something to do with the earthquakes may be valid.

JOHN FOSS (*Bell Communication Research, Morristown, N.J.*): Several years back studies were performed at the California Institute of Technology using radiation dating of rocks. By these means the investigators were able to look backward about two thousand years along the San Andreas fault, and to estimate past earthquake activity in that area. I realize that the San Andreas fault is much better defined than are the faults in the East, but I would like to know whether similar studies are planned for the eastern sector.

STEVEN WESNOUSKY (*Memphis State University, Memphis, Tennessee*): Concerning the first question about whether earthquakes occur where they've happened in the past I would say that geologically that's true. But where earthquakes occur at a specific site, for example, along the San Andreas fault, earthquakes may occur on average every one, two, or three hundred years. So, in the span of our lives, it's not likely that a large earthquake is going to rupture the exact same segment of the San Andreas fault twice. Going one step further we know that California is characterized by many faults, and each of the names that was given for sites of earthquakes corresponds to a fault that produced a large earthquake historically. In the span of our lifetime, it seems unlikely that those faults will generate another large earthquake at the same site.

Rather, future earthquakes are likely to be centered on the other faults which have not ruptured historically. Within the context of the eastern United States, we know that New Madrid and Charleston were very big earthquakes and, in turn, have been the faults to most recently release a large amount of accumulated strain. As a consequence these fault zones may presently represent a small seismic hazard. A real question is: Are there other structures in the eastern United States that are sitting there primed to produce a large earthquake? That's a question we're trying to answer.

In terms of the question of studies such as that on the San Andreas fault which try to recognize past prehistoric earthquakes and obtain some idea of how frequently earthquakes occur, there will be a very interesting paper presented in a later session by Steve Obermeier related to the Charleston area. Investigators are presently trying to look at the prehistoric record of earthquakes in the New Madrid area, and I will touch on that in a later session as well.

HAMILTON: To further elaborate the last question: the study referred to in California was the work by Kerry Sieh along the San Andreas fault. And, of course, a major difference in the East is that we do not identify surface faulting in the same way that we do in the West. Probably the most similar feature that we have would be the Meers fault in southern Oklahoma. Attempts have been made to look at earthquake recurrence in the area of the Meers fault, but generally techniques of that nature cannot be applied in the East simply because we don't see the faults at the surface.

MIKE UZUMERI (*University of Toronto, Toronto, Canada*): I am the Chairman of the Canadian National Committee on Earthquake Engineering, which develops the Canadian code.

Professor Whitman's comments were very interesting and I would like to address one of them.

If I recall correctly, Professor Whitman indicated that reduced ductility requirements in lower risk areas may result in lower safety. I believe that statement is a very important one, but taken by itself, it causes me difficulty. In Canada, as well as in the eastern part of the United States, one has to vary the so-called "ductility." In other words, the quality of detailing the structural system will vary with the area. If this is not done, the result is not economically sound. It is unreasonable to expect to have the same level of ductility or the same ability to deform and sustain the deformation inelastically everywhere across the country. So, in the Canadian code, we have moved away from the word "ductility." Instead we are trying to directly link a set of provisions in material specifications (concrete code or otherwise) to the use of the K factor (or R factor) in a given risk zone. This way we are saying to the engineer that if he is to benefit from use of a reduction factor, he must follow the corresponding structural details. This also deals with the wind-versus-earthquake problem that Professor Whitman mentioned. That is a very important point because, as he indicated, many people, including those in Canada who do wind-versus-earthquake comparisons, say that wind controls when the wind forces are higher although they have used a K factor equal to 1 in earthquake analysis. Therefore, they say, one can forget about the earthquake!!

Those two points are very important. We've tried to address them in Canada, by linking the reduction factors directly to the design and detailing rules of the material codes.

JOE WALLACH (*Atomic Energy Control Board, Ottawa, Ontario, Canada*): Dr. Hamilton, your connections with the Zobacks might help you to answer my question. You have shown a uniform northeast trend of maximal horizontal stress, which we've all recognized, and every one of us would accept the notion that it is related to plate tectonics. If it's tied somehow to spreading off the mid-Atlantic ridge, which appears to have a northwest trend, I'm curious as to why we have the northeast compressive stress.

HAMILTON: I'll try to answer that, but somebody can jump in if he or she feels I am going astray.

There are two mechanisms that have been proposed to account for the stresses within the crust. One is the push that the crust gets at the ridge line where the new crust is formed. And the other is the drag that would be taking place on the bottom of the crust as it moves through the fluid material underneath. It is clear that the relative plate motion should be more or less perpendicular to the direction of the ridge, assuming that the material is formed at the ridge and then moves away in a perpendicular direction.

So, if you follow the ridge as it goes into the North Atlantic you see that it splits, and part of it would go up through Iceland and at least an older version would cut to the northwest. And if you compute absolute plate motions, taking into account the movement of all the plates around the globe, you end up with an absolute plate motion toward the southwest, and that's what correlates with the stress direction.

Does anyone wish to correct me on that?

JOHN ADAMS (*Geological Survey of Canada, Ottawa, Ontario, Canada*): We are all misled by maps that are on Mercator projections. If you look at a globe, in fact, the mid-Atlantic ridge extends through the Arctic Ocean—it's actually north of Alaska. So, if you like, on a Mercator map of North America, the ridge is pushing from the upper right corner.

Also, according to our current knowledge, the Labrador ridge has not been active for fifty million years.

HAMILTON: So part of what I said was wrong. What really has to be taken into account is that the ridge actually goes through the Arctic Ocean, and so it's from that direction that the motion would be coming down.

UNIDENTIFIED SPEAKER: Are any of these computations available? Are there open file reports or any publications on this? I would certainly appreciate information on the overall plate motion or the computations of the net movement towards the southwest. A number of technical papers have been published on that subject, but the most helpful one was prepared by Allan Cox and some of his students at Stanford, which provides a sort of handbook on plate tectonics. I believe that the American Association of Petroleum Geologists published it.

WALLACH: Peter Basham and John Adams spoke about seismicity in Canada, but I would like to mention a far broader program in eastern Canada, the acronym of which is MAGNEC—the Multi-agency Group for Neotectonics in Eastern Canada. Basically MAGNEC's intent is to deal with some of the questions that Leonardo Seeber and others have raised. We need to know about the geology in order to more realistically estimate seismic risk. The lead agency in this is the Geological Survey of Canada, and other members include the Atomic Energy Control Board and Ontario Hydro. There are other participating provincial agencies within Ontario, one of them being the Ontario Geological Survey and the other the Ontario Center for Remote Sensing. From Quebec, there is Hydro Quebec to some extent, and provincial agencies from New Brunswick and Nova Scotia are now on board.

Not only are we trying to come up with more realistic estimates of seismic risk, but we're also attempting to bring together geological as well as seismological data. One of the things that we're looking at, and it does cause problems, is deformation in the unconsolidated sediments because those are the youngest geological strata that we have available to us. But we have a problem there: when we see deformation, it's most tempting for some of us—for people like me—to see it as proof of a paleoseismic event. But, in fact, we have to be very careful about glacial dynamics themselves and ice. In any case, though, we have an integrated program.

Another thing to point out is the phenomenon called offset bore holes—geological evidence that was mentioned on a large scale in Missouri. We see in Ottawa offset bore holes on newly created reverse faults exposed along fresh sites. The movements are reverse and the trend of the surface along which the movement takes place is northwest, which is entirely consistent with the stress orientation.

Another point to make is the presence of a quarry in the Ottawa area where we have pop-ups or buckles on the quarry floor. There are many of these, and lots of them trend northwest, which is again consistent with what we could expect from a northeast-oriented principal compressive stress. But there are others also on that floor that trend northeast or approximately northeast. They are mutually perpendicular. Focal mechanisms have been brought up already as an indicator of stress orientation and rock response. They trend northwest and northeast, with reverse motions indicated by these things.

Offset glacial striations now have been identified; actually they were initially identified back in the 1880s in the Province of New Brunswick, and Ken Burke at the University of New Brunswick is going to be doing some work on these. We have seen the displacements: they are along northeast trending structures and the striations have been displaced in a reverse sense. Much detail stress measurement work has been done by Duncan McKay of Ontario Hydro; he deserves much credit for what we know about stresses in parts of eastern Canada.

The point of all of this is that there are at least two principal compressive stress directions: the northeast one, which is clearly predominant, and a northwest one, which seems to come up quite often. So the picture isn't quite as simple as we would like it to be, although the northeast stress trend is still predominant.

In summary, I would like you to be aware of MAGNEC's program, and if anyone wants more information, I would be very happy to provide it. We are also looking to get into a cooperative effort with our American colleagues in order to exchange information and establish information bases.

HAMILTON: The point you make about not everything fitting into this overall stress pattern is one that Mark Zoback mentioned. There are plenty of examples, particularly in the northeast, where very shallow seismicity is not, in many cases, consistent with this overall pattern. But I think that Dr. Zoback would argue that many of the large earthquakes and the deeper seismicity do fit.

Would anyone else care to comment on that particular point?

RENÉ LUFT (*Simpson, Gumpertz and Heger, San Francisco, Calif.*): Among the interface areas between seismologists and engineers are the seismic hazard studies and the resulting seismic hazard maps. I'd like to pose a question: would the work that has been done in the Northeast regarding plate tectonics and crustal stresses in any way modify the basic methodology that is used for seismic hazard studies in the Northeast? Specifically, if one looks at many of the studies done, the point zones, line zones, and other concepts come from the type of earthquakes we associate with California. I wonder whether the new studies that have been done over the last ten years would somehow modify the fundamentals of seismic hazard studies for the Northeast.

ADAMS: The Canadian experience indicates rather amorphous clusters of earthquakes collapsing into zones; we don't see them collapsing into lines. We do see some elongation in these zones, which I think is real and represents a series of *en echelon* faults which are being activated. The 1985 National Building Code of Canada used the Cornell-McGuire method, which involves drawing a box around the earthquakes and using the known seismicity to determine the earthquake return rates. This method is ideal for those zones. What I'm not so sure about is what you do with other earthquake source models which may imply large earthquakes where none are known. Should you have another box around these zones—places we think earthquakes may happen but haven't yet? That's much more problematic. But the basic idea of the Cornell-McGuire method is sound. We know where the earthquakes are. We can draw boxes around them, and we can determine the earthquake rates. This is not the full answer, though.

ALLIN CORNELL (*Stanford University, Stanford, Calif.*): When you look at seismic hazard assessment for a specific site, you have to remember that you are looking at an assessment of seismicity that is localized both in time, the next fifty years, let's say, and space, the 50 to 100 kilometers around the site you're interested in. The likelihood that earthquakes occur in that window in time and space is probably best estimated by those that have occurred in the last fifty to one hundred years in the same location. That simple estimate does not rule out the fact that we observe in North America both surprises in space and surprises in time. It does not rule out cyclic-type events with very long return periods. But what it says is that the likelihood that those surprises occur in that small time and space window, and hence appear to have a big impact on your estimated seismic risk, is probably not very great, comparatively speaking.

MICHEL BRUNEAU (*Morrison Hershfield Ltd., Toronto, Ontario, Canada*): Dr. Basham indicated that large earthquakes in eastern North America will not necessarily occur in the same geographical region where they have been observed to occur in the past. This is well illustrated by the Magnitude 7 Nahanni earthquake, which occurred in a region of low seismicity, where the seismic requirements for buildings are the same as in Toronto. Following this reasoning, for regions said to be of low seismicity, we are left with only two options. One is to do nothing and simply wait for major earthquakes to occur: a solution that does have some wisdom, in a sense. The other option is to upgrade the level of the design earthquake for those regions. In this perspective, I would like to know what Dr. Basham's approach would be to setting the design earthquake for regions of low seismicity.

BASHAM: I can only respond by saying that I don't think it is economically practical to protect every structure against the remote large earthquake. If we took all of the earthquakes we know historically in eastern North America and established a rate at which they occur and then assume for the next fifty years that they are going to be completely at random and estimated the seismic hazard based on that, that hazard at any particular point on the map will not give you protection against any significant earthquake. It's throwing everything into one pot and saying we don't know anything. Computing a hazard at moderate probabilities, we get very little protection.

On the other hand, we cannot assume that a large earthquake is going to occur very close to a particular site in the next fifty years; from what we know the probability of that happening is so low that we would not impose that as an economic design requirement on various structures.

So, in my view, there are various ways to do seismic hazard estimates that account for various types of models of how the future earthquakes will occur. But even with those we are not going to provide protection at any arbitrary site against a large earthquake. The bottom line probably is to make sure that our decision-makers are aware that we are saying that their site could experience a large earthquake that is not being accommodated in the building design. When the policy-makers get that message from the earth scientists, it's up to them to decide what to do about it.

For a very critical facility, however, you can impose a very conservative design. But for a common facility, you have to recognize the very low probability that it will experience a very damaging earthquake. The social response to that goes far beyond the area dealt with by the earth scientist.

Earthquake Hazards:
A Political Perspective

EDWARD J. WOODHOUSE

Department of Science and Technology Studies
Rensselaer Polytechnic Institute
Troy, New York 12180

From observing several decades of intense controversy over technological risks, political scientists are beginning to understand the ingredients of successful policy-making about risks, as well as some of the obstacles to it. This paper reviews part of what we have learned, with special attention to nuclear power plants, biotechnology, and toxic chemicals. It suggests how the key insights might be applied to the regulation of earthquake hazards, and argues that researchers and other relevant professionals have a responsibility to take a more political approach if they want their analyses to lead to improvements in society's protections against earthquake damage.

FACING UP TO INEVITABLE UNCERTAINTIES

Policy-making ordinarily has to proceed in the face of considerable uncertainty about the probability and severity of risks. Moreover, the efficacy of mitigation strategies often cannot be known at the outset. Therefore, major errors are quite likely, and some form of trial, error, and error correction is an inescapable part of decision-making. The only choices facing earthquake researchers, risk analysts, citizens, and policy-makers are whether to explicitly acknowledge that uncertainty, and whether to face up to its consequences.

In contrast, many people still believe that scientists can produce clear and universally accepted "answers" if they just try hard enough. As a former Senator from Rhode Island put it,

> . . . [W]e have had a lot of gobbledegook from both sides of this question, a lot of verbiage that sometimes is hard for the average citizen (or Senator?) to understand. I wish at some time somebody would come before this committee and tell the committee categorically a nuclear reactor is safe or isn't safe, so that the public will know exactly where it stands.[1]

Senators are not alone. Risk professionals flee from uncertainty by tightly bounding the range of hazards they will examine, typically ignoring some crucial ones. Thus, toxicologists generally ignore synergisms among chemicals. The same effect is achieved by focusing only on what can be quantified, as the nuclear engineering community did in staking its hopes on fault-tree analyses of reactor safety instead of bargaining with the critics to find out what would be broadly acceptable. Professionals also retreat into enclaves, conducting dialogues almost solely with others who share common assumptions, jargon, and ways of approaching problems. This seems to have happened to the numerous nuclear physicists and engineers who knew there were alternatives to the giant light-water reactors that have caused so much conflict throughout the U.S. and Europe; but it took until the late 1980s for the information to filter down to the rest of us, by which time it was too late to change course.[2]

Administrative agencies and courts likewise retreat from uncertainty. Increasingly, across diverse areas of environmental law and other hazard domains, it has become impermissible for agencies to reach decisions that are merely plausible. They are expected to undertake exhaustive, court-like administrative procedures that comprehensively review the scientific and other evidence available on a pending action, and rigorously back up whatever decision they reach. European regulators, in contrast, tend to have much greater discretion, and therefore have taken action against many more chemicals than have U.S. agencies, and they have done this by bargaining with industry and environmental groups.[3]

Journalists, policy-makers, and ordinary people also flee from uncertainty. Psychologists have discovered numerous processes by which humans level and sharpen information to make it less fuzzy; discrepant information tends to be heavily discounted or rejected altogether.[4] When survey researchers ask people their opinion on almost any subject, the rarest answer is "I don't know."

But uncertainty—and hence, disagreement—seldom can be avoided in any realm of political life, because there are good reasons for it. One is that many risks simply cannot be calculated very exactly. In the case of lead, for example, in spite of extended attention over a long period, researchers cannot agree on how much lead (if any) the human body can tolerate without physiological changes.[5] Atmospheric researchers have been studying the possibility of a greenhouse effect for more than three decades, and still cannot agree on the probability or severity of it.[6] Most European nations refused to go along with the U.S. restrictions on ozone-depleting fluorocarbons in the late 1970s because there was "no proof"; they waited for the dramatic ozone hole over Antarctica.[7] Much the same can be said of nuclear power plant safety and many other health/safety/environment issues. The range of legitimate disagreement can be narrowed, but often not far enough to be of much use to policy-makers. Even for debates we think of as settled, such as that over smoking and lung cancer, inspection of the scientific literature often reveals a lively minority viewpoint.[8]

In addition to factual uncertainties, interpretations of risks require controversial value judgments, on which disagreement is certain. Should a new hazardous waste disposal facility be sited far away from cities, for example, to minimize the number of people potentially exposed if problems occur? Or should it be located near cities, so those who benefit from activities generating the waste also bear a fair share of the risks? Should risk regulations aim to fully protect the most vulnerable members of the public, or should they aim to protect the average person? How much weight should be given to semi-"objective" professional expertise compared with citizen perceptions? Several thoughtful, logical opinions can be given to each of these questions, none of which is obviously conclusive.[9]

In sum, there is no way to avoid uncertainty—and, hence, disagreement—in regulating risky endeavors. The wise policy-maker, citizen, or scientist/engineer will accept this fact, and prepare to act intelligently without waiting for certainty and consensus. This is the first lesson of good decision-making on technology.

STRATEGIES USED IN OTHER AREAS OF RISK

Given all the uncertainty and the consequent inevitability of error, how has decision-making been conducted in various areas of risk regulation where errors are potentially too catastrophic to tolerate? Obviously not always very well. But when decision-making has succeeded, it turns out to have followed a set of (often implicit) strategies for proceeding intelligently in the face of uncertainty. Professionals interested in improving earthquake protection regulation may be able to benefit from understanding these strategies.

In risk arenas from the stock market to regulation of nuclear power, humans have gradually been evolving a considerable repertoire of tactics for coping with high uncertainty and potentially catastrophic consequences. These can be distilled into five general strategies. When taken together as an integrated approach to decision-making, these strategies constitute a system of "sophisticated trial and error." It improves on ordinary trial and error by taking out insurance against uncertain hazards, guarding against the worst potential consequences of error, while actively preparing to learn from whatever errors do occur.[10]

First, even in highly uncertain endeavors, we can usually foresee and protect against some of the worst risks. Homeowners do not have to calculate the likelihood of a house burning down; the mere possibility is enough to warrant taking out insurance as an *initial precaution* against catastrophic loss. Through much the same rationale, civilian nuclear reactors in the United States have heavily reinforced containment buildings, which can prevent most of the radioactivity released during an accident like that of Three Mile Island from entering the environment. Those in charge of the Chernobyl reactor, in contrast, unwisely relied on impeccable performance.[11] The tactics would be different for other types of problems, but the basic idea is to take some kind of initial precautions rather than merely hoping for the best.

A related strategy is to *err on the side of caution,* as the U.S. did in regulating fluorocarbons before there was direct evidence that ozone depletion was occurring. The American Chemical Society complained that proposed legislation would constitute "the first regulation to be based entirely on an unverified scientific prediction."[12] But Congress and the Environmental Protection Agency (EPA) banned most aerosol fluorocarbon sprays, even though few other nations were doing so. Another aspect of proceeding cautiously is to put the burden of proof on advocates of risky activities. Until the 1947 Federal Insecticide, Fungicide, and Rodenticide Act, the government had to take each instance of unsafe pesticide practice to court, and prove the danger. Manufacturers now must conduct testing to demonstrate that their products are not unduly dangerous.

A third strategy seeks to promote quicker learning about problems than would happen with ordinary trial and error, by *testing the risks.* After bad experiences with DDT, vinyl chloride and PCBs, the 1976 Toxic Substances Control Act (TSCA) required chemical companies to notify the EPA prior to manufacturing a new chemical, and gave the agency authority to require extensive information and testing. Similarly, instead of waiting to measure actual depletion of ozone, laboratory researchers simulated atmospheric conditions hypothesized to catalyze ozone destruction, and helped to verify the theory.

Actively *preparing to learn from experience* is a fourth strategy. Well-designed trial and error is not purely reactive: it carefully monitors initial trials and prepares to learn from error. Failure to do so allows problems to be repeated and get worse, as when the Nuclear Regulatory Commission (NRC) had a chance to prevent the Three Mile Island accident by learning from similar problems with stuck valves in another reactor. But the NRC had no procedure for identifying and acting quickly on the important information in the thousands of reports submitted to it each year. In contrast, the EPA now has a program for monitoring and reappraising new chemicals when they are put to significantly new uses that may be more dangerous than was the case with their original purposes.

Priority-setting is a fifth strategy used in conjunction with advance testing and deliberate learning from experience. Hazards research and regulation usually is so complex that there are only two choices: be haphazard or set priorities. Thus, since the 60,000 chemicals used by industry cannot all be tested any time soon, TSCA created an Interagency Testing Committee to identify a few chemicals each year that are in relatively widespread use, are close analogues of chemicals known to be carcinogens, or otherwise are considered suspicious and under-emphasized.[13] But this is an exception; more typical, unfortunately, is current work on the greenhouse effect. There, tens of millions of dollars are being thrown at carelessly "prioritized" atmospheric chemistry and physics research, while crucial action-oriented research on reforestation, land-use planning for protection of coastal property, and regional water conservation is being neglected.[14]

In sum, whereas ordinary trial and error merely hopes for the best and corrects when failures become apparent, sophisticated trial and error restructures decision-making to make it more appropriate for dealing with high uncertainty and potentially catastrophic consequences. Decision-makers protect against catastrophe by taking more stringent initial precautions than are really expected to be necessary. They then actively set out to reduce uncertainty about the magnitude and probability of the hypothesized risks, through prioritized testing and careful monitoring of early trials. If uncertainty can be substantially narrowed, the initial precautions then are revised. Precautions can be relaxed when threats prove less serious than feared, enhanced when threats prove more dangerous than initially suspected, and otherwise refined to make them more effective and perhaps less costly.

This was more or less the procedure scientists voluntarily applied in the mid 1970s to protect against the unknown dangers of early biotechnology research, known then as "recombinant DNA research." There was a voluntary moratorium on potentially risky research, while a regulatory strategy was worked out through the National Institutes of Health. Six classes of especially risky experiments were prohibited altogether, and precautions were adopted for the others, varying in stringency according to the anticipated degree of risk. Special laboratory facilities were intended to prevent bacteria from escaping; and intentionally enfeebled strains were used for most of the research, so that even if bacteria escaped they would have great difficulty surviving. These precautions were more stringent than many scientists believed necessary. Worst-case experiments then sought to determine whether researchers might accidentally create virulent new organisms. As uncertainty was reduced, more experiments were allowed at lower levels of containment; by the early 1980s, most of the containment requirements were dropped, and no type of experiments remained altogether prohibited.[15] This system has not been followed quite as well now that commercial biotechnology is being introduced, but the early period stands as an exemplar of how humans can intelligently proceed in the face of high uncertainty and potentially grave consequences of error.

Application to Earthquakes

These strategies are very much in line with reforms that many earthquake professionals have been advocating. Rather than hoping for the best, as most easterners now do, it would make sense to take initial precautions of various kinds: avoiding building on some sites, designing new buildings to better resist shaking, and protecting utility and other lifelines. How cautious to be is a political judgment, of course, on which there is no analytic answer. And some geographical areas have higher risk profiles than others, and presumably would want to be more cautious. So one of the high-priority tasks is to develop a spectrum of policy options, with cost estimates for various areas, so that informed debate can occur on the "How safe is safe enough?" issue.

In the research phase, priority setting is necessary also. If the goal is to refine scientific understanding of earthquakes, then the relevant disciplines clearly are on the right track; enormous progress has been made in the past several decades, and there is every reason to believe that further refinements will be coming along regularly. If, however, the goal is to help society protect itself against the financial, social, and psychological damage caused by earthquakes, it is questionable whether much has been accomplished in the past two decades. So researchers and professionals need to decide where their priorities are.

If actually effecting change in earthquake regulations is to be made a priority, then the Federal Emergency Management Agency's (FEMA's) overall research program will need to go even further than it has in emphasizing work that will actually be helpful to ordinary citizens, interest groups, and policy-makers. Further refinements in seismic maps will not do it. What decision-makers need pertains more to the costs of the potential damage, and the options available for mitigating such damage. I return to this issue in the next section.

EARTHQUAKE PROFESSIONALS AND POLITICS

In principle, then, strategies are available to substantially reduce the financial and human costs of a major earthquake. But the strategies are not being applied because few people believe they need to be, except (perhaps inadequately) in California. As discussed in the first section, although it is uncongenial for technical people to admit it, no "scientific" argument can demonstrate that the hazards posed by earthquakes are sufficiently grave to deserve more of society's scarce resources. Nor, in all likelihood, will there ever be a scientific consensus on the probability or severity of future earthquakes for any particular region, at least in sufficient detail for policy-makers. Scientists and engineers, moreover, can offer only the vaguest notions about the economic, social, psychological, and political costs of the damage. While better "guesstimates" of these could be offered by economists and other relevant professionals, they would fall over a wide range. The costs of taking steps now to mitigate the damage likewise are not fully knowable: what, for example, is the cost of refusing to allow development of a piece of prime real estate in an especially vulnerable area? Finally, there is no conclusive, analytic answer to the question, "How risk-averse should we be?"

The essence of this paper's argument, however, is that policy-making can proceed just fine without analytic certainty on these points. What the society needs is a *political* judgment about whether to invest in further earthquake protections, and if so, which ones. At present, there is virtually no debate on the subject. Why not? Earthquake researchers and design professionals typically point the finger at the public and/or policy-makers. But is this a constructive approach? We know that the failure to invest in earthquake protections does not stem from a literal lack of funds, nor a lack of concern about risks. Tens of billions of dollars have been expended on mitigation of risks from nuclear power, and several billion dollars are spent annually on protection from other types of hazards (not counting enormous expenditures for life and property insurance). So people are concerned with at least some types of hazards—even morbidly concerned—and the society spends enormous sums as a result.

That leaves three main possibilities:

1. People are ignorant that there is a risk of serious damage from eastern earthquakes;
2. They believe the probability of such damage to be trivial, or at least not worth additional investment;
3. They want to abate the hazard, but don't know how.

Should it be up to citizens to actively seek information about earthquake risks? Perhaps. But there are so many issues in a technological society that no one can pay attention to them all. So, realistically, most citizens (and policy-makers) depend largely on ideas that have been elevated onto the priority agenda, as defined by the media and by those who supply the media with stories and claims for priority attention. Hence, any substantial reduction in the society's ignorance about earthquake hazards will have to come from someone who knows something about the issue, and has an incentive to push for change. It is hard to see who that would be if not earthquake and design professionals, and perhaps insurance companies.

If earthquake professionals were to accept this, what then? The first rule of American politics is that interest groups are crucial intermediaries and stimulants for policy-making. While FEMA and the building industry associations can do something, they are not enough. If there is to be substantially better earthquake protection in the next decade or two, it is imperative to get interest groups working on the issue. It is conceivable that environmental groups can be brought into the fray: for example, if there is a credible case to be made that an earthquake would seriously disrupt hazardous waste dumps. Organizations of educators, hospital administrators, or consumer groups that deal with public utility commissions are other possibilities. None of these is interested in earthquake hazards per se; but each has concerns that bear on some aspect of it. No one other than earthquake professionals is likely to be in a position to forcefully point out that connection.

Once groups take up the cause, they will need volunteer expert assistance. A little help can go a long way, and the expertise does not necessarily have to be from forefront researchers to be of use. To illustrate the poverty of the technical expertise typically available, when the Three Mile Island reactor was initially being licensed, the intervenor group had to make do with a single expert witness, while the utility company had several dozen. So a few geophysicists or design professionals could be extremely helpful, especially in deliberations that occur at the local level. This can be taken a step further by actually becoming part of the board (or even the audience) of relevant local commissions.

Next is the task of developing a rhetorical stance that is likely to win attention. The rhetoric need not be flamboyant, but it must somehow raise the salience of the issue and override normal presumptions against costly new restrictions. New earthquake protections could be couched as a form of "insurance," and also as a type of long-term "cost-effectiveness," by estimating and publicizing the loss "potential to exposed books of insured business."[16] This could appeal to the businessmen who typically exert heavy influence in local government, while not offending either liberals or conservatives. Good rhetoric also (gently) paints the opposition as foolhardy, as by pointing to instances where people hoped for the best and ended up regretting it; ozone depletion is an obvious possibility, but so are local fires, floods, droughts, or leaking hazardous waste dumps. Because humans tend to think in very concrete and personal images, moreover, vivid illustrations of what happens in an earthquake—"an ambulance rendered useless beneath the ruins of its carport, collapsed overpass structures"—may win more converts than will more abstract analyses.[17]

Indirect action also needs to be considered. One obstacle to improved building design, Beavers argues, is that undergraduate engineering students at many colleges are not required to learn dynamic analysis.[18] Since accrediting procedures for engineering schools can exert a powerful influence on the curricula, it is not inconceivable over a period of years that revisions in the curriculum of civil engineers could be achieved. Improvements may also be needed in ongoing professional education of "many design professionals and others directly involved with the building community who have never been taught seismic design."[19]

While there still are not nearly enough journalists interested in scientific and technological issues, their numbers have multiplied rapidly in recent years; and a growing number of newspapers and magazines have special sections on the subject. Because the cadre of science journalists is relatively small, it would not take an outlandish effort to educate a substantial fraction of such journalists regarding earthquake hazards. Subsequent media treatments of earthquakes probably would be more sympathetic toward the need for improved regulation.

Finally, political campaigns require funding. While earthquake research funds come overwhelmingly from government, lobbying funds cannot. Earthquake education funds conceivably might come from government, but probably will not. So it will be important to develop an alternative source, which probably means an existing foundation or a new one set up explicitly for the purpose by a wealthy donor. Someone in the large network of professionals interested in the subject knows someone who knows someone who might help. Whether that potential donor is ever actually contacted and persuaded to act remains to be seen.

CONCLUSION

Overall, then, contemporary thinking by earthquake professionals is generally compatible with political scientists' emerging understanding of the appropriate strategies for regulating risky technologies, with two exceptions. First, there is a misguided tendency to believe that society will wake up to earthquake dangers if only researchers can reduce the remaining uncertainties about the timing and severity of future New Madrid or other eastern earthquakes. In fact, it is virtually inconceivable that physical science research could conclusively answer the questions faced by policy-makers, nor is it necessary to do so to proceed intelligently in tightening earthquake protection regulations.[20]

Secondly, earthquake professionals tend to underemphasize the political elements in risk regulation: Like it or not, effective earthquake policy will come about only through a partisan process in which interest groups and professionals stick their necks out and argue for erring on the side of caution. Atmospheric scientists have been doing it for the ozone and greenhouse issues; the Union of Concerned Scientists and the American Physical Society (among others) did it for nuclear power; and the relevant scientific community acted in concert to take precautions in early biotechnology research.

Design professionals and earthquake researchers who believe, as citizens, that policy should be erring on the side of caution, need to help raise funds, stimulate interest group activity and media attention, serve on local boards, and otherwise turn a portion of their energies toward the policy-making process if they want to alter the status quo. Likewise, FEMA and other funding sources for earthquake research need to stimulate policy-oriented research and civic education, not just further studies of seismicity.

In sum, good policy-making requires a willingness to enter into the political process, and employment of intelligent strategies in the face of the uncertainty that is an inevitable and permanent part of policy-making on risk. Earthquake policy is poised for that leap, but taking it depends on the commitment of concerned professionals.

NOTES AND REFERENCES

1. PASTORE, J. O. 1973. *In* The Status of Nuclear Reactor Safety. U.S. Congress, Joint Committee on Atomic Energy. Washington, DC.
2. MORONE, J. G. & E. J. WOODHOUSE. 1989. The Demise of Nuclear Power? Lessons for Democratic Control of Technology. Yale University Press. New Haven, CT.
3. BRICKMAN, R. *et al.* 1985. Controlling Chemicals: The Politics of Regulation in Europe and the United States. Cornell University Press. Ithaca, NY.
4. FISCHHOFF, B. *et al.* 1984. Acceptable Risk. Cambridge University Press. New York.
5. COLLINGRIDGE, D. & C. REEVE. 1986. Science Speaks to Power: The Role of Experts in Policy Making. St. Martin's. New York.
6. For an overview on the greenhouse effect, see chapter 6 of Morone, J. G. & E. J. Woodhouse. 1986. Averting Catastrophe: Strategies for Regulating Risky Technologies. University of California Press. Berkeley, CA.
7. For further details on ozone depletion, see Morone and Woodhouse, 1986, chapter 5.
8. COLLINGRIDGE & REEVE, 1986, chapter 10.
9. HISKES, A. L. & R. P. HISKES. 1986. Science, Technology, and Policy Decisions. Westview. Boulder, CO; JOHNSON, D. G. 1985. Computer Ethics. Prentice-Hall. Englewood Cliffs, NJ; SHRADER-FRECHETTE, K. S. 1983. Nuclear Power and Public Policy: The Social and Ethical Problems of Fission Technology, 2nd ed. D. Reidel. Boston, MA.
10. For elaboration of these strategies, see MORONE and WOODHOUSE, 1986, especially chapter 7.
11. WILSON, R. 1986. Chernobyl: Assessing the accident. *In* Issues in Science and Technology 3: 21–29.
12. TANNENBAUM, J. A. 1978. Fluorocarbon battle expected to heat up. *In* Wall Street Journal. January 19: 38.
13. WOODHOUSE, E. J. 1985. External influences on productivity: EPA's implementation of TSCA. Policy Studies Rev. 4: 497–503.
14. On the failure to set priorities in greenhouse research, see MORONE and WOODHOUSE, 1986, chapter 6.
15. KRIMSKY, S. 1982. Genetic Alchemy: The Social History of the Recombinant DNA Controversy. MIT Press. Cambridge, MA; MORONE and WOODHOUSE, 1986, chapter 4.
16. FRIEDMAN, D. G. 1987. Uses of earthquake information in an insurance operation. *In* Proceedings from the Symposium on Seismic Hazards, Ground Motions, Soil-Liquefaction and Engineering Practice in Eastern North America, October 20–22, 1987. K. H. Jacob, Ed.: 515–517. National Center for Earthquake Engineering Research. Buffalo, NY.

17. STATTON, T. C. 1987. Some thoughts regarding the state of earthquake mitigation in the Eastern U.S. *In* Jacob, 1987: 513–514.
18. BEAVERS, J. E. 1985. Current practices in earthquake preparedness and mitigation for critical facilities. *In* Societal Implications: Selected Readings: 2/1–2/7. Building Seismic Safety Council. Washington, DC.
19. DILLON, R. M. 1985. Development of seismic safety codes. In Societal Implications: Selected Readings: 3/1–3/8.
20. For another perspective on the politics of seismic hazards abatement, see SYLVES, R. T. 1987. Political, social, and economic factors to be addressed in developing a program to abate seismic hazards to electric power lifelines. *In* Abatement of Seismic Hazards to Lifelines: Papers on Political, Social, and Economic Issues. Federal Emergency Management Agency. Washington, DC.

Quantitative Ground-Motion Estimates[a]

DAVID M. BOORE

United States Geological Survey
Menlo Park, California 94025

INTRODUCTION

Estimates of ground motion in eastern North America (ENA) are hampered by a lack of data. Because of this, most estimation schemes have been based, at least in part, on data obtained from the more seismically active and better-instrumented western North America (WNA). In view of the different geologic structures and tectonic settings in the two regions, however, it is not clear *a priori* that data from WNA can be exported to ENA. A theoretical model has been developed in the last few years that does not require WNA data for the estimation of ground motions in ENA. This model constructs the ground motion from filtered random Gaussian noise, for which the filter parameters are determined by a seismological model of both the source and the wave propagation. Many of the parameters in the model can be determined from independent seismological investigations in ENA or in other regions with similar tectonic characteristics. Crucial ingredients of this model are the shape of the radiated energy spectrum and the relation between the seismic moment of the earthquake and the corner frequency of the spectrum (usually referred to as the "scaling law"). In two recent papers, predictions of ENA ground motions using a scaling law characterized by a constant stress parameter were in reasonable agreement with the sparse observed data at rock sites, most of which come from earthquakes near moment magnitude 4.5. Subsequent to these studies, however, two variable-stress scaling laws have been proposed. All the laws lead to similar motions for moderate earthquakes (M = 4.5), but for larger events the estimates diverge (the range in predicted motions is a factor of 6 for an M = 7.5 earthquake, the estimates from the constant stress scaling being between the other two). The new model can be used to gain insight into similarities and differences between ground motions for ENA and WNA earthquakes. The calculations suggest that the motions would be the same at distances within about 60 km for ground motions with frequencies less than about 10 to 15 Hz. At greater distances the differences in regional attenuation of seismic waves leads to larger ground motions in ENA (although these motions would probably not cause much damage), and at higher frequencies the generally more competent rock in ENA will allow higher frequencies to propagate to the site than would be the case for typical rock sites in WNA.

STOCHASTIC MODEL

The stochastic model for estimating ground motions is best described with reference to FIGURE 1; more complete descriptions can be found in papers published by Boore,[1] Boore and Atkinson,[2] and Toro and McGuire,[3] and the refer-

[a] The research reported here was funded by the U.S. Nuclear Regulatory Commission.

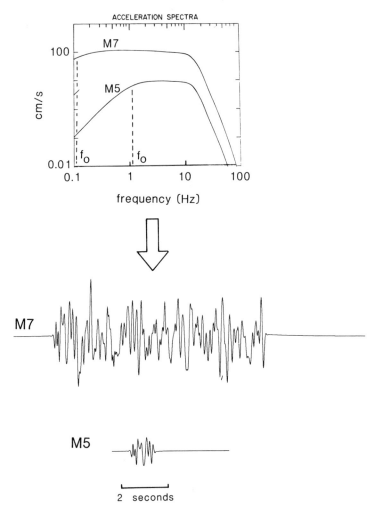

FIGURE 1. Basis of stochastic model. Radiated energy described by spectra in *upper part* of figure is assumed to be distributed randomly over a duration equal to the inverse of the lower frequency corner (f_0). Time series is one realization of a random process. The levels of the low-frequency part of the spectra are directly proportional to the logarithm of the seismic moment and thus, by definition of Hanks and Kanamori,[15] to the moment magnitude **M**. Unless specifically noted otherwise, all magnitudes in this paper refer to moment magnitude.

ences therein. The spectra of the radiated motion is specified by a seismological model (in this case, the ω^2 spectrum with constant stress parameter), and the motion is assumed to be spread out in a stochastic manner over a specified duration (in this case, the inverse of the corner frequency). The peak motions can be determined either by a suite of time-domain simulations or, more conveniently, by random process theory. Boore and Atkinson[2] (hereafter referred to as BA87) used this model with a 100-bar stress parameter to predict pseudo-relative velocity spectra (*PSV*) as well as peak acceleration for hard-rock sites in eastern North America (ENA). Note that the spectra and time series shown in FIGURE 1 have not been sketched by hand; they were computed for the magnitudes shown.

SOURCE SCALING

Crucial to the ground-motion predictions is the specification of the source spectrum. This usually comprises two parts: the shape of the spectrum, and the way in which the spectral corners depend on seismic moment (this dependence is usually referred to as the "scaling law"). At the time BA87 was written, the only scaling law offered for ENA earthquakes was that of Nuttli[4] (referred to hereafter as N83). BA87 considered this scaling law, but found it to lead to ground motions significantly lower than the few available recordings (mainly for earthquakes around M4.5). BA87 found that scaling with a constant stress parameter of 100 bars gave reasonable predictions of the available data. Since publication of BA87, Nuttli *et al.*[5] (referred to hereafter as N87) and Boatwright and Choy[6] (referred to hereafter as B&C) have published scaling laws for ENA and intraplate earthquakes. FIGURE 2 compares the magnitude scaling of the high-frequency level of

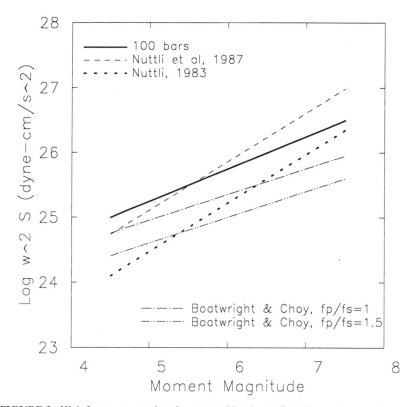

FIGURE 2. High-frequency acceleration spectral levels predicted by various scaling laws.

$(2\pi f)^2$ times the moment rate spectrum (in effect this is the acceleration spectrum normalized such that the long period level of the corresponding displacement spectrum is equal to the seismic moment) for the various scaling laws.

Boatwright and Choy[6] used teleseismic P waves to derive their source spectra. In order to apply their results to the radiation of S waves, I have assumed that the shape of the source spectra is the same for both P and S waves, with a specified ratio between the P- and S-wave corner frequencies. A frequency-independent multiplicative factor accounted for the radiation pattern and seismic velocity differences. The high-frequency spectral level of the S wave depends on this

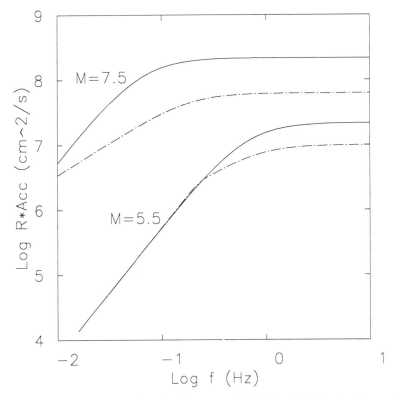

FIGURE 3. Distance-corrected acceleration spectra for 100-bar (*solid line*) and Boatwright and Choy (*dot-dash line*) scalings. The ratio of P-wave to S-wave corner frequencies was 1.0.

multiplicative factor and on the ratio of corner frequencies. In most of the computations presented here, the corner frequency ratio has been taken to be unity. This produces a result that is consistent with Choy and Boatwright's derivation of high-frequency S-wave spectral amplitudes from their P-wave scaling.[7] For comparative purposes, FIGURE 2 also contains results for a corner frequency ratio of 1.5.

From FIGURE 2 alone, we can see that all other things being equal, the N83 scaling will lead to much lower ground motions than the 100-bar scaling, and that for earthquakes less than about M 6.5 the revised scaling given by N87 will lead to motions similar to those given by the 100-bar relation used by BA87. The B&C

relation with a corner frequency ratio of 1.5 leads to low motions, as will their relation with a ratio of 1.0 (except for events near M = 4.5).

The spectral shapes assumed for the constant stress (100 bar) and B&C scalings, for earthquakes of magnitudes 5.5 and 7.5, are shown in FIGURE 3. The B&C spectra are characterized by two corner frequencies. The higher of the two corners is determined by a least-square line fit to B&C observations of corner frequency versus moment magnitude; the lower corner is found by intersection of the low-frequency part of the spectrum, determined by moment, and a line of unit slope between the observationally determined high-frequency corner and spectral level.

COMPARISON WITH ENA GROUND MOTIONS

FIGURE 4, patterned after Figure 7 in BA87, shows the predicted PSV as a function of distance for the various scaling laws. The symbols are available data from rock sites in ENA. Coincidentally, the N87 and B&C (with corner frequency ratio of unity) laws give virtually identical motions (see the previous figure also). It would be difficult to choose between the 100-bar, N87, and B&C laws based on a comparison of the predictions to the observations.

The scaling with magnitude of PSV for the various scaling laws, at a fixed distance of 10 km, is shown in FIGURE 5. It is clear that although the 100-bar, N87, and B&C laws lead to similar motions at M = 4.5 (see previous figure also), the predictions diverge for the large earthquakes of most importance for engineering design. Data from these large events are needed to distinguish between the models. B&C based their study on teleseismic recordings of earthquakes greater than M = 5, and so preference should perhaps be given to their law. On the other hand, the 100-bar relation gave a reasonable prediction of m_{Lg} values for larger ENA earthquakes,[2,8] and Somerville et al.[9] found an average stress drop of 100 bars in their analysis of teleseismic recordings of large earthquakes in eastern North America.

COMPARISON WITH NAHANNI EARTHQUAKE DATA

The M = 6.5 December 23, 1985, Nahanni, Northwest Territories earthquake was recorded on three nearby SMA stations. For several reasons, including similarities of types of faulting and geologic conditions, the event is considered by Wetmiller et al.[10] to be representative of events that might occur in ENA. The strong motion recordings then take on particular significance for the estimation of seismic hazard in ENA. The aftershock locations and instrument locations are shown in FIGURE 6, which was taken from Weichert et al.[11]

Observed and predicted pseudo-relative velocity spectra at the three stations are shown in FIGURE 7. The PSV at station 1 was computed using the first 7 seconds of record, thereby eliminating the large burst of energy late in the record (this prominent burst of energy is not seen on the records from the other stations and may not be representative of the overall fault rupture). The open circles are predictions using equations in BA87, with two distance measures—closest distance, and distance to the center of aftershocks; the ×'s are predictions from Joyner and Boore's analysis of records from western North America.[12] The agreement between observed and predicted values is reasonable, especially in view of

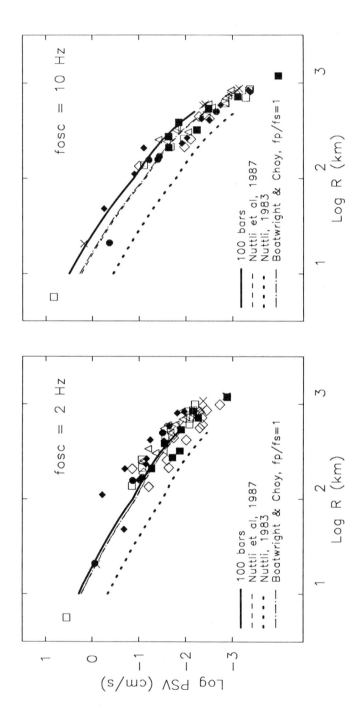

FIGURE 4. Comparison of observed (*symbols*) and predicted (*lines*) 5 percent damped pseudo-velocity response spectra (*PSV*) for 2- and 10-Hz oscillators. Data have been normalized to M = 4.5 by means of scaling obtained from the theoretical predictions using 100-bar scaling. (See Figure 7 in Boore and Atkinson[2] for key to symbols.)

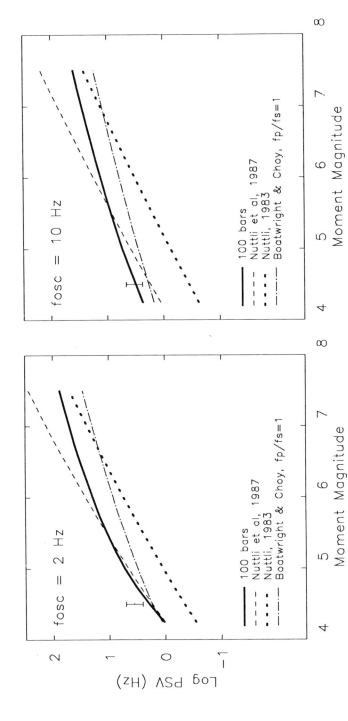

FIGURE 5. Scaling of *PSV* with moment magnitude (M) at 10-km distance for 2- and 10-Hz oscillators. The scant data suggest *PSV* values indicated by the bar at M = 4.5.

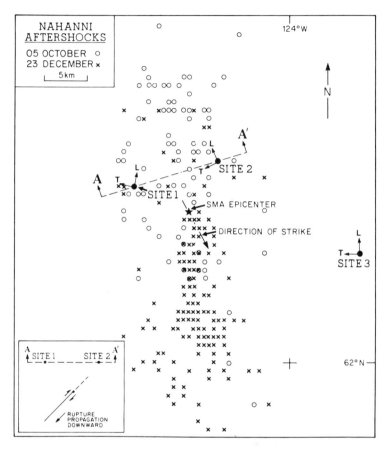

FIGURE 6. Station locations of SMA instruments (*solid circles*) and aftershocks of Nahanni earthquakes. (From Weichert *et al.*[11] Reprinted by permission.)

the large scatter of individual observations that exists in any attempt to predict inean values of ground motion.

The comparison in FIGURE 7 was between *PSV* derived from observations and previously published predictions of the *PSV*. The relation between the data and theoretical predictions using B&C's scaling, as well as the 100-bar stress parameter model, are shown in FIGURES 8 and 9. The comparison is in terms of Fourier amplitude spectra (*FAS*) rather than *PSV* (the relative agreement of observations and predictions should be similar for both quantities, however). The observations are given by heavy lines (where, as with *PSV*, only the first 7 seconds were used in computing the *FAS* for station 1). In FIGURE 8, the predictions used the distances and radiation pattern terms used by Choy and Boatwright.[7] The predictions in FIGURE 9 use my estimates of distances (closest distance and distance to center of aftershocks) and the radiation pattern term used in BA87. B&C give a better fit to the observations when using their parameters, and as expected from the comparison in FIGURE 7, the 100-bar prediction is reasonable for stations 1 and 2 when using the BA87 parameters. B&C, however, consistently have a better fit to the station 3 acceleration spectrum than that given by the 100-bar model. Some cautions are necessary before drawing any firm conclusions from these comparisons. First, the BA87 predictions are for the mean value of motion, and the motions from any one earthquake could be systematically higher or lower than the mean value. Second, radiation pattern and geometrical spreading terms, as well as directivity and partition of energy into various components, affect the overall amplitude level, but are somewhat uncertain, especially close to large faults. For this reason, a firm conclusion that one scaling model is better than another should not be reached solely on the basis of the Nahanni data.

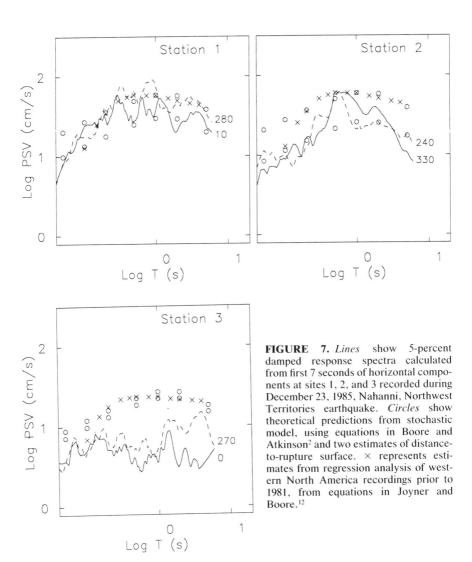

FIGURE 7. *Lines* show 5-percent damped response spectra calculated from first 7 seconds of horizontal components at sites 1, 2, and 3 recorded during December 23, 1985, Nahanni, Northwest Territories earthquake. *Circles* show theoretical predictions from stochastic model, using equations in Boore and Atkinson[2] and two estimates of distance-to-rupture surface. × represents estimates from regression analysis of western North America recordings prior to 1981, from equations in Joyner and Boore.[12]

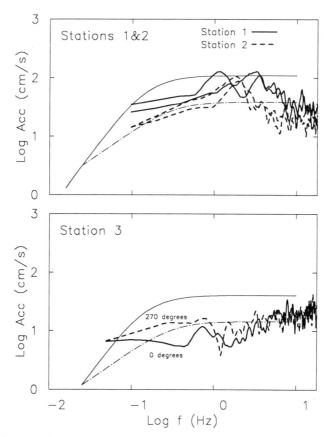

FIGURE 8. *Heavy lines* show acceleration from first 7.5 seconds of horizontal components at sites 1, 2, and 3 recorded during the 23 December 1985 Nahanni earthquake. *Light lines* show estimated spectra, using 100-bar model (*solid line*) and Boatwright and Choy model (*dash-dot*: P- to S-wave corner frequency ratio = 1.0). Distances and radiation patterns are from Choy and Boatwright.[7]

COMPARISON OF PREDICTIONS IN EASTERN AND WESTERN NORTH AMERICA

I used the stochastic model to make a comparison of *PSV* at four oscillator periods for the ENA model of BA87 and the WUS model in Boore.[13] The results are shown in FIGURE 10. The two models differ in several ways: in ENA the crustal velocity is slightly higher, f_m is much higher (50 Hz versus 15 Hz), little amplification is expected at hard-rock sites, and a higher stress parameter is required to fit the meager data (100 bars versus 50 bars). These differences are reflected in a complicated way in the comparison shown in FIGURE 10. For example, compared to the WUS, for the 20-Hz *PSV* the ENA motions are considerably higher because of the higher f_m, and for the 5-Hz *PSV* the ENA motions are lower because of the higher crustal velocity and lack of an amplification effect. This

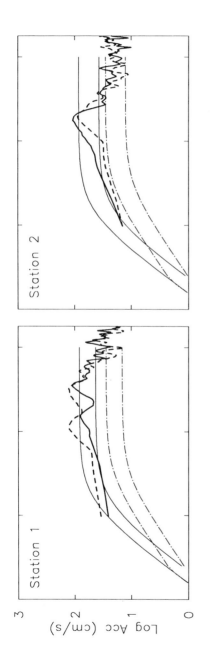

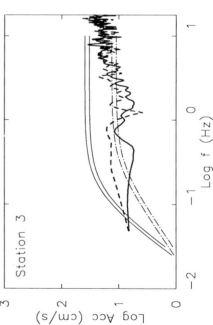

FIGURE 9. Observed and predicted acceleration spectra from 23 December 1985 Nahanni earthquake. Predictions used my estimates of distances and the radiation pattern correction of Boore and Atkinson.[2] The *upper* and *lower curve* of each pair used the closest distance to the rupture surface and distance to the center of the rupture surface, respectively. The *solid curves* used the 100-bar model and the *dot-dashed curves* used the Boatwright and Choy scaling.

amplification effect is frequency-dependent, being near unity for frequencies less than about 1 Hz, and therefore the 1-Hz *PSV* again shows that ENA motions are higher (because of the higher stress parameter). Finally, the 0.2-Hz *PSV* is not very sensitive to the stress parameter, since the corner frequencies of all but the M = 7 event are higher than the oscillator frequency, and therefore the ENA and WUS motions are similar.

ESTIMATION OF RESPONSE SPECTRA FROM PEAK MOTIONS

A common way of estimating *PSV* is to multiply peak acceleration and/or peak velocity by appropriate factors and to plot the resulting levels on log-log paper (Newmark and Hall;[14] most U.S. and Canadian building codes). When applied to the stochastic model calculations (FIG. 11), this method produces reasonable estimates of *PSV* for magnitude 5 events but overestimates *PSV* for larger earthquakes by as much as a factor of two, for frequencies less than about 10 Hz. For all size events the method consistently underestimates the high-frequency part of

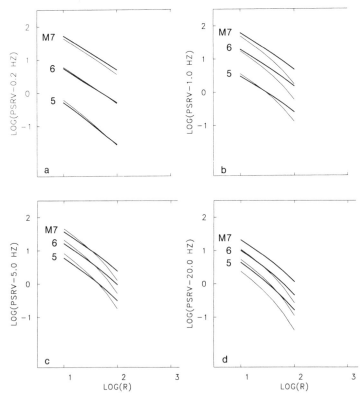

FIGURE 10. log *PSV* versus log *r* for 5-percent-damped oscillators at four frequencies (**a, b, c, d**). In each figure, results are shown for moment magnitudes 5, 6, and 7. The ENA and WUS predictions are given by *heavy* and *light lines,* respectively.

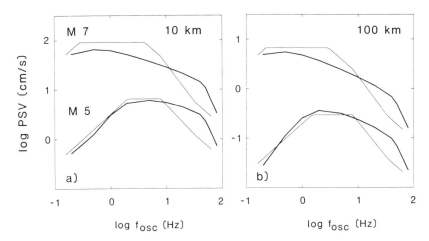

FIGURE 11. Predicted 5-percent-damped pseudo-velocity response spectra (*PSV*) at 10 and 100 km as a function of oscillator frequency (f_{osc}). Predictions from the stochastic model are shown by the *heavy curves*. The *light lines* were obtained from the peak velocity and acceleration predicted from the model by applying median amplification factors from Table 1 in Newmark and Hall,[14] with faring to the high-frequency asymptote beginning at 8 Hz and ending at 33 Hz (see Figure 5 in Newmark and Hall[14]). (From Boore and Atkinson.[2] Reprinted by permission.)

the spectrum, by up to a factor of 3, even though the peak accelerations used in the method were those computed for the ENA ground-motion model. This is in large part due to the much higher frequency content in the theoretical motions (controlled by the choice of $f_m = 50$ Hz) than exist in the WNA data from which the amplification factors and faring frequencies published by Newmark and Hall[14] were derived. The comparison emphasizes the need to predict expected response spectra directly, rather than through the use of standard spectral shapes. Simply put, the frequency content and attenuation of ENA ground motions preclude the use of standard spectral shapes developed empirically from western data. In particular, the enrichment of ENA motions in higher frequencies means that the relationship between maximum ground acceleration and maximum acceleration response is not a simple constant in the frequency range of engineering interest (less than 30 Hz). The acceleration response spectra, S_a, will increase with increasing frequency until frequencies near f_m (50 Hz) are approached.

ACKCNOWLEDGMENTS

I thank Jack Boatwright, Tom Hanks, and Bill Joyner for comments on this paper.

REFERENCES

1. BOORE, D. M. 1987. The prediction of strong ground motion. *In* Strong Ground Motion Seismology. M. Ö. Erdik & M. N. Toksöz, Eds.: 109–141. D. Reidel. Dordrecht.

2. BOORE, D. M. & G. M. ATKINSON. 1987. Stochastic prediction of ground motion and spectral response parameters at hard-rock sites in eastern North America. Bull. Seismol. Soc. Am. **77:** 440–467.
3. TORO, G. R. & R. K. MCGUIRE. 1987. An investigation into earthquake ground motion characteristics in eastern North America. Bull. Seismol. Soc. Am. **77:** 468–489.
4. NUTTLI, O. W. 1983. Average seismic source-parameter relations for mid-plate earthquakes. Bull. Seismol. Soc. Am. **73:** 519–535.
5. NUTTLI, O. W., D. S. BOWLING, J. E. LAWSON, JR., & R. WHEELER. 1968. Some aspects of seismic scaling and the strong ground motion of the eastern Missouri earthquake of January 12, 1984. Seismol. Res. Lett. **58:** 53–58.
6. BOATWRIGHT, J. & G. L. CHOY. 1988. Acceleration source spectra for large earthquakes in North Eastern North America. *In* Proceedings of the Workshop on Earthquake Ground-Motion Estimation in Eastern North America. EPRI NP-5875:16–1—16–16. Electric Power Research Institute, Palo Alto, CA.
7. CHOY, G. L. & J. BOATWRIGHT. 1988. Teleseismic analysis of the Nahanni earthquakes in the Northwest Territories, Canada. Bull. Seismol. Soc. Am. **78:** 1627–1652.
8. ATKINSON, G. M. & D. M. BOORE. 1987. On the m_N, M relation for eastern North American earthquakes. Seismol. Res. Lett. **58:** 119–124.
9. SOMERVILLE, P. G., J. P. MCLAREN, L. V. LEFEVRE, R. W. BURGER & D. V. HELMBERGER. 1987. Comparison of source scaling relations of eastern and western North American earthquakes. Bull. Seismol. Soc. Am. **77:** 347–365.
10. WETMILLER, R. J., R. B. HORNER, H. S. HASEGAWA, R. G. NORTH, M. LAMONTAGNE, D. H. WEICHERT & S. G. EVANS. 1988. An analysis of the 1985 Nahanni earthquakes. Bull. Seismol. Soc. Am. **78:** 590–616.
11. WEICHERT, D. H., R. J. WETMILLER & P. MUNRO. 1986. Vertical earthquake acceleration exceeding 2 *g*? The case of the missing peak. Bull. Seismol. Soc. Am. **76:** 1473–1478.
12. JOYNER, W. B. & D. M. BOORE. 1982. Prediction of earthquake response spectra. U. S. Geol. Surv. Open-File Rep. **82–977,** El Cerrito, CA.
13. BOORE, D. M. 1983. Stochastic simulation of high-frequency ground motion based on seismological models of the radiated spectra. Bull. Seismol. Soc. Am. **73:** 1865–1894.
14. NEWMARK, N. M. & W. J. HALL. 1982. Earthquake Spectra and Design. Earthquake Engineering Research Institute.
15. HANKS, T. C. & H. KANAMORI. 1979. A moment magnitude scale. J. Geophys. Res. **84:** 2348–2350.

Implications of Some Important Strong-Motion Recordings of Mid-Plate Earthquakes

DIETER H. WEICHERT

Geological Survey of Canada
Pacific Geoscience Centre
Sidney, British Columbia, Canada V8L4B2

INTRODUCTION

Two earthquakes of unprecedented magnitude for that region occurred in the North Nahanni River area of the Northwest Territories of Canada in October and December 1985. The earthquakes had moment magnitudes 6.6 and 6.8, surface-wave magnitudes 6.6 and 6.9, and body-wave magnitudes of 6.5 and 6.4, respectively. The Geological Survey of Canada (GSC) installed strong-motion accelerographs in the epicentral region immediately after the occurrence of the first large earthquake and captured important records from the 23 December earthquake. These exhibited peak horizontal accelerations of 1.25 g and peak vertical accelerations that were beyond the limits of the recording medium, but must have exceeded two times the acceleration of gravity (g).

The Nahanni earthquakes occurred in an area of relatively low seismicity, the so-called "Mackenzie" earthquake source zone of the source zone model developed by Basham *et al.*[1] On the basis of the short historical record, the Mackenzie source zone had been assigned an upper-bound magnitude of 6.0, which was significantly exceeded by the 1985 earthquakes. The implications of these surprise earthquakes for hazard estimation in the Canadian Cordillera have been discussed by Wetmiller *et al.*[2]; implications for eastern Canada are discussed in this paper.

The seismicity in the Mackenzie (see FIGURE 1) and the enclosed smaller Richardson Mountains source zone, with higher activity and magnitudes, has parallels in seismic source regions in eastern Canada and the northeastern United States. There, similarly diffuse patterns of earthquakes are observed in the Northern Appalachian seismic zone and clusters of higher activity, some with large earthquakes, occur elsewhere, for instance in the Charlevoix source zone.

In this paper it is argued that the seismotectonic environments in the Northeastern Cordillera and the Northern Appalachian regions are sufficiently similar so that Nahanni-type surprise earthquakes should also be seriously considered in the seismic zones of eastern North America (ENA) at an appropriately low probability. A short review of the geologic and tectonic setting and the faulting mechanisms of the Nahanni earthquakes is accompanied by references to eastern analogues. Earthquake recurrence statistics of the Mackenzie and the Northern Appalachian source zone are shown to be similar. Comparisons of the intensity pattern of the Nahanni events with the similarly sized 1925 Charlevoix earthquake in ENA are given as circumstantial evidence that the Nahanni observations are relevant for the East. Finally, the strong ground motions recorded from the December event are reviewed and it is concluded that they are unlikely to be site effects.

TECTONIC ENVIRONMENT

FIGURE 1 shows the location and relative level of seismicity of the Mackenzie region in the context of the total Canadian seismicity. While the activity along the Pacific plate boundary is high and well defined, the earthquakes within the continent and mostly surrounding the Canadian shield occur in diffuse low-activity clusters. The Mackenzie seismicity is at least 800 km distant from the tectonically well-understood earthquake zones of the circum-Pacific belt and has no obvious causal connection.

Magnitude 6 and 6.5 earthquakes have occurred in the Beaufort Sea and in the Richardson Mountains, about 500 km northwest of the current activity in the Mackenzie Mountains, but within the diffuse cluster that includes the Nahanni events, the largest previous historical earthquakes have only been near magnitude 5. It is conceivable that as the Nahanni aftershock series continues, it may eventually resemble the clustered activity in the Richardson Mountains near the 1940 and 1955 M6.5 and M6.6 earthquakes. This would imply that for long-term hazard estimation such clusters around relatively recent large earthquakes must not be separated from the tectonically similar or indistinguishable surrounding zone.

The Nahanni earthquakes occurred in the Mackenzie Plain, part of the Mackenzie Fold Belt of the northeastern Cordillera. This fold belt underwent various degrees of deformation of the Proterozoic and Paleozoic sedimentary rocks during the late Cretaceous to early Tertiary Laramide Orogeny, which was the main deformational episode in the region. There are several mapped surface faults in the area,[3] but none of these seems to have been involved in this earthquake.

Contemporary stress indicators[4] suggest that the crustal rocks in this region are subjected to high horizontal compressive stress in NE-SW direction, similar to that found in most of northeastern North America.

This tectonic regime, especially the compressive stress, has similarities in regions east of the old North American craton, in the Appalachian fold belts and St. Lawrence valley, where in principle it should also be capable of producing similar-size earthquakes (i.e., up to M7), unless sound arguments against this assumption can be found.

Adams and Basham[5] have recently discussed the eastern source zones of the Canadian hazard model. They conclude that the larger significant earthquakes in continental Canada have been associated with the Paleozoic rift system of the St. Lawrence and suggest that the relatively uniform and shallower seismicity in the Appalachian source zone may not be capable of supporting maximum earthquakes larger than M6.

EARTHQUAKE RECURRENCE RELATION AND HISTORICAL ACTIVITY

The Mackenzie seismic zone was characterized by Basham et al.[1] as a zone of "background" seismicity. It includes some clustered activity, with magnitudes as large as 5.3. The upper bound or maximum magnitude earthquake was set at 6.0,

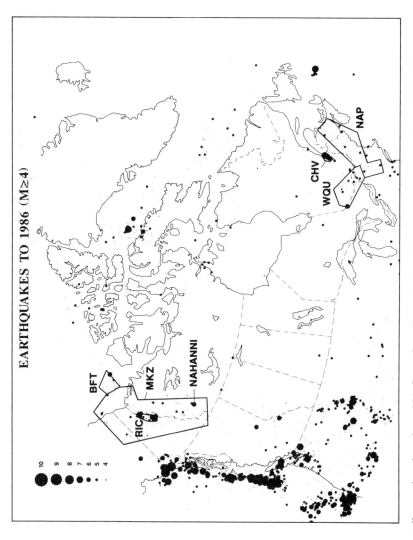

FIGURE 1. Canadian earthquakes above M4. Some Canadian seismic model source zones are outlined: MKZ, MacKenzie: RIC, Richardson Mountains; BFT, Beaufort Sea; NAP, Northern Appalachian zone; CHV, Charlevoix; WQU, Western Quebec.

but the rates of the larger events were poorly defined because of the short time interval of observation in this thinly populated area. Periods of complete reporting for the northern Cordillera are short, with magnitude 6 or larger only complete since the 1920s, and magnitude 5 since about 1950.[6] In eastern Canada, Basham *et al.* recognized a similarly large zone of rather uniform seismicity throughout New Brunswick, Maine, New Hampshire and Vermont and they called this the Northern Appalachian zone (NAP). The southern boundary was arbitrarily drawn to include the seismicity beyond the Canadian border that might affect Canadian territory, but to exclude the large number of historic events that are catalogued for the Boston region during the time of early settlement. The zone was extended to the southeast just far enough to include the seismicity of southeastern New York

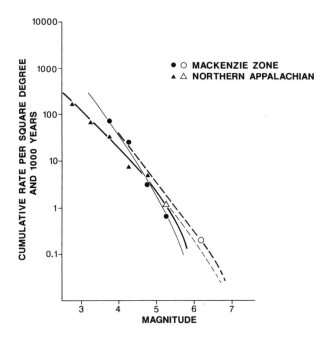

FIGURE 2. Earthquake recurrence rates for the MacKenzie and the Northern Appalachian source zones. *Open symbols* represent earthquakes that occurred after the original calculation of recurrence rates.

along the Hudson river, but was terminated north of the Ramapo fault system. Thus, the boundaries of this NAP zone in the United States were arbitrary ones, not intended to prejudge conclusions of U.S. researchers. Since the larger historic earthquakes in the Northern Appalachian zone have magnitudes near 5, the maximum magnitude adopted for the zone was 6.0, using similar reasoning as in the Mackenzie zone.

FIGURE 2 compares the recurrence rates of the Mackenzie and the Northern Appalachian seismic zones, normalized to one thousand years and to one square degree. Confidence limits on the rates are not shown, but the major distinction between the two zones is the length of their earthquake records of about 300 years. At the magnitude threshold of potentially damaging M5 earthquakes the activity level in the two zones is essentially the same. It is not easy to decide how to include the Nahanni swarm activity into the recurrence relations. Except for the larger events of 1985/86, which have their own aftershock swarms, all events are essentially aftershocks with the attendant doubts about their statistical treatment.[7] Moreover, if the Richardson Mountains activity of the last half century is assumed to be a temporary swarm (i.e., aftershocks of the 1940 and 1955 earthquakes), the recurrence relation has to be drastically redefined. The open circle in FIGURE 2 represents the Nahanni series. The plotting position in magnitude is shown as an M6.5 within the original 0.5 groupings (i.e., greater than 6.25), but should perhaps be 0.5 units higher; the associated rate is equally subjective. The maximum magnitude for the zone clearly has to be increased to M7 as suggested by the dashed curve.

In the Northern Appalachians, the 1982 earthquake series in the Miramichi area of New Brunswick occurred as a surprise for eastern Canada at that time, but with magnitudes M 5.7 and 5.4, it can still be easily accommodated in the earlier estimated recurrence statistics of NAP with maximum magnitude 6. This is shown in FIGURE 2 by the open triangle, which represents the large 1982 events; the large number of small aftershocks does not increase the steepness of recurrence relation significantly.

If, however, in the interest of conservatism the arguments in this paper are accepted, the maximum magnitude should be increased to M7; FIGURE 2 then gives an approximate hazard rate of one Nahanni-type event per square degree in 100,000 years. While negligible for the Cordilleran hinterland, this rate becomes appreciable when multiplied by the exposure and population density of, say, the Newark basin.

FAULT MECHANISMS

The mechanisms of the two large Nahanni earthquakes were determined from P-wave first motions, analysis of surface-wave spectra, teleseismic p waves, and interpretation of aftershock distributions.[3] Rupture surfaces strike 175° with uncertainties of about 30°; rake angles are 90° with uncertainties of about 20°. The October fault rupture, with an essentially identical epicenter, dips toward the west at about 34°, while the December rupture dips at a shallower 25° to the west.

The aftershock zone defining the active area was determined from data obtained during field experiments and data from the Canadian seismograph network.[8] Aftershocks occurred over an area about 15 by 50 km, but could not be associated with any of the mapped surface faults, nor with their inferred extension at depth. Depending on the fraction of aftershock area considered, the slip may have been between 1 and 2 m.[8] No evidence for surface rupture has been found, in particular not on the Iverson thrust fault that runs right through the epicentral area. Aftershock depths located from field data ranged from 3 km to 12 km. The fault ruptures appear to have started in the sedimentary rocks and propagated downward, taking the rupture into the crystalline rocks of the Precambrian craton.

The mechanisms, including the strike-slips, are consistent with the ENE compressive stress field observed in the general area. The situation in ENA is the same. Focal depths of the Nahanni events are intermediate between those in the Charlevoix sources zone and the Miramichi swarm of 1982, although it is possible that the 1925 Charlevoix earthquake may have been as shallow as 10 km.[15] The lack of surface rupture for an event of this size has its parallels in both the Charlevoix and the Miramichi areas, and perhaps even in the significantly larger 1886 Charleston earthquake.

MERCALLI INTENSITIES

The value of Mercalli intensity observations and patterns for engineering design is limited, but the similarities of the Nahanni observations with those from the 1925 Charlevoix earthquake appear relevant and are therefore reviewed here.

Neither of the two Nahanni earthquakes caused any significant damage, as the immediate epicentral area is uninhabited. Several cases of cracked plaster were reported in single-storey, wood-frame buildings in the closest communities of Wrigley (120 km) and Fort Simpson (160 km). The felt area of the October event was estimated as about 1.5 million km², judged from the macroseismic data from the south and east. The December felt area may have been as much as 50 percent greater—more than 2 million km²—on the basis of areas inside the intensity IV contour (for the later event the outer felt contour was not fully defined).[3]

Because the epicentral area was devoid of human habitation and engineered structures, it is not possible to establish accurate epicentral intensities. One occupied cabin at a distance of 60 km experienced intensity V. Extrapolating inward from the extent of the intensity V contours we believe that maximum epicentral intensity could have been about IX.

There were many small slope failures and one major rock avalanche in the epicentral region. Landslides are formally rated as intensity X on the modified Mercalli intensity scale, but significant mountain rock-slope failures have been observed at intensities as low as VI.[9] Slope failures may therefore not be the most useful for establishing intensity.

The distribution of intensities shows a strong elongation in the NW-SE direction, parallel to the strike of the Cordillera; intensity IV was reported to at least 1000 km to the southeast of the epicenters, but to only about 500 km to the west and less to the east. This asymmetry of intensity patterns in western Canada has been noted previously[10] and is thought to be due to more efficient propagation of L_g parallel to the structural trends in the Cordillera. A comparison[14] with the isoseismals of the 1925 St. Lawrence earthquake shows a closely similar asymmetry, with the long axis of the 1925 event parallel to the St. Lawrence river.[11] Along the major axis, the Nahanni event was felt towards the southeast about 1800 km into the United States, the same distance as the 1925 event was toward the southwest. This is in agreement with an estimate of M6 3/4 for the 1925 event by Street and Turcotte,[12] the same magnitude as the larger Nahanni event but slightly smaller than the older Canadian estimate for 1925.[11]

STRONG GROUND MOTIONS

World Record Peak Acceleration

After the October earthquake, three 3-component SMA-1 accelerographs were installed in the epicentral area with separations of 10 to 20 km.[13] Sites 1 and 2 are about 11 km apart, and at about 10 km hypocentral distance from the 23 December initial rupture; site 3 was about 25 km to the east. The record of that more distant site showed no unusual features: the strong shaking lasted about 10 seconds with peaks reaching about 0.2 g. In the early part of the records the ground motion was similar at the closer two sites, having peak accelerations of 0.4 and 0.5 g in the horizontal directions at frequencies between 1 and 3 Hz.

Later into the earthquake, accelerations at site 1 increased, showing distinct subevents near 3, 6, 8 and 9 seconds. One upward acceleration pulse near 9 seconds exceeded the limits of recording film and thus was not directly measured. Conservative interpolation estimates this peak at about 2 to 2.3 g.[13] FIGURE 3

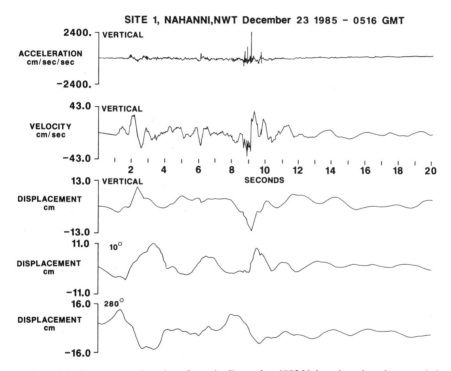

FIGURE 3. Strong ground motions from the December 1985 Nahanni earthquake recorded at site 1. Acceleration and velocity are shown only for the vertical component. A 10-second highpass filter has been applied to the traces.

shows the vertical acceleration and velocity, and the three components of displacement for this site, filtered above 10 sec.

Permanent displacement does not appear to be recoverable; however, during the first S arrival the total displacement was about 0.3 m toward 70° and upward at 25°; this corresponded well with the earlier quoted mechanism solution. The later displacement pulse corresponding to the large 9-sec acceleration spikes is very similar to the first pulse in shape, but smaller in the horizontal direction, and larger and inverted on the vertical.

Site Investigations

The credibility of the strong ground motion at site 1 has obvious implications for building codes and engineering design, since all components of this particular record exceed the design spectrum for the highest seismic zone in the National Building Code of Canada, for periods shorter than 0.3 to 0.4 seconds.[2] The possibility of some site effect has been considered, especially since the record at the approximately equidistant site 2 shows very little energy comparable to the site 1 energy around 9 seconds.

During an instrument service visit to the area in June 1987, a careful visual inspection of all sites was made. Partial excavation with pick and shovel convinced me that instrument 1 was firmly bolted to an outcrop of solid bedrock. The site was on top of a gently undulating plateau, about 30 to 50 m from a 100- to 150-m escarpment that here marks the Iverson Thrust. Apart from possible topographic amplification due to this nearby 50 to 75 percent slope, site 1 looked sound, with no nearby loose rock or debris. No sign of surface rupture was recognized on the Iverson Thrust.

Statistical evidence against unusual site amplification at site 1 is obtained from a comparison of peak acceleration amplitudes from 66 aftershocks recorded at both sites 1 and 2. On the average, at site 2 almost 30 percent higher accelerations were recorded, after application of a correction for nominal epicentral distance. This empirical attenuation over the first few tens of kilometers is approximately as distance$^{-2.5}$.[14] A similarly strong attenuation with exponent 2 to 2.5 is obtained at 10 to 20 Hz from analysis of the response spectra of the 23 December event for sites 2 and 3.

Origin of Strong Motion

Special site effects appear to be excluded as causes of the extreme acceleration peaks in the later part of the site 1 record; instead, these strong motions must originate from either very shallow asperities, or, less likely, from deeper asperities on the main fault plane with radiation pattern that favors site 1 strongly.[14]

The three instruments were triggered by the longitudinal wave; the arrival of the shear wave was recorded at all three sites and allows triangulation of the point of rupture initiation, that is, the hypocenter. It appears to lie several kilometers south of the site 1–2 axis at a depth of about 6 km, perhaps slightly closer to site 2. This is consistent with the hypocenter determination from other network data.

The shallow focus and the aftershock distribution indicate that the rupture of this thrust fault propagated down to the west.[8] Thus, the rupture would move closer to site 1. We already know that the 23 December event ruptured on a shallower fault plane than the 5 October event; it is therefore not unreasonable that some of the 23 December subevents ruptured on even shallower splay faults, thus coming much closer to site 1. The three or four subevents exhibit slightly decreasing S-P intervals, with the last and shortest perhaps 0.5 sec, corresponding to about a 4-km travel path. With the earlier noted attenuation of peak acceleration, the large acceleration would be perfectly compatible with a 4-km distance. The period of the strongest acceleration spike is about 120 msec, implying an asperity size of a few hundred meters.

The noted downward displacement during this late event is not readily explained with the above assumption, but might suggest rupture on an orthogonal plane to the main fault plane; this has been suggested by Wetmiller et al.[3] on the basis of other considerations.

CONCLUSIONS

The Nahanni earthquakes occurred in geological structures and in a contemporary stress field that extends throughout most of eastern North America and stretches as far west as the eastern Cordillera. For conservative seismic hazard evaluation, larger earthquakes than hitherto observed will clearly have to be considered in the immediate area of the recent earthquakes. More importantly, these earthquakes should also be considered typical of earthquakes that must be expected in the active zones of eastern Canada and the United States throughout the Appalachian zone where the causative stress field and the normalized seismic activity are almost indistinguishable from those of the Mackenzie zone. In particular, the December Nahanni earthquake matched the 1925 St. Lawrence earthquake in magnitude, estimated epicentral intensities and macroseismic intensity patterns, or indeed may have approached the Charleston earthquake.

Having occurred in a fold belt not fundamentally different from the Appalachian system, the Nahanni events seem to negate the view that such large intraplate earthquakes must be causally connected and restrained to the rift systems of eastern North America. The implications are that Nahanni-type extreme, short-duration accelerations should be expected almost anywhere in eastern North America, regardless of the existence of nearby recognized potential surface faults, not only in the Charlevoix zone or other places where such earthquakes have actually occurred in the past few hundred years, but almost at random in regions with a relatively low level of seismic activity, such as the Northern Appalachians, and perhaps including the New York region.

SUMMARY

In 1985, two earthquakes with $M_s6.6$ and 6.9 occurred in a hitherto only moderately seismic mid-plate environment, in the Nahanni River area of the Canadian Northwest Territories. They exceeded the stipulated maximum magnitude for the region and could not have been predicted on the basis of historical seismicity patterns. Accelerographs installed after the first earthquake recorded ground motions lasting about 12 seconds with peak horizontal accelerations reaching 0.4 to 0.5 g early in the record. In addition, at one recording site several seconds of extremely strong motion appeared late in the record, with the vertical component exceeding 2 g. This motion seems to have been caused by the rupture of asperities within 4 to 6 km of the site. Response spectra exceed the nominal design spectrum for the highest Canadian seismic zone at periods shorter than 0.4 seconds. Since the seismotectonic environment, the compressive stress field, and source mechanisms are similar to those in some eastern Canadian and U.S. regions, it is argued that similar surprise earthquakes should also be expected there and that the Nahanni earthquakes must be considered as representative design earthquakes for engineered structures in the active seismic zones of eastern North America.

ACKNOWLEDGMENTS

I appreciate discussions with colleagues and especially Bob Horner, at the Geophysics Division and the Pacific Geoscience Centre of the Geological Survey of Canada.

REFERENCES

1. BASHAM, P. W., D. H. WEICHERT, F. M. ANGLIN & M. J. BERRY. 1982. New probabilistic strong seismic ground motion maps of Canada: A compilation of earthquake source zones, methods and results. Earth Phys. Br. Open File Rep. **82–33**.
2. WETMILLER, R. J., P. W. BASHAM, D. H. WEICHERT & S. G. EVANS. 1987. The 1985 Nahanni earthquakes: Problems for seismic hazard estimates in the northeast Canadian Cordillera. Proceedings of the Fifth Canadian Conference on Earthquake Engineering, Ottawa: 695–703.
3. WETMILLER, R. J., R. B. HORNER, H. S. HASEGAWA, R. G. NORTH, M. LAMONTAGNE, D. H. WEICHERT & S. G. EVANS. 1988. An analysis of the 1985 Nahanni earthquakes. Bull. Seismol. Soc. Am. **78**: 590–616.
4. ADAMS, J. 1987. Canadian crustal stress database: A compilation to 1987. Geol. Surv. Canada Open File Rep. **87–1622**.
5. ADAMS, J. & P. BASHAM. 1988. The seismicity and seismotectonics of eastern Canada. *In* Geological Society of America "Decade of North American Geology" series, Vol. GSMV-1; Neotectonics of North America. (D. B. Slemmons, E. R. Engdahl, D. Blackwell, D. Schwartz & M. Zoback, Eds. Submitted for publication.
6. LEBLANC, G. & R. J. WETMILLER. 1974. An evaluation of seismological data available for the Yukon Territory and Mackenzie Valley. Can. J. Earth Sci. **11**: 1435–1454.
7. WEICHERT, D. H. 1980. Estimation of the earthquake recurrence parameters for unequal observation periods for different magnitudes. Bull. Seismol. Soc. Am. **70**: 1337–1346.
8. HORNER, R. B., R. J. WETMILLER, M. LAMONTAGNE & M. PLOUFFE. 1988. Aftershock studies of the Nahanni earthquakes. In preparation.
9. KEEFER, D. K. 1984. Landslides caused by earthquakes. Geol. Soc. Am. Bull. **95**: 406–421.
10. ROGERS, G. C., R. M. ELLIS & H. S. HASEGAWA. 1980. The McNaughton Lake earthquake of May 14, 1978. Bull. Seismol. Soc. Am. **70**: 1771–1786.
11. SMITH, W. E. T. 1967. Basic seismology and seismicity of eastern Canada. Seism. Series of the Dominion Observatory 1966-2, Ottawa.
12. STREET, R. L. & F. T. TURCOTTE. 1977. A study of northeastern north American spectral moments, magnitudes, and intensities. Bull. Seismol. Soc. **67**: 599–614.
13. WEICHERT, D. H., R. J. WETMILLER & P. MUNRO. 1986. Vertical earthquake acceleration exceeding 2g? The case of the missing peak. Bull. Seismol. Soc. Am. **76**: 1473–1478.
14. WEICHERT, D. H. & R. B. HORNER. 1987. The Nahanni earthquakes. *In* the Proceedings of the Symposium on Seismic Hazards, Ground Motions, Soil-Liquefaction and Engineering Practice in Eastern North America. National Center for Earthquake Engineering Research. Buffalo, NY.
15. EBEL, J. E., P. G. SOMERVILLE & J. D. McIVER. 1986. A study of the source parameters of some large earthquakes in northeastern North America. J. Geophys. Res. **91**: 8231–8247.

Earthquake Source and Ground-Motion Characteristics in Eastern North America[a]

PAUL SOMERVILLE

Woodward-Clyde Consultants
566 El Dorado Street
Pasadena, California 91101

EARTHQUAKE SOURCE SCALING RELATIONS

The use of synthetic seismograms to simulate strong ground motions is becoming an increasingly important approach to the evaluation of ground motions for seismic design.[1-3] This approach is especially pertinent in regions such as eastern North America where earthquakes are relatively infrequent and strong ground-motion recordings are correspondingly sparse. The simulation methods entail the use of a source model to specify the level of high-frequency radiation from the source in relation to its seismic moment. This may be done for a given seismic moment by specifying the source duration (or equivalently, the spectral corner period), which then determines the stress drop.

At present there are two principal uncertainties in the description of source characteristics required for the use of these methods in eastern North America. These uncertainties relate to whether earthquakes in eastern North America follow a constant or nonconstant stress drop scaling relation, and whether their average stress drops are similar to or higher than those of earthquakes in other regions such as western North America. Unresolved, these issues give rise to uncertainty in the estimation of strong ground-motion characteristics using simulation methods. It is therefore important to construct a well-constrained source scaling relation for earthquakes in eastern North America, and to compare it with the scaling relation for earthquakes in western North America.

Scaling relations were constructed from the source parameters of earthquakes from eastern North America, other continental interiors, and western North America. The other continental interior events were used to provide constraints on source scaling for the larger magnitudes, which are sparse in eastern North America. The seismic moments and source durations used in constructing the scaling relationships were estimated from the time-domain modeling of body waves.[4-6] The parameters of the eastern North American events are given in TABLE 1.

The source scaling relation for each regional category is given by a linear relation between the logarithm of seismic moment and the logarithm of source duration. Plots of seismic moment against source duration for eastern North America and western North America are shown in FIGURES 1 and 2, respectively. The error estimates for each measurement are shown in order to indicate the weights given to each point in the least-squares fit. Also shown are scaling rela-

[a] This work is part of a program being conducted by the Electric Power Research Institute.

TABLE 1. Seismic Moments, Source Durations, Depths, and Lg Magnitudes in Eastern North American Earthquakes

No.	Region	Date	Depth	m_{Lg}	M_0	U	DUR	U
1	Charlevoix	1925/ 3/ 1	10.0	7.0	0.22E+27	4.0	5.0	1.2
2	Timiskaming	1935/11/ 1	10.0	6.3	0.51E+26	1.7	5.0	1.2
3	Charlevoix	1939/10/19	8.0	5.6	0.10E+25	5.0	1.5	1.3
4	Ossippee	1940/12/20	10.0	5.7	0.12E+25	5.0	0.8	1.3
5	Missouri	1963/ 3/ 3	15.0	4.7	0.11E+24	1.5	0.5	1.5
6	Baffin Bay	1963/ 9/ 4	7.0	—	0.16E+26	2.1	3.0	1.4
7	Missouri	1965/10/21	4.0	4.8	0.90E+23	1.5	0.5	1.2
8	Illinois	1968/11/ 9	25.0	5.5	0.13E+25	1.5	0.7	1.7
9	Maine	1973/ 6/15	6.0	5.0	0.62E+23	1.5	0.8	1.6
10	Quebec	1979/ 8/19	6.5	5.0	0.15E+24	1.5	0.9	1.3
11	Kentucky	1980/ 7/27	13.5	5.2	0.48E+24	1.9	1.1	1.2
12	New Brunswick	1982/ 1/ 9	7.0	5.8	0.13E+25	1.5	0.8	1.3
13	New York	1983/10/ 7	7.0	5.2	0.25E+24	1.5	0.6	1.5

NOTE: Mo: seismic moment in dyne · cm; DUR: source duration in seconds; U: standard deviation factor of preceding value.

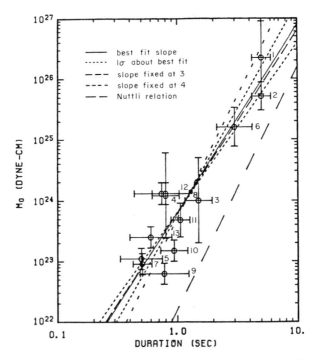

FIGURE 1. Source scaling relations for eastern North America. The median values and standard deviation factors of seismic moment and source duration are given in TABLE 1, and the scaling relations are given in TABLE 2.

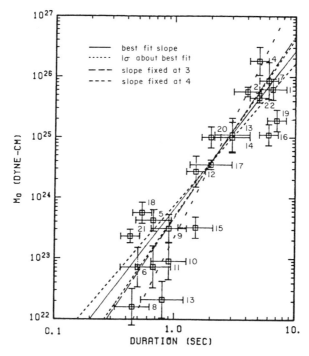

FIGURE 2. Source scaling relations for western North America. The scaling relations are given in TABLE 2.

TABLE 2. Scaling Relations: $\text{Log}_{10}(\text{Moment}) = \text{Intercept} + \text{Slope} \times \text{Log}_{10}(\text{Duration})$

Region	Intercept	Slope[a]	Standard Deviation of Slope	Degrees of Freedom	Sums[b]	Stress Drop
ENAM	23.81	3.09	0.34	11	8.7	
	23.80	3				120
CINT	23.53	3.09	0.72	9	16.2	
	23.61	3				75
ENAM &	23.77	2.85	0.17	22	25.8	
CINT	23.66	3				85
WNAM	23.80	2.60	0.29	20	48.8	
	23.67	3				90
ENAM, CINT (Nuttli,[7] 1983)	22.20	4		11	109.4	

NOTE: ENAM: eastern North America; CINT: other continental interiors; WNAM: western North America.

[a] Upper values: slope unconstrained; lower values: slope constrained to 3 (constant stress drop).

[b] Weighted sum of squares of residuals.

tions showing the best-fit line, lines of one standard deviation in slope, and lines of slope 3 (corresponding to constant stress drop) and 4 (corresponding to Nuttli's[7] scaling relation). The scaling relations obtained using the least-squares fitting method of York[8] are summarized in TABLE 2. The uncertainty in slope is represented by its standard deviation about the mean.

Comparison of Slopes of Scaling Relations

The slopes of the scaling relations for the different regions are summarized in TABLE 2 and FIGURE 3. The slopes of the scaling relations for eastern North America, other continental interiors, and these two sets combined are consistent with a slope of 3 and inconsistent with a slope of 4. This indicates that the source characteristics of eastern North American earthquakes are consistent with constant stress drop scaling, and inconsistent with the nonconstant stress drop scaling proposed by Nuttli.[7] The slope of the scaling relation for western North America was found to be 2.6, which is somewhat inconsistent with the value of 3, but very inconsistent with the value of 4. This indicates that the source character-

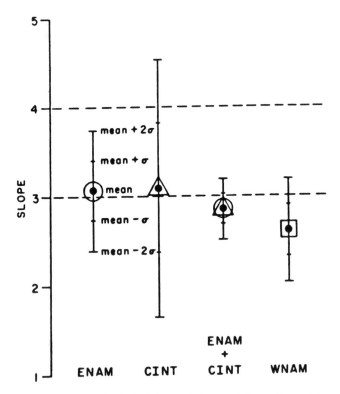

FIGURE 3. Means and standard deviations of slopes of the scaling relations. ENAM: eastern North America; CINT: other continental interiors; WNAM: western North America.

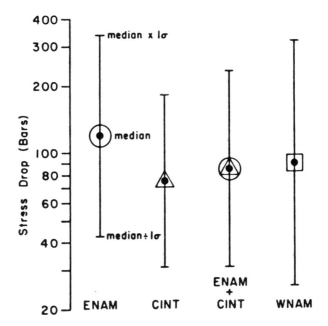

FIGURE 4. Median values and standard deviation factors of stress drops assuming constant stress drop scaling. ENAM: eastern North America; CINT: other continental interiors; WNAM: western North America.

istics of western North American earthquakes are somewhat inconsistent with constant stress drop scaling.

Comparison of Stress Drops

The source characteristics of earthquakes in the three regions are most conveniently compared using stress drop as a means of describing the relation of seismic moment to source duration. Uncertainties in the estimation and interpretation of stress drop include limitations in the model used to calculate stress drop, uncertainties in the seismic moment and source duration values used in the calculation, and ambiguities in the definition of stress drop of events having complex ruptures. In order to facilitate the comparison of stress drops, it is simplest to assume constant stress drop so that each population can be represented by a single average stress drop value. As we have seen, the scaling relations for eastern North America, other continental interiors, and the combination of these two sets are consistent with constant stress drop, whereas the western North American events are somewhat inconsistent with this assumption.

The mean and standard deviations of the stress drops of the three populations are listed in TABLE 2 and illustrated in FIGURE 4. The eastern, western and other continental interior events have median stress drops of 120, 90 and 75 bars, respectively. The differences between these median values are not statistically significant, and all are consistent with a median stress drop of 100 bars.

Comparison of Source Characteristics

The focal depths of large earthquakes in eastern North America, estimated within an accuracy of a few kilometers using depth phases, are confined to the upper crust as in western North America. The source functions of these events as observed teleseismically[5] also show a range of complexity that is comparable to that of western North American earthquakes. However, subtle differences may exist in the variability of source parameters in each population. When the number of degrees of freedom are compared with the weighted sum of squares of residuals in TABLE 2, it appears that individual events in western North America show significant departures from a single source scaling relation, while for eastern North America and other continental interiors, individual events are consistent with a single relation.

The uniformity in source properties of eastern events may be explained by the model of Kanamori and Allen,[9] which relates static stress drop to average repeat time. Although their data set does not contain information for any eastern North American earthquakes, they suggest that earthquakes in eastern North America have very long repeat times, and expect their stress drops to be high. According to this model, the relative uniformity in stress drops of eastern events may be due to their uniformity in repeat time.

Kanamori and Allen[9] demonstrated that the static stress drops of western North American events span a wide range and that they show a tendency to increase with earthquake repeat time in accordance with their model. This wide variation in repeat times may be responsible for the wide range in stress drops of western events used in the present study. This variability indicates that it is inappropriate to group western events into a single category for the purpose of evaluating source characteristics.

These considerations notwithstanding, western events are commonly grouped together for the purpose of evaluating strong ground motion characteristics. This practice has been justified because differences in ground motion between different kinds of events have not been clearly established, although they may exist. The combination of western events in this study rests on a similar justification. When this is done, the median stress drop of the western events is found to be not significantly different from that of the eastern events. While it may be true, as Kanamori and Allen[9] state, that a factor of 5 difference in average stress drop is commonly seen between earthquakes with short and long repeat times, this difference is not observed between the median values for western and eastern North American earthquakes.

The principal differences in the source properties of eastern and western North American earthquakes are the greater variability of the source characteristics of the western events, and the tendency of their stress drops to decrease slightly with seismic moment. These differences may be interpreted by means of a model in which the largest stress drops in each region are controlled by the maximum shear strength of fault zones. The maximum strength is realized when the whole rupture surface is a large asperity or contains a dense distribution of asperities. These conditions appear to be independent of earthquake size in eastern North America, giving rise to relatively uniform source characteristics and constant stress drop scaling. However, as earthquake size increases in western North America, the proportion of the rupture surface that consists of asperities decreases on some faults, causing a corresponding decrease in stress drop. This gives rise to a scaling relation in which stress drop decreases slightly with earthquake size, and to a great degree of variability in the source characteristics of earthquakes in western North America.

Implications for Strong Ground-Motion Estimation

Detailed analysis of the uncertainties in the scaling relations has allowed two important issues concerning the source scaling of earthquakes in eastern North America to be addressed. First, the source characteristics of large earthquakes in eastern North America and other continental interiors are consistent with constant stress drop scaling, and are inconsistent with nonconstant scaling models such as that of Nuttli.[7] Second, the stress drops of large earthquakes in eastern North America and other continental interiors are not significantly different from those of earthquakes in western North America, and have median values of approximately 100 bars. This result is constrained by seismograms whose frequency content does not extend very far above the corner frequency. Projection of this result to the higher frequencies of engineering interest using simple spectral models suggests that those aspects of strong ground motion amplitudes that are attributable to the earthquake source may also not be significantly different between eastern and western North America.

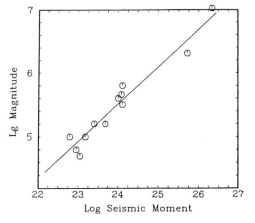

FIGURE 5. Independently determined values of seismic moment and m_{Lg} for twelve eastern North American earthquakes listed in TABLE 1. The least-squares linear fit to the data is given by the equation $m_{Lg} = 0.59$ $\log M_0 - 8.6$.

However, differences in ground motion amplitudes are expected to arise from differences in crustal structure, scattering, anelastic absorption, and characteristic site conditions between eastern and western North America. The influence of crustal structure, focal depth and other parameters on ground motion attenuation in eastern North America has been demonstrated by Barker et al.[3]

The seismic moment estimates of eastern North American earthquakes given in TABLE 1 can be used to construct an empirical relationship between seismic moment and Lg magnitude. The table contains twelve earthquakes whose seismic moments span the range from 6×10^{22} to 2×10^{26} dyne-cm and the m_{Lg} range from 4.7 to 7.0. Using the standard deviation factors of seismic moment given in TABLE 1 and a standard deviation of 0.3 for m_{Lg}, the least-squares procedure of York[8] yields the relation

$$m_{Lg} = 0.59 \log M_0 - 8.6$$

This relation, shown in FIGURE 5, allows earthquake size in the eastern North American catalog to be quantified in terms of seismic moment, thereby facilitating the use of simulation methods for ground-motion estimation in seismic hazard studies.

WAVE PROPAGATION MODELING OF STRONG GROUND-MOTION ATTENUATION

An important element in the estimation of seismic hazard in eastern North America is the accurate characterization of ground motion attenuation. The empirical data set is quite sparse in this region, particularly within 200 km of the source. Typical approaches to estimating ground-motion attenuation include scal-

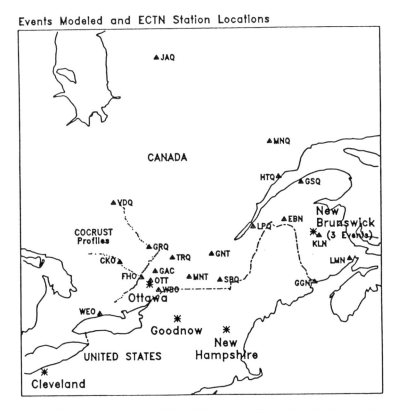

FIGURE 6. Map of the northeastern United States and adjacent Canada showing the seven earthquakes modeled in this study and the locations of the Eastern Canada Telemetered Network (ECTN) stations. Also shown as *dotted lines* northwest of Ottawa are the lines recorded during the COCRUST refraction survey.

ing and combining these empirical data and solving for the best-fitting one- or two-segment monotonically decreasing function of horizontal range. In this study, we use semiempirical synthetic seismogram techniques in order to model the ground motion attenuation for seven earthquakes in eastern North America. We find that along specific azimuths from a particular event, the ground motions result from the interference of upgoing, diving and postcritically reflected crustal S waves. The resulting attenuation curves contain far more detail than one or two simple linear trends. Combining data from different azimuths, or from events at different depths or in different crustal structures tends to obscure the details of attenuation.

The data modeled were digitally recorded on the Eastern Canada Telemetered Network (ECTN) from earthquakes in the northeastern United States and adjacent Canada (FIG. 6). For this paper, we will concentrate on the 1983 Ottawa earthquake and one of the aftershocks of the 1982 New Brunswick earthquake. Information on other events modeled, as well as further details and illustrations, may be found in Barker et al.[3]

Ground-Motion Attenuation in the Grenville Province near Ottawa

The detailed crustal velocity structure near Ottawa was obtained by modeling P and S waves from a nearby crustal refraction survey.[10] The locations of the refraction lines are shown as dotted lines northwest of Ottawa in FIGURE 6. The resulting crustal velocity structure model appropriate for the Grenville province near Ottawa is shown on the left side of FIGURE 7. It consists of a smooth crustal gradient and a total crustal thickness of 39.4 km.

The left side of FIGURE 8 shows a profile of the vertical velocity records from ECTN stations within 200 km of the 1983 Ottawa earthquake plotted with an S-wave reduction velocity. Superimposed are the travel times for a source at a depth of 15.5 km in the Grenville velocity structure model. The right side of FIGURE 8 includes a profile of semiempirical synthetic seismograms computed using generalized ray theory, including the upgoing direct S ray and diving rays that turn at each of the layer interfaces beneath the source. The response was computed along a line N45°E from the surface using the mechanism of Wahlstrom.[11] This was convolved with an empirical source function obtained by windowing and scaling the OTT record (at 20 km) to produce the seismograms shown. Comparing the two profiles, we see that although the observed data were recorded at a variety of azimuths, the arrival times and relative amplitudes of the S waves are well modeled. With the exception of unmodeled scattering and the development of Lg, which elongate the wavetrains recorded beyond 100 km, the character of the observed records is adequately described by the synthetic seismograms.

One of the benefits of waveform modeling with generalized ray theory is the ability to decompose the synthetic seismograms to obtain the responses for specific ray sets. FIGURE 9 includes such a decomposition, in which for a number of ranges, the top trace is due to the upgoing direct S wave (S_{up}), the second trace is due to diving S waves that turn within the crustal gradient (S_{div}), the third trace includes the reflection from the Moho and rays that turn in the upper mantle ($S_M S$), and the bottom trace is the total synthetic seismogram. At 20 km, the response is due entirely to S_{up}. At 80 km, S_{div} constructively interferes with S_{up}, and is beginning to appear as a late arrival. By 100 km, S_{div} destructively interferes with S_{up}, and $S_M S$ has reached its critical angle and is of comparable amplitude to S_{up}. At 120 km, S_{up} is now smaller in amplitude than S_{div}, but the peak amplitude is now due to $S_M S$. From 160 km to 200 km, $S_M S$ continues to account for the peak amplitude, but approaches S_{div} in time. The Moho head wave (S_n) may be seen as the first arrival at 200 km. The predicted peak ground-motion amplitude thus results from the interaction and interference of different S-wave phases, each of which sample different features of the crustal velocity structure. This interference is a sensitive function of range.

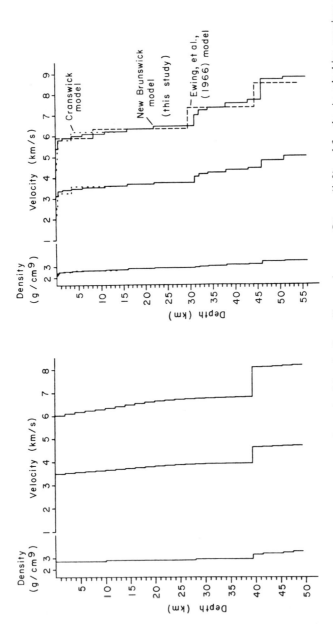

FIGURE 7. Crustal velocity structure models determined for the Grenville province near Ottawa (*left*) and for the Appalachian province near Miramichi, New Brunswick (*right*).

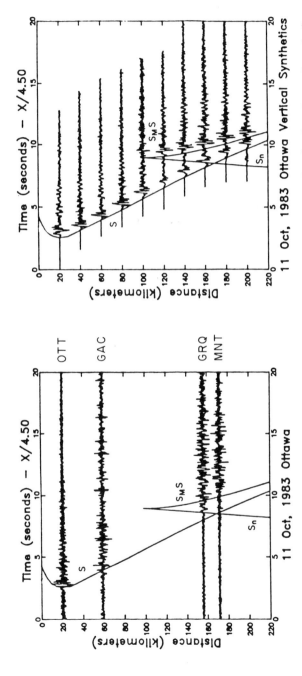

FIGURE 8. (*Left*) Vertical velocity recordings from ECTN stations within 200 km of the 1983 Ottawa earthquake. Superimposed are the S-wave travel-time curves for a source at 15.5-km depth in the Grenville velocity structure model. (*Right*) Semiempirical, synthetic vertical velocity seismograms for a profile running N45°E from the Ottawa earthquake.

The left side of FIGURE 10 shows a comparison of the observe and synthetic peak vertical velocity measured in the time domain as a function of epicentral range for the Ottawa earthquake. Observed ground motions are plotted as open symbols, while synthetic ground motions are plotted as closed symbols. Beyond about 200 km, the peak observed ground motions are due to Lg, while our synthetic seismograms include only crustal S waves. At a range of about 200 km, we expect crustal body waves and Lg waves in the observed data to have comparable

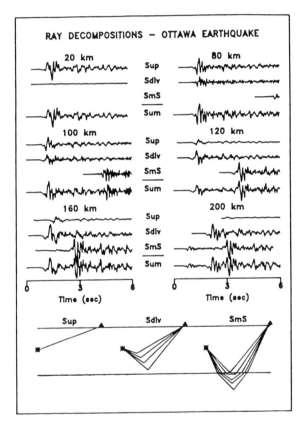

FIGURE 9. Ray decompositions of the semiempirical synthetic seismograms for the Ottawa earthquake. For each range, the *top trace* includes only the upgoing, direct S wave (S_{up}); the *second trace* includes diving S waves that turn within the crustal gradient (S_{div}); the *third trace* includes the S wave that reflects from the Moho and rays that turn within the mantle (S_MS); and the *bottom trace* is the total synthetic seismogram.

amplitudes. Although computed along a single azimuth, the synthetic peak values appear to adequately interpolate the observed values between 20 and 200 km. Also indicated as dashed lines are the peak synthetic velocities obtained when only the upgoing S, diving S, or S_MS rays are included. The level of the upgoing S amplitude is almost constant due both to the source depth and to the rotation of the S wave such that an increasing amount of the S energy appears on the vertical component as the incidence angle becomes more horizontal. The diving S wave

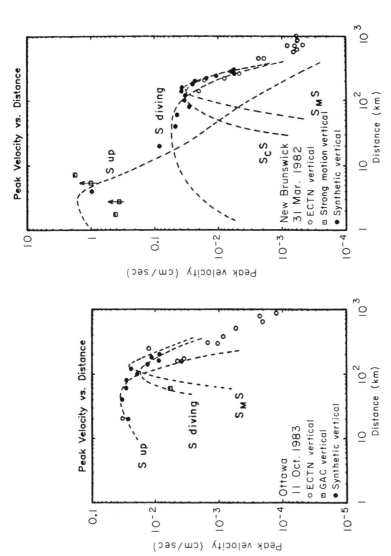

FIGURE 10. (*Left*) Attenuation of observed (*open symbols*) and synthetic (*solid symbols*) peak vertical velocity for the Ottawa earthquake. The ground motion at station GAC has been corrected to resemble ECTN data. Also shown on the left is the attenuation of peak synthetic velocity when only specific sets of rays are included. (*Right*) Attenuation of observed and synthetic peak vertical velocity for the March 31, 1982 New Brunswick aftershock. Strong-motion recordings from within 10 km (*open boxes*) have been corrected to resemble ECTN data. Also shown on the left is the attenuation of peak synthetic velocity when only specific sets of rays are included.

becomes important beyond about 60 km, but arrives simultaneously with the upgoing S, causing complicated interference phenomena. Between 100 and 120 km, the Moho-reflected S wave becomes postcritical and accounts for the largest ground motion to beyond 200 km.

Ground-Motion Attenuation in the Appalachian Province near Miramichi

The structure model developed for the Appalachian province is shown on the right side of FIGURE 7. It was obtained by fitting simultaneously the arrival times of P and S waves from three principal aftershocks of the 1982 Miramichi, New Brunswick earthquake recorded on the ECTN. The resulting model has a mid-crustal discontinuity and a total crustal thickness of 46 km. Semiempirical synthetic seismograms were computed for the March 31, 1982 aftershock using generalized ray theory for the source mechanism of Nabelek[12] at a depth of 3.5 km. An empirical source function was obtained by windowing the vertical S wave from a strong-motion recording 7.2 km from the source, integrating to velocity and correcting to the ECTN instrument response. The resulting synthetic profile along a N60°W azimuth is compared with ECTN recordings within 300 km in FIGURE 11. Once again, we see that the timing and relative amplitudes of the S-wave phases within 200 km are well modeled. Beyond 200 km, when arrivals can be interpreted from within the P coda, the arrival times are well matched, but the relative amplitudes suggest that the gradient beneath the mid-crustal discontinuity may be slightly too strong. Overall, with the understanding that scattering and the onset of Lg beyond 200 km are not modeled, the character of the observed waveforms has been adequately modeled by the semiempirical synthetic seismograms.

FIGURE 12 includes a ray decomposition of the synthetic seismograms for the New Brunswick aftershock. The ray sets are comparable to those of FIGURE 9, except that rays that reflect from the midcrustal discontinuity, or turn in the gradient between this discontinuity and the Moho are denoted S_CS. At 20 km, S_{up} and S_{div} arrive at the same time and with comparable amplitudes. The result is a complicated interference of the two. By 60 km, S_{up} is diminished, S_{div} accounts for the peak amplitude, and S_CS is beginning to appear. At 120 km, S_CS has passed its critical angle and contributes the peak amplitude, although S_{div} is significant. At 180 km S_MS has passed its critical angle, and the total synthetic seismogram shows a complicated interference of S_{div}, S_CS and S_MS. At 240 km and beyond, the head waves (S_n and S*) may be seen in front of S_{div} and the reflected part of S_CS, but the latter still control the peak amplitudes.

The comparison of the observed (open symbols) and synthetic (solid symbols) peak vertical velocities for the New Brunswick aftershock is shown on the right side of FIGURE 10. The observed ground motions at ranges less than 10 km are measured from strong-motion recordings which have been corrected to resemble ECTN velocity records. Once again, the observed ground motions beyond about 200 km reflect Lg waves, which are not included in the synthetic seismograms. The synthetic ground motions interpolate the observations between 10 and 200 km quite well, while defining the detailed shape of the attenuation curve over these ranges. Also shown are the synthetic values of peak velocity obtained if only S_{up}, S_{div}, S_CS or S_MS are included. At a 3.5-km depth, the March 31 aftershock is much shallower than the Ottawa earthquake (15.5 km), so the peak ground motion of the direct, upgoing S wave decays much more rapidly with epicentral range. The peak ground motion between 40–100 km is controlled by the diving S wave that turns within 5–10 km beneath the source. Postcritical reflections from the mid-crustal

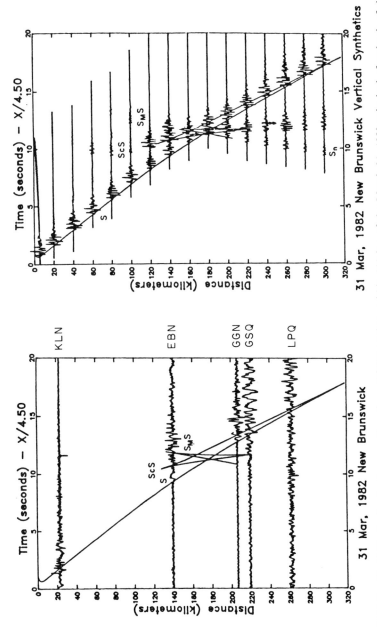

FIGURE 11. (*Left*) Vertical velocity recordings from ECTN stations within 300 km of the March 31, 1982 principal aftershock of the New Brunswick earthquake. Superimposed are the S-wave travel-time curves for a source at 3.5-km depth in the Appalachian crustal velocity structure model. (*Right*) Semiempirical vertical velocity seismograms for a profile running N60°W from the March 31, 1982 New Brunswick aftershock.

and Moho discontinuities combine with the upper-crustal turning ray to increase peak amplitudes between 120–160 km and to define the attenuation until Lg dominates. Clearly, the ground-motion attenuation within about 200 km is a complex interference of S waves from four distinct raypaths, each having its own depth- and structure-dependent rate of attenuation.

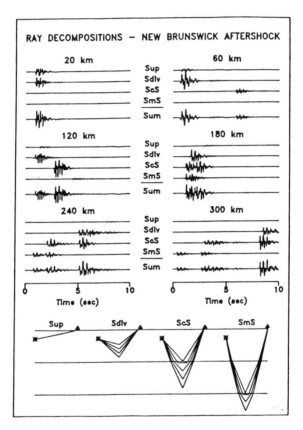

FIGURE 12. Ray decompositions of the semiempirical synthetic seismograms for the New Brunswick aftershock. For each range, the *top trace* includes only the upgoing, direct S wave (S_{up}); the *second trace* includes diving S waves that turn within the crustal gradient (S_{div}); the *third trace* includes the S wave that reflects from the midcrustal discontinuity and rays that turn between the depth and the Moho (S_CS); the *fourth trace* includes similar reflections from the Moho and rays that turn within the mantle (S_MS); and the *bottom trace* is the total synthetic seismogram.

Sensitivity of Ground-Motion Attenuation to Crustal Structure and Source Depth

We have shown through semiempirical synthetic waveform modeling of sparse observations that peak ground motions in the northeastern United States and adjacent Canada are controlled by (in order of increasing range) the direct upgoing

S wave, the diving S wave that turns within the crust, postcritical reflections from mid-crustal and Moho discontinuities and, finally, Lg. Each of these wave types has its own form of attenuation, and the total ground motion attenuation is the result of the complex interaction and interference of these phases.

We have developed crustal velocity models for the Grenville and Appalachian provinces, which differ primarily in crustal thickness and the existence or lack of a mid-crustal discontinuity. By comparing synthetic ground motions for events of similar depth and source mechanism but located within different velocity structure, we may isolate the effect of gross crustal structure on the shape of the attenuation curves. Of the events modeled by Barker *et al.*,[3] the 1983 Goodnow, New York earthquake occurred in the Grenville structure whereas the January 11 aftershock of the New Brunswick earthquake occurred in the Appalachian structure. Otherwise, each was located at the same depth and both had reverse faulting mechanisms. The left side of FIGURE 13 includes a comparison of peak vertical acceleration for these events, in which we have scaled the peak ground motions of the Goodnow event to the moment of the New Brunswick aftershock. The Grenville ground motions (shown as solid circles) reflect large direct S at 20 km, low amplitudes at 40–80 km as direct and diving S interfere, and a slight increase in amplitude beyond the S_MS critical range at 100 km. The Appalachian ground motions (open circles) much more variable in the 20–100 km range, reflecting the complex interaction of direct and diving S in the presence of the mid-crustal discontinuity. It is apparent from this comparison that ground motion attenuation within 200 km of the source is highly sensitive to the gross features of the crustal velocity structure between the source and receivers.

On the right side of FIGURE 13, we illustrate the effect of source depth by comparing synthetic peak accelerations for the Ottawa earthquake (15.5 km, open circles) and for the Goodnow earthquake (7.0 km, solid circles). Both occurred in the Grenville structure and both had shallow reverse mechanisms. We have scaled the Goodnow ground motions to the moment of the Ottawa source. The Ottawa ground motions are controlled by the upgoing S wave out to about 80 km and the attenuation curve is quite broad. For the Goodnow event, upgoing S interferes with diving S between 40 and 100 km, causing peak ground motions to drop much more rapidly than those of a deeper source. This interaction of the direct, upgoing S wave and the diving S wave that turns within the crust appears to be the most important factor in determining ground motion attenuation within 100 km from the source. Further, this interaction appears to be most sensitive to source depth.

Implications for Strong Ground-Motion Estimation

These results indicate that ground motion attenuation within a given region (crustal structure) may be described by a family of characteristic curves that depend primarily on the source depth within that structure. When strong motion data from regions with contrasting crustal structures, or from events having different focal depths, are combined into a single data set, the detailed characteristics of regional attenuation relations are smeared out, leaving a data set having broad scatter that is most reasonably fit using a smooth attenuation curve. If the purpose of the attenuation curve is to predict ground motions over a very wide region such as eastern North America, the use of such a curve seems appropriate. However, if the purpose of the attenuation curve is to predict ground motions at a given site or within a given region, it seems more appropriate to use an attenuation curve that reflects the wave propagation characteristics of that region. This

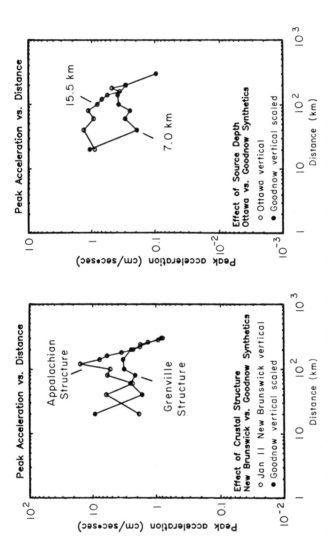

FIGURE 13. (*Left*) A comparison of synthetic peak vertical acceleration for an event in the Appalachian crustal structure (January 11 New Brunswick aftershock, *open circles*) and an event in the Grenville crustal structure (Goodnow, *solid circles*). The Goodnow data have been scaled to the moment of the New Brunswick event. (*Right*) A comparison of synthetic peak vertical acceleration for two events in the Grenville crustal structure at different depths. The Ottawa earthquake (*open circles*) was at 15.5-km depth, while the Goodnow earthquake (*solid circles*) was at 7.0-km depth. The Goodnow data have been scaled to the moment of the Ottawa event.

should produce a reduction in the uncertainty in ground motion attenuation in the region. Our study has shown how empirical data and simulation methods that explicity embody wave-propagation effects can be used toward this end.

SUMMARY

From waveform modeling of body waves, we have found that earthquakes in eastern North America and other continental interiors have constant stress drop source scaling, and stress drops whose median value of about 100 bars is not significantly different from that of western North American events. This suggests that differences in ground motions between eastern and western North America may be mainly due to differences in crustal structure, scattering, absorption (Q), and local site conditions. Analysis of empirical ground-motion data in the east using theoretical seismograms indicates that ground-motion amplitudes are controlled (in order of increasing range) by the upgoing direct S wave, the diving S wave, postcritical reflections from the mid-crustal and Moho discontinuities, and finally Lg. The shape of the ground-motion attenuation curve is controlled by the relative amplitudes and interference of these rays, and so is very sensitive to crustal structure and source depth.

ACKNOWLEDGMENTS

Many of the results of these studies were predicted by Don Helmberger, and we are grateful for his guidance. Jeff Barker, Roy Burger, Vicki LeFevre, and Jim McLaren participated in the work. These studies are part of a program being conducted by the Electric Power Research Institute under the direction of Dr. J. Carl Stepp to develop seismic wave attenuation functions for the eastern United States.

REFERENCES

1. BOORE, D. M. & G. M. ATKINSON. 1987. Stochastic prediction of ground motion and spectral response parameters at hard-rock sites in eastern North America. Bull. Seismol. Soc. Am. **77**: 440–467.

2. BURGER, R. W., P. G. SOMERVILLE, J. S. BARKER, R. B. HERRMANN & D. V. HELMBERGER. 1987. The effect of crustal structure on strong ground motion attenuation relations in eastern North America. Bull. Seismol. Soc. Am. **77**: 420–439.

3. BARKER, J. S., P. G. SOMERVILLE & J. P. McLAREN. 1987. Modeling of Ground Motion Attenuation in the Eastern North America. NP-5577. Electric Power Research Institute. Palo Alto, CA.

4. SOMERVILLE, P. G. 1986. Source Scaling Relations of Eastern North American Earthquakes. NP-4789. Electric Power Research Institute. Palo Alto, CA.

5. EBEL, J. E., P. G. SOMERVILLE & J. D. McIVER. 1986. A study of the source parameters of some large earthquakes in eastern North America. J. Geophys. Res. **91**: 8231–8247.

6. SOMERVILLE, P. G., J. P. McLAREN, L. V. LeFEVRE, R. W. BURGER & D. V. HELMBERGER. 1987. Comparison of source scaling relations of eastern and western North American earthquakes. Bull. Seismol. Soc. Am. **77**: 322–346.

7. NUTTLI, O. W. 1983. Average seismic source-parameter relations for mid-plate earthquakes. Bull. Seismol. Soc. Am. **73**: 519–535.

8. YORK, D. 1966. Least squares fitting of a straight line. Can. J. Phys. **44:** 1079–1086.
9. KANAMORI, H. & C. R. ALLEN. 1986. Earthquake repeat time and average stress drop. *In* Earthquake Source Mechanics, Geophysical Monograph 37, (Maurice Ewing Series 6): 227–235. American Geophysical Union. Washington, DC.
10. MEREU, R. F., D. WANG, O. KUHN, D. A. FORSYTH, A. G. GREEN, D. MOREL, G. G. R. BUCHBINDER, D. CROSSLEY, E. SCHWARTZ, R. DUBERGER, C. BROOKS & R. CLOWES. 1986. The 1982 COCRUST seismic experiment across the Ottawa-Bonnechere graben and Grenville Front in Ottawa and Quebec. Geophys. J. **84:** 491–514.
11. WAHLSTROM, R. 1987. The North Gower, Ontario, earthquake of 11 October, 1983: Focal mechanism and aftershocks. Seismol. Res. Lett. **3:** 65–72.
12. NABELEK, J. 1985. The January 9, 1982, New Brunswick, Canada earthquake. *In* A Study of New England Seismicity with Emphasis on Massachusetts and New Hampshire: Technical Report Covering 1982–1984. Earth Resources Laboratory, Massachusetts Institute of Technology. Cambridge, MA.

General Discussion: Part II

KLAUS JACOB, CHAIRMAN (*Lamont Doherty Geological Observatory, Palisades, N.Y.*): To open discussion let me ask you a question, Dr. Somerville: Am I correct in saying that stress drop in the eastern United States is on average comparable or not distinguishably higher than in the West?

PAUL SOMERVILLE (*Woodward-Clyde Consultants, Pasadena, Calif.*): That's correct.

JACOB: Now let me press you a little bit harder on this, for those earthquakes where we have probably the best observations in defining the source size, such as the examples addressed by Leonardo Seeber on the Goodnow earthquake. Would you still maintain your position or do you think this is a case where scientists may quibble?

SOMERVILLE: First, I'd like to point out that our study covered the magnitude range of about 5 to 7, which is the magnitude range of engineering concern. And, of course, the Goodnow and the New Brunswick earthquakes were in our data set, but at the lower end of it. I don't rule out the possibility that something different may happen for smaller earthquakes.

My view is that the maximum strength of the crust in eastern North America is similar to that in western North America, and that maximum strength is somewhere on the order of 100 bars when an asperity breaks. The fact that the stress drops of large earthquakes in western North America tend to decrease with earthquake size can be explained by the suggestion that in the West, while fault asperities exist, they are separated by large intervals of weak fault zone. But when an earthquake happens, it usually happens because an asperity has to be broken, and when that breaks it breaks with about the same stress drop in the West as when an old fault in the East breaks.

JACOB: Would you reiterate why we even worry about the high stress drop?

SOMERVILLE: Because, using simple spectral scaling models, the peak acceleration is proportional to the two-thirds power of the stress drop. So, for example, Professor Kanamori has said that perhaps there's a factor of 5 difference; that eastern events have a stress drop that may be five times larger than that in the West. That would translate into peak accelerations three times higher in the East than the West.

JACOB: For the high-frequency motions?

SOMERVILLE: For frequencies above the corner of the event. For everything about 1 Hz, Kanamori would have to say that the peak accelerations would be three times higher. But I don't think that's borne out in the data, and I don't think that when you look at eastern earthquakes and compare them with western earthquakes in the manner that we have done, you see that difference.

UNIDENTIFIED SPEAKER: Your second conclusion suggested that the source mechanism details are not important, that it's crustal depth and depth of earthquake and crustal structure. How does this jibe with concerns in western earthquakes about asperities, asperity locations, and so on having a significant impact upon at least the near-source accelerations?

SOMERVILLE: That comment referred to the focal mechanism, to the orientation of the fault plane, in other words, whether it was a strike slip or a reverse or a normal fault. It wasn't referring to these complications of the source related to asperities and so forth.

But on that subject, looking at these older larger events, like the Timiskaming, Charlevoix, Ossipee earthquakes, a study was done by John Ebel and me and another individual where we looked, admittedly teleseismically, at rather long-period information. But we saw the same kinds of things that we see for western events. One event was very simple. The next event was complex, with maybe a couple of asperities. And we couldn't really distinguish that kind of behavior from western events that sometimes are simple and other times are complex. So, systematically we don't see any difference between eastern events and western events in terms of their source characteristics. But, as I said, the data that we're using are rather limited to low frequencies.

The reason we are simulating strong ground motions for eastern North American earthquakes is because we don't have strong-motion records from large earthquakes in the East. And we simply don't know very much about the high-frequency behavior, certainly of larger events, in the East. So, we're trying to gain whatever information we can by other means.

JACOBO BIELAK (*Carnegie-Mellon University, Pittsburgh, Pa.*): From a practical point of view, I think that the idea of having synthetic accelerograms should be emphasized more than it has in the past. It is very important to get peak values of acceleration, displacement, velocity, and so on. But as we get into inelastic design and response it is also very important to have synthetic accelerograms that allow us to do further analysis and to better understand the behavior of structures. Clearly one value is not sufficient when we have to know the complete history.

JACOB: Dr. Weichert, you showed us accelerations of 200% *g*. Are they really of consequence? Or, since you have not seen them on the second strong-motion instrument, is it a totally erratic localized feature that is so close to the recording site that it is just important within a few hundreds of meters?

DIETER WEICHERT (*Pacific Geoscience Centre, Sidney, B.C., Canada*): No. To start with, the strong event is clearly visible on site 2. It's also visible on site 3.

JACOB: I'm not sure whether that was an earthquake.

WEICHERT: On site 3, 30 kilometers away? In fact, if you look carefully, it is seen even in the displacement on site 3. But the point is, if I take peak accelerations from these three sites for aftershocks and I plot all 66 of them, which I have available at the moment, I find the peak acceleration in that first 20 kilometers attenuates as distance to the −2½. It attenuates very rapidly.

With this attenuation I don't need a rupture closer than 4 kilometers to site 1. If it moved 4 kilometers under site 1, I'm getting that size of acceleration. That would just be another splay fault.

It was new to me to hear that someone said he had placed the second event even shallower than the first one. I showed it at the same angle. If it was shallower, this would confirm the idea of a splay fault or an imbricate fault. And the solution, the source mechanism solution of the second event is also shallower. So, it may have broken the way I indicated on the splay fault, and there may be more of these splay faults. All I need is something 4 kilometers deep, as against 6 kilometers. I would estimate the asperity size on the order of a few hundred meters, not 5 kilometers, as someone else had in his solution, but a few hundred meters at a depth of 4 or 5 kilometers.

JACOB: Let's sort that out. You present the point of view of an earth scientist who is interested in the rupture process. Should engineers be interested in it? How large is the area over which the high accelerations can be found? We cannot exclude the fact that in a very small area something unusual is going to happen occasionally. The question is: Should we worry about such high accelerations in a hazard sense or ultimately in a risk sense?

WEICHERT: In a general hazard sense, perhaps not. But if you look at critical structures, you should know what you can come up against. It is not just a high-frequency spike that does not do anything to the response spectrum, but it does raise the response spectrum significantly. So, the high peak accelerations measured are quite visible there.

JACOB: How do you explain the rapid fall-off with distance that you just described for the high frequencies? For what frequencies is that high fall-off?

WEICHERT: Which high frequencies? The large spike is about 8 Hz.

JACOB: You described a very rapid fall-off with distance.

WEICHERT: I can't explain that. But I see this strong attenuation near our array in the west of Canada; the same $-2\frac{1}{2}$ exponent over the first 10–20 kilometers, peak acceleration going down faster than the distance squared; $-2\frac{1}{2}$. Call it 2 if you like, but uncertainty is no more than $\pm.3$.

JACOB: Do you attribute that to a Q or any other things? That's attenuation!

WEICHERT: Yes, that's attenuation.

JACOB: Attenuation in an unelastic or scattering sense . . .

WEICHERT: I would think so because it's the very shallow layers in which it is being attenuated. Also, it's not a constant frequency that attenuates there; the higher frequencies go down very quickly. At the more distant sites you see slightly lower frequencies. So, perhaps the higher frequencies go even faster.

Implications of Eastern Ground-Motion Characteristics for Seismic Hazard Assessment in Eastern North America

GAIL M. ATKINSON

125 Dunbar Road S.
Waterloo, Ontario, Canada N2L 2E8

INTRODUCTION

There has been much recent work on the nature of ground motion in eastern North America (ENA).[1-3] This has led to a new understanding of ENA ground motions, and in particular how they differ from better-observed western North American (WNA) ground motions.[4] These differences in characteristics, combined with those in the occurrence rates of earthquakes, have important implications for seismic hazards in eastern North America.

The purpose of this paper is to review the salient features of ENA ground motions and their implications for seismic hazard in simple engineering terms. Illustrative examples show typical expected response spectra for probability levels relevant to building codes and critical facilities. Conclusions are drawn as to how seismic hazard in ENA differs from that in the more active western regions.

CHARACTERISTICS OF ENA EARTHQUAKE OCCURRENCE

Ground-Motion Characteristics

Understanding of the generation and propagation of seismic ground motion in ENA has advanced significantly in the past five years because of improvements in the empirical database, coupled with advances in theoretical modeling techniques. Until five years ago, the ground-motion database for ENA consisted almost entirely of qualitative observations, such as maps of Modified Mercalli Intensity (MMI) from historical earthquakes. At present, the database contains about a hundred quantitative recordings, including peak ground accelerations, velocities, and response spectra. These records were gathered during recent moderate (magnitude > 4) earthquakes in New Brunswick, New Hampshire, New York, Ontario and Ohio. For some of these events there are strong-motion recordings of the motions at distances close to the earthquake source,[5-7] while other events were recorded by digital seismographic networks at distances of 100 to 1000 km.[8] Ground-motion data show that ENA earthquakes have more energy at high frequencies than do western events. Western earthquakes typically exhibit a rapid decrease of ground-motion amplitudes at frequencies above approximately 15 Hz,[9,10] whereas amplitudes for ENA events do not decline rapidly until frequencies of about 40 Hz or greater.[1] The high-frequency cutoff (often labelled f_{max}) may be a site effect[9]; the relatively high f_{max} values in ENA could be attributed to the competent crustal conditions, which allow propagation of high-frequency energy. Alternatively, f_{max} may be a source effect.[11] Because peak ground accelerations increase with increasing high-frequency content, higher values are observed in ENA than in WNA, for records at the same magnitude and distance.[1] This fact was not known prior to the recording of ENA earthquakes at near-source distances.

The quantitative database also shows that eastern ground motions decay more slowly with distance than do their western counterparts. This corroborates earlier MMI observations, which indicated that eastern earthquakes are felt at very large distances. The slower eastern attenuation is attributed to the relatively stable and unfractured crust.

The available ground-motion data have been well matched by a theoretical model.[1-3] The model, referred to as the stochastic model, has its origins in the work of Hanks and McGuire,[12] who showed that high-frequency ground motions can be treated as band-limited Gaussian white noise. The spectral amplitudes are determined by a seismological model of the source spectrum, filtered by regional attenuation properties. In the stochastic model, differences between eastern and western ground motions arise primarily from different f_{max} values, and from differences in anelastic attenuation. Different source properties and crustal constants are also factors. There is now reasonable confidence in ground-motion relations derived using the model, although its validity for large magnitude earthquakes in ENA is still subject to ongoing debate concerning the underlying seismic source model.[2,4,13]

To illustrate the key differences between eastern and western ground motions, FIGURE 1 shows random horizontal acceleration time histories of simulated events of moment magnitude (M) 5, at a hypocentral distance (R) of 10 km, for both regions. The simulations are based on the stochastic model, using the methodology of Boore.[10] Note the much higher frequency content of the eastern records, and the higher peak ground accelerations.

The higher peak ground acceleration (pga) values for eastern earthquakes do not necessarily imply a greater damage potential for most structures, because the eastern peak ground accelerations are carried by frequencies higher than those to which most structures respond. FIGURE 2 illustrates the point by showing smoothed median response spectra (horizontal component) for events of M 5 and 7 at $R = 10$ and 100 km, for eastern and western events (from the relations of Boore and Atkinson[2] and Joyner and Boore,[14] respectively). The plots show the pseudo-acceleration (psa) values as a function of natural vibration frequency, for 5% of critical damping. The psa values are the maximum acceleration that a simple oscillator would experience during the ground motion. This method of depicting ground motions brings out their engineering significance, because many analysis methods for the response of structures use as their input the "design" psa value for the natural frequency of the structure. (There may be more than one natural frequency if several modes of vibration are possible; the natural frequency of the structure is determined by analyzing its stiffness and geometry.) Because the psa values are directly related to structural response, higher psa values imply greater damage potential.

The response spectral plots show that at distances close to the earthquake source, motions will be similar in the east and west for frequencies less than 10 Hz. This implies that the damage potential of the events would be similar for structures with vibration modes in this frequency range, such as most buildings, major dams, and nuclear power plant reactor buildings. For frequencies above 10

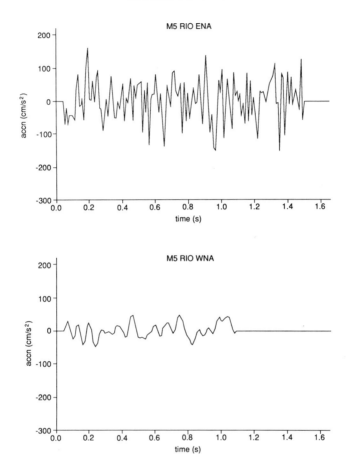

FIGURE 1. Simulated time histories for M5 earthquake at $R = 10$ km for ENA and WNA.

Hz, psa values are higher in the east as a consequence of the higher f_{max} values. This may imply greater damage potential for high-frequency structures such as small concrete dams and some nuclear power plant equipment; it also suggests that higher vibration modes may be important for many structures.

At large distances, the response spectra of eastern earthquakes are significantly higher than the equivalent western ones at all frequencies because of the effects of differing attenuation rates. This implies that eastern earthquakes may cause damage (or at least be strongly felt) for greater distances than their western counterparts. Historical observations corroborate this prediction.

Earthquake Occurrence Rates

Another important factor in comparing seismic hazards for different regions is differences in the occurrence rate of significant events. FIGURE 3 shows that although damaging earthquakes (say M > 5 to 6) are by no means restricted to

WNA, they are certainly more frequent there, particularly in western California. In ENA, events tend to be more diffuse, correlate poorly with individual faults, and are relatively infrequent. There is considerable variability in seismicity patterns in ENA, with some areas (e.g., Charlevoix, Quebec) being very active, some areas (e.g., the Canadian prairies) nearly aseismic, and other areas (e.g., the Appalachians and much of the St. Lawrence River valley) being moderately active.

As a broad generalization, one might say that earthquakes of any given magnitude are about 10 to 20 times more frequent in the active areas of California than in the moderately active regions that typify much of ENA. Of course, this implies greater hazard in the west than the east. As an interesting aside, however, note that this does not necessarily imply greater *risk*, since risk is also a function of consequence; it may be that the consequences of a large earthquake are more severe in many eastern areas because of the high population density. This paper does not attempt to treat this issue, and is restricted to comparing *hazards* only.

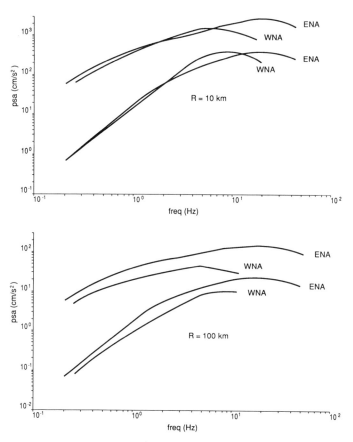

FIGURE 2. Median response spectra (5% damped) for M5 and M7 earthquakes at $R = 10$ and $R = 100$ km for ENA and WNA.

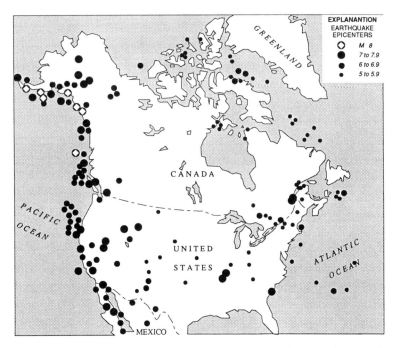

FIGURE 3. Epicenters of large (M > 6) historical earthquakes in North America. M > 5 events since 1930 also plotted for region east of 110° W. (After Page and Basham.[15])

Another factor to consider when comparing eastern and western hazards is that maximum possible magnitudes (M_x) are likely higher in many active western regions than for most of ENA. Extended fault sources near the California coast would be capable of larger earthquakes than would areas in the stable craton, for example. However, large events (M > 7) have occurred in parts of ENA, often those underlain by ancient rift systems such as the St. Lawrence River. This suggests that for many ENA situations M_x values may be similar to those for typical western fault systems, although the occurrence of such events may be less frequent.

SEISMIC HAZARD IMPLICATIONS

In this section, the implications of the differences in earthquake characteristics between ENA and WNA are considered. To illustrate the significance of the effects for seismic design of typical structures, a simple probabilistic hazard analysis is performed to obtain expected psa values for probability levels of 0.002 (500-year "return period") and 0.0001 (10,000-year return period) per annum (p.a.). The 0.002 p.a. values are often used as the design basis in building codes for conventional structures, whereas the 0.0001 p.a. values are often used for critical facilities. The expected psa values for these probabilities are obtained by the Cornell-McGuire[16-18] method. The method integrates contributions to the exceedence probability of a specified psa value, at a given natural frequency, by summing over all possible magnitudes and distances. The input parameters to the calculations are the regional ground-motion relations, and the earthquake occurrence rates and maximum magnitudes for all defined seismic source zones or faults.

For demonstration purposes, a large homogeneous source zone is assumed, about 200 km by 200 km, with the site located in the middle. Ground-motion relations (for 5% damped horizontal psa) are assumed to be as given by Boore and Atkinson[2] for ENA, and by Joyner and Boore[14] for WNA. The relations are for rock sites. (The effects of fault rupture length and different distance definitions are ignored for this illustration.) For the eastern example, earthquake occurrence rates within the zone are chosen so as to be fairly typical of moderately active areas. Accordingly, it is assumed that the rate for M5 is N5 = 0.02, and the recurrence relation has slope b = 0.9 (also a typical value). For the western example, it assumed that N5 = 0.4, and b = 0.9, typical of an active western area.

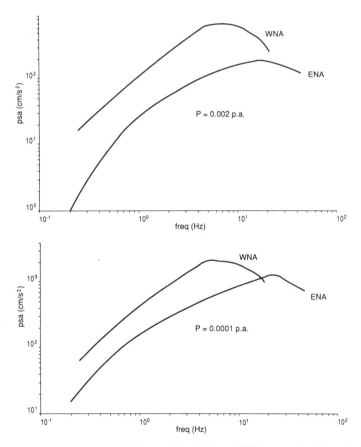

FIGURE 4. Response spectra (5% damped) for ENA and WNA examples for probabilities of 0.002 and 0.0001 per annum.

These parameters mean that in the eastern zone an M5 event would occur once in 50 years on average, an M6 once in 400 years, and an M7 once in 3200 years. In the western zone an M5 quake would occur on average once in 2 years, an M6 event once in 20 years, and an M7 once in 160 years. For both regions it is assumed that the maximum possible magnitude is 7.25.

FIGURE 4 shows the results of the hazard analysis, for probability levels 0.002 and 0.0001 p.a., for both the eastern and western cases. The results apply to rock sites. The important feature to note is the difference in the shape of the expected ENA spectra relative to that of the WNA case. Expected psa values for the two cases are similar for frequencies above 10 Hz, because the high-frequency content of eastern earthquakes is offset by the lower occurrence rates. For lower frequencies, the hazard is much greater in WNA because of the higher occurrence rates. Note that the differences at low frequencies would be even more pronounced if a lower M_x value for the eastern region, relative to the western region, had been assumed.

The difference in shape of the expected spectra has important implications. First, any attempt to scale spectra to an index parameter such as pga would have to follow very different rules of thumb in the east than those devised for WNA, or the result would be gross overconservatism for low frequencies. Standard spectra such as those developed by Newmark, Blume and Kapur for NRC Regulatory Guide 1.60 should not be used. In any event they are unnecessary since expected spectra can easily be calculated without resort to scaling by peak ground-motion parameters.

Another consequence of the difference in shape is that it implies a fundamental difference in the nature of the hazard in ENA. For structures with significant high-frequency modes of vibration, the hazard may be just as great in much of ENA as for active areas of California. For structures with longer periods, the hazard will be much less in ENA. Thus generalizations about earthquake hazards derived from California experience cannot simply be "transported" to the east on a scaled-down level. The types of structures at risk, and the types of damage to be expected, are potentially different. Note also that soil response, which has not been considered here, may cause further significant differences.

CONCLUSIONS

The nature of seismic hazard in ENA differs from that in WNA due to the combined effects of differences in the character of ground motion, and in recurrence rates. Expected spectra for a constant probability level, which are a good indication of appropriate seismic design parameters and/or damage potential, have different shapes in the two regions. High-frequency structures in many parts of ENA face a hazard comparable to that in many seismically active areas of California, whereas the hazard for low-frequency structures in ENA is relatively modest.

SUMMARY

The salient features of eastern North American (ENA) ground motions are reviewed, and their implications for seismic hazard demonstrated by an illustrative example. Recent data show that eastern ground motions are enriched in high frequencies. For seismic hazard evaluation, both the characteristics of the ground motions and the occurrence rates of earthquakes are important. When the differences in eastern and western ground motions are combined with those in occurrence rates, the implications for seismic hazard can be seen. The shape of response spectra for a constant probability is quite different for a typical ENA site than it is for a typical California site. High-frequency structures in many parts of ENA face a hazard comparable to that in many seismically active areas of California, whereas the hazard for low-frequency structures in ENA is relatively modest.

REFERENCES

1. ATKINSON, G. 1984. Bull. Seismol. Soc. Am. **74:** 2629–2653.
2. BOORE, D. & G. ATKINSON. 1987. Bull. Seismol. Soc. Am. **77:** 440–467.
3. TORO, G. & R. McGUIRE. 1987. Bull. Seismol. Soc. Am. **77:** 468–489.
4. BOORE, D. & G. ATKINSON. 1987. Proceedings of the EPRI Workshop on Strong Ground Motion Predictions in ENA. Palo Alto, CA.
5. WEICHERT, D., P. POMEROY, P. MUNRO & P. MORK. 1982. Earth Phys. Branch Open File Rep. **82-34.** Ottawa, Ontario, Canada.
6. CHANG, F. 1983. NUREG/CR-3327. U.S. Nuclear Regulatory Commission. Washington, DC.
7. BORCHERDT, R. 1986. U.S. Geol. Surv. Open File Rep. **86-181.**
8. ATKINSON, G. 1985. Earth Phys. Branch Open File Rep. **85-5.** Ottawa, Ontario, Canada.
9. HANKS, T. 1982. Bull. Seismol. Soc. Am. **72:** 1867–1879.
10. BOORE, D. 1983. Bull. Seismol. Soc. Am. **73:** 1865–1894.
11. PAPAGEORGIOU, A. & K. AKI. 1981. Proceedings of the U.S. Geol. Surv. Nuclear Regulatory Commission Workshop on Strong Motion, Lake Tahoe.
12. HANKS, T. & R. McGUIRE. 1981. Bull. Seismol. Soc. Am. **71:** 2071–2095.
13. BOORE, D. 1988. This volume.
14. JOYNER, W. & D. BOORE. 1982. Proceedings of the 51st Annual Convention of Structural Engineering Association (California).
15. PAGE, R. & P. BASHAM. 1985. U.S. Geol. Surv. Bull. **1630.**
16. CORNELL, C. 1968. Bull. Seismol. Soc. Am. **58:** 1583–1606.
17. McGUIRE, R. 1976. U.S. Geol. Surv. Open File Rep. **76-67.**
18. McGUIRE, R. 1977. Int. J. Earthquake Eng. Struct. Dyn. **5:** 211–234.

Issues in Strong Ground-Motion Estimation in Eastern North America[a]

ROBIN K. McGUIRE AND GABRIEL R. TORO

Risk Engineering, Inc.
5255 Pine Ridge Road
Golden, Colorado 80403

INTRODUCTION

The estimation of quantitative characteristics of strong earthquake ground motion in eastern North America (ENA) is difficult. No instrumental records are available from destructive shaking in the region, and there are perceived differences between shaking in ENA and in California, thereby invalidating the use of empirical equations from California (where data are abundant) in ENA. Quantitative estimation of ground motion is a critical step in the mitigation of risk for existing facilities and in the selection of design criteria for new facilities. Therefore in the absence of empirical observations of strong shaking we must evaluate and use relevant theories, low-amplitude seismographic observations, and data from other regions to deduce what might be the range of strong ground-motion characteristics in ENA. Correct decisions (in the sense of optimal decisions under uncertainty) about seismic-risk mitigation will be made only if available alternatives are judged on their scientific merits and if current uncertainties on ground-motion characteristics are reported honestly. This paper summarizes the current scientific issues causing uncertainty in ground-motion estimation for ENA, and evaluates methods of ground-motion prediction on their ability to incorporate and represent current alternative viewpoints.

AVAILABLE DATA

While there are no instrumental records of strong shaking in ENA, there are abundant assessments of Modified Mercalli Intensity (MMI) values from earthquakes in the region. These data (and the descriptions on which they are based) document in a qualitative way that damaging earthquakes have occurred in ENA. MMI data are of critical importance in establishing the locations and sizes of preinstrumental shocks in the region, but are of very limited value in making estimates of strong ground motion (for reasons discussed below in the section entitled ENGINEERING ESTIMATES OF GROUND MOTION).

Instrumental records of ground motion in ENA are in fact quite abundant. TABLE 1 summarizes the available data for ENA and related areas (those for which the seismic source and attenuation characteristics might be similar to ENA). These are wide-bandwidth data that are generally available to the scientific

[a] This research is part of a program conducted by the Electric Power Research Institute.

TABLE 1. Available Ground-Motion Records in Eastern North America and Related Areas

Description	Magnitude	No. of Records	Instrument	Distance	Soil Condition	Comments
U.S. earthquakes and explosions, 1962–1967	m_b 3.0–4.8	~250	LRSM	200 to 1000 km	Varied	Three-component analog records
Blue Mountain Lake, NY, July–August 1973	M_L 2.8	3	SMA-1	Near-source	Bedrock or buried boulder	Shallow events
Mississippi Valley earthquakes, 1975–1976	m_b 4.2–5.0	9	SMA-1	10 to 100 km	Mississippi embayment	Most records on earth dams
Monticello Reservoir, 1978	M_L 2.8	3	SMA-1	Near-source	Alluvium	Shallow, induced events
Canada, eastern U.S. events, 1980–1986	m_b 4.1–5.6	60	ECTN	100 to 1000 km	Rock	Vertical component digital data
New Hampshire, January 19, 1982	m_b 4.8	12	SMA-1	10 to 100 km	Alluvium	Most records from earth dams
New Brunswick, March and May 1982	m_b 4.0 and 4.8	5	SMA-1	3 to 7 km	Shallow alluvium and rock	Three-component digital records
U.S. earthquakes 1983–1986	m_b 3.2–5.1	~140	RSTN	200 to 1000 km	100-m boreholes in rock	Recorded on massive foundation
Painesville, Ohio 1986	m_b 4.9	1	SMA-3	18 km	Shale	
Nahanni, NWT, Canada, December 1985	M_s 5.0–6.9	17	SMA-1	Near-source	Rock	Most records at epicentral distances of less than 10 km
Gazli, USSR, May 17, 1976	m_b 6.2 M_s 7.0	1	Accelerograph	Near-source	Sandstone and clay	Three-component optical record at Karakyr Point

community. Not included are analog seismograph records from instruments with peaked response, and data from aftershock and earthquake swarm studies that are not readily available (these are predominantly from $m_b \leq 3.5$ earthquakes). As TABLE 1 indicates, the data are primarily from small earthquakes at regional distances (100 to 1000 km). For this reason the accelerograph records from the Nahanni and Gazli events are important: if the crustal stress release causing these earthquakes was similar to what might be expected in ENA, the records give empirical evidence of the associated, near-source ground motions. This issue is important especially in the case of the Nahanni events, which were well recorded from near-source to teleseismic distances.

Additionally the extent to which some records have been affected by underlying soil conditions is unclear, particularly for accelerograph data from Mississippi valley earthquakes and the Miramichi and New Hampshire shocks. To the extent that soil effects cannot be removed from a record, the usefulness of that record to predict ground motions for other geologic conditions (in particular, rock) is limited. The issue of soil effects in ENA is discussed more fully below.

THE SOURCE SPECTRUM

The most direct method to estimate earthquake ground motion is to start with a representation of energy released at the source. The energy can then be attenuated to the site, and any relevant soil effects can be added, to obtain an estimate of ground motion at a particular site for a specified earthquake. Both attenuation and soil effects are discussed in subsequent sections.

The Fourier spectrum of shear waves caused by slip displacement on a fault can generally be modeled by a shape proposed by Brune.[1,2] In this model, the Fourier spectrum of acceleration increases with frequency as f^2 up to some corner frequency f_0; above that spectral amplitudes are constant and proportional to $M_0 (2\pi f_0^2)$, where M_0 is the seismic moment of the earthquake. This simple representation of the source energy, coupled with relationships between the seismic moment, stress drop, source size, and corner frequency provides a convenient means of estimating the energy released as shear waves during earthquakes.

More complicated representations of the source spectra have been synthesized for some California earthquakes.[3] However, in spite of these more complicated spectral shapes, the simple Brune model of seismic shear wave spectra has been shown to provide adequate engineering estimates of strong ground motion in California, where there are sufficient data for evaluation.[4-6] This spectrum has also been used to estimate ground motion in ENA,[7-9] although most of the observations available for comparison with predictions are low-amplitude ground motions. Thus the simple Brune model of the source spectrum can be considered adequate for engineering predictions of ground motions, recognizing that the spectrum from any particular earthquake might deviate from it to some degree.

In the Brune model, seismic moment M_0, stress drop $\Delta\sigma$, and corner frequency f_0 are related by:

$$f_0 = \left(\frac{\Delta\sigma\beta^3}{8.44 \, M_0} \right)^{1/3} \qquad (1)$$

The spectral amplitude above the corner frequency usually controls ground motions of engineering interest. This amplitude is proportional to $M_0^{1/3} \, \Delta\sigma^{2/3}$; the root-mean-square (rms) amplitude (assuming that duration is the reciprocal of corner frequency) is proportional to $M_0^{1/6} \Delta\sigma^{5/6}$. Thus, the stress drop that drives the high frequency is an important parameter for the prediction of ground motions of engineering interest.

The stress drop obtained by observing f_0 from ground-motion spectra, or using a source radius based on aftershocks inverting Equation 1, and calculating $\Delta\sigma$ as a function of f_0 and M_0 is a *static* stress drop, a measure of the average stress across the faulting surface before the earthquake, minus the average stress after the earthquake. Static stress drops for large earthquakes are typically in the range of 1 to 100 bars. The stress drop that drives the high frequencies of ground motion is not well understood, but is more like an average dynamic stress drop across the faulting surface. This stress drop can be inferred from measurements of root-mean-square acceleration, a_{rms}, from accelerograph records, and is found to be 50 to 100 bars in California.[4,5,10] This distinction is important since the dynamic stress drops that drive high-frequency strong ground motion can be quite different from the static stress drop for the same event.

Several studies have examined the difference in stress drops between plate-margin and intraplate earthquakes. Liu and Kanamori[11] have studied the relationship between M_s and m_b for various earthquakes and conclude that stress drops in intraplate regions are higher than for plate margins, by a factor of about 3. The stress drops driving these observations are more like dynamic stress drops. Using more indirect methods, Kanamori and Allen[12] conclude that for a given magnitude, earthquakes with longer repeat times have shorter fault lengths (and thus higher stress drops) than do events in more active regions. Scholz et al.[13] come to a similar conclusion, deriving a factor of 6 difference in stress drops between plate-margin and intraplate events. These inferred differences are relevant to static rather than dynamic stress drops.

Nuttli[14,15] compared relationships between m_b, M_S, and M_0 for intraplate earthquakes and concluded that stress drops increase as $M_0^{1/4}$ (or equivalently, corner frequency varies as $M_0^{-1/4}$). More recently, Nuttli (personal communication, 1987) has revised his earlier source scaling model based on $M_0 - f_0$ data from ENA and Japan. In this new scaling relationship the stress drop still increases as $M_0^{1/4}$, but the seismic moment corresponding to a corner frequency of 1 Hz is 3×10^{23} (the corresponding moment in the 1983 model was roughly 10^{22}).

Somerville et al.[16] used time-domain modeling of teleseismic body waves to estimate the source parameters of thirteen ENA moderate and large earthquakes. They concluded that stress drops do not depend on source size, and have an average value of 100 bars.

The issue of stress drops of earthquakes in ENA cannot be unequivocally resolved with data because of scatter. FIGURE 1 illustrates one comparison of observations of m_{bLg} and M_0 and predictions using a constant stress drop of 100 bars and Nuttli's[17] increasing stress drop. Either interpretation is allowed by the data, and the issue of whether and how stress drops in ENA are different from those in California likely will not be resolved until several moderate earthquakes are well recorded in ENA.

In addition to issues of stress drop, ENA earthquakes are observed to generate higher amplitudes at high frequencies than California earthquakes. This may be an effect of higher frequency energy generated at the source, or of higher frequency being transmitted more efficiently through the earth's crust (including through near-surface rocks and soils). Whatever the cause, significant energy at frequencies up to 40 Hz is often observed at near-source distances in ENA. The energy at these frequencies has little effect on the response of ordinary engineering structures, but can result in records with high peak accelerations. For this reason the direct prediction of response spectrum amplitudes avoids issues of scaling spectral shapes to a high-frequency peak acceleration.

ATTENUATION OF GROUND MOTION

The manner in which ground motion attenuates with distance is important in two respects. At distances less than 100 km attenuation is critical to estimate amplitudes of strong shaking. At farther distances (100 to 1000 km) the attenua-

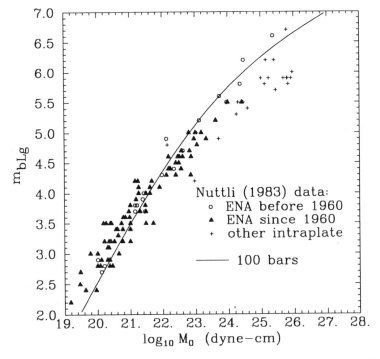

FIGURE 1. Theoretical relationships between m_{Lg} and M_0 compared to data. (After Toro).[9]

tion of ground motion must be accurately modeled in order to predict seismograph responses and compare those predictions with observations, and to model magnitude measurements from specified source spectra. At the closer distances (less than 100 km) empirical evidence in California indicates that $1/R$ geometric decay of body-wave amplitudes is accurate. This is also a reasonable assumption in ENA. Modeling of multiple wave arrivals for ENA earthquakes by Burger et al.[18] and Barker et al.[19] indicates that while individual waves may attenuate faster or slower than $1/R$, the composite representation of amplitude decay is adequately represented by $1/R$ (especially when considering the average over possible source depths and focal mechanisms).

TABLE 2. Eastern North American Q Models Reported in the Literature

Model	Investigators
$Q = 1300 f^{0.38}$	Dwyer et al.[21]
$Q = 900 f^{0.20}$	Hasegawa[22]
$Q = 982 f^{0.376}$	Gupta and McLaughlin[23]
$Q = 500 f^{0.65}$	Shin and Herrmann[20]
$Q = 540 f^{0.414}$	Atkinson and Sinclair[24]

At distances beyond 100 km Lg waves dominate high-frequency ground motions in ENA. The frequency-domain amplitudes of these surface waves decay as $R^{-1/2}$; their time-domain amplitudes decay as $R^{-5/6}$. Synthetic modeling has confirmed these attenuation rates.[20]

The duration of shaking at near-source distances is usually taken to be equal to the source duration T_0. When surface waves dominate the motions, dispersion increases the duration of motion by $0.05\,R$, where R is distance in kilometers. This effect has been confirmed by examining seismographic data.[8] Anelastic attenuation in ENA is more problematic. This attenuation takes the form of a factor $\exp(-\gamma R)$, where $\gamma = \pi f / Q\beta$, f is frequency, Q is quality factor, and β is shear-wave velocity. Depending on the investigator and the data set, various Q models have been reported in the literature for ENA; some of these are listed in TABLE 2. The effect of these different models is large (a factor of five in ground-motion amplitude) at 10 Hz, as illustrated in FIGURE 2. The model by Shin and Herrmann[20] gives results intermediate to the others.

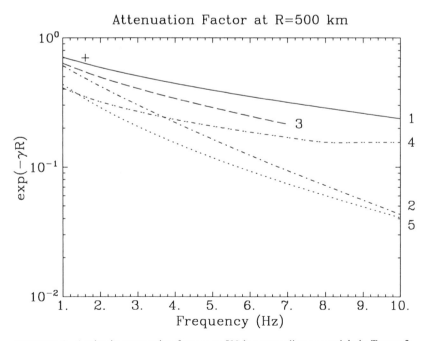

FIGURE 2. Anelastic attenuation factors at 500 km according to models in TABLE 2.

TABLE 3. Reported Amplification Factors (for Response or Fourier Spectra) of Deep Soil/Rock Based on Analyses of Empirical Data

Reference	10 Hz	5 Hz	3 Hz	2 Hz	1 Hz
Campbell[26]	1.0[a]	1.0[a]	NR	2.0[b]	2.0[b]
Campbell[27c]	1.0[a]	1.0[a]	NR	1.7–1.9[b]	1.7–1.9[b]
Bernreuter et al.[25]	0.8	1.1	1.3	1.5	1.7
Joyner and Fumal[28]					
Generic Soil:	0.95	0.98	1.1	1.4	1.9
V_s = 200 m/s	1.0	1.0	1.35	1.9	2.9
V_s = 500 m/s	1.0	1.0	1.05	1.3	1.8
Rogers et al.[29]					
Mean	NR	3.4	3.4	3.3	3.3
Range	NR	2.8–5	2.8–5	1.6–5	1.5–5
Toro and McGuire[8]	1.0	1.3	1.75[d]	1.8	2.1

NOTE: NR = Not reported.

[a] Value based on reported ratio for peak horizontal acceleration.

[b] Value based on reported ratio for peak horizontal velocity.

[c] Values shown are maximum reported (soil depth > 6 km) for equations unaltered by special conditions in Utah.

[d] Average of values reported at 2 Hz and 4 Hz.

SURFICIAL GEOLOGY

Estimating the effects of soils on ground motions is important both for the prediction of motion on soils during future earthquakes and in understanding the effects that soils have had on the records currently available. To avoid a site-specific dynamic analysis at every site, requiring site-specific soil properties, it is common to designate several soil categories and to estimate the average effects of soils in each category.

There is some consistency among empirical results obtained in California for deep soils (thicker than, say, 30 m). TABLE 3 lists reported amplification factors (ratios of Fourier spectra or response spectra) as a function of frequency. The general trend is for a factor of unity (i.e., no amplification) at 10 Hz, and a factor of two at 1 Hz, with intermediate values between these two frequencies. (An exception is the study by Rogers et al.,[29] who used microseismic recordings of distant nuclear shots to derive amplification factors.) Shallow soils are more problematic. Both empirical and theoretical studies[25] indicate that shallow soils can amplify high frequencies of ground motion, but it has not been demonstrated that this effect can be accurately quantified with a few generic categories. This is a particular problem in assessing the strong motion data from Miramichi, which were obtained on shallow soils. The records may not be representative either of deep soil sites or rock sites. In a similar vein, the records from the New Hampshire earthquake were obtained mostly at earth dam sites (including crest, abutment, and downstream sites) that undoubtedly modified the ground motions from rock conditions. The estimation of ground-motion effects for geologic conditions such as these is not easy in general, and it may be especially difficult in ENA for sites of unconsolidated glacial till overlying competent, high-velocity bedrock.

ENGINEERING ESTIMATES OF GROUND MOTION

To estimate strong ground motion for engineering purposes, we classify available procedures into three groups. These are reviewed here.

Calibrated theoretical methods use a mathematical representation of the source spectrum and ground-motion attenuation based on theory and seismological observations to derive the form of equations to estimate ground motion. The parameters in these equations are then chosen so that the absolute values of estimates match available observations. In this way low-amplitude instrumental data for small-magnitude earthquakes and large distances can be used to calibrate the equations, and extrapolations to close distances, large magnitudes, and high ground-motion amplitudes can be made with some degree of confidence. Herrmann and Nuttli[30] were the first to use methods of this type; they predicted peak acceleration and velocity. Other recent investigators[7-9,31] have also used these methods to predict response spectrum amplitudes. These methods are the most preferred, because they can explicitly account for characteristics of ENA events (higher stress drops, higher amplitudes at high frequencies) as compared to California events. The effects of uncertainties in seismological interpretations (e.g., of stress drop versus magnitude) can be obtained by application of these models with different assumptions. FIGURE 3 illustrates the agreement between one application of a model of this type and ENA data (see Toro[9] for definition of the symbols), including Nahanni data.

Semi-theoretical methods (also called *semi-empirical methods*) make the assumption that, in the near-source region, ground motion characteristics for ENA

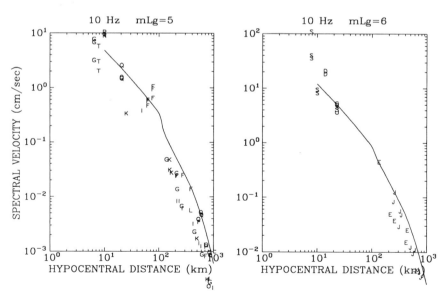

FIGURE 3. Estimates of 10-Hz response spectrum velocity and data from ENA; see Toro[9] for definitions of symbols.

earthquakes will be identical to those for California earthquakes of the same magnitude. This would follow if the properties of the energy release during dynamic faulting were the same for earthquakes in the two regions. Estimates of ground motions near the source are based on empirical observations in California. At farther distances the near-source predictions are reduced using a geometric attenuation (which is region-independent) and an anelastic attenuation appropriate for ENA. In this manner the empirical, near-source estimates are modified by theory in ENA to account for the different attenuation in that region. Early work of this type was conducted by Nuttli,[32] Campbell,[26] and Algermissen *et al.*[33,34] More recently, ground motion records obtained in the near-source regions of ENA earthquakes indicate substantially more energy at high frequencies than do comparable records in California. This may reflect the generation of high-frequency energy by ENA faults (due perhaps to more small-scale heterogeneities) or may reflect greater ability of the earth's crust and surficial rock to transmit these frequencies in ENA. In either case the assumption of similarity between California and ENA near-source ground motions likely is not valid; any difference in high-frequency content will affect response spectrum amplitudes at those high frequencies, and will have a particularly large effect on peak acceleration. For this reason, and because stress drops hypothesized to be different from those in California cannot be handled by these methods, these methods are not preferred.

Finally, *intensity-based methods* use a qualitative intensity scale, typically the MMI scale, to estimate ground-motion characteristics. In this method the dependence of MMI on earthquake size and source-to-site distance is quantified using equations calibrated with MMI observations from historical ENA earthquakes. From another region (typically California, where abundant instrumental records are available), correlations are developed between MMI and quantitative measures of ground motion (response spectrum amplitudes and peak acceleration and velocity). The two sets of equations are then combined, often by simple mathematical substitution, to obtain equations predicting ground motion as a function of earthquake size and distance in ENA. An exhaustive summary of studies that use these substitution methods is given in Bernreuter.[25] One major difficulty with these methods is that both the equation predicting MMI in ENA, and the equation converting MMI to ground-motion amplitudes, are statistical correlations, not deterministic relationships. Mathematical substitution of one equation for the other leads to biased results.[35] A way to avoid these problems is to develop predictive statistical equations with the same independent variables for both steps,[36] but this is difficult to accomplish and is almost never done in practice. Even if it were, it would still require the assumption that, for example, MMI = VIII is the same ground motion in ENA as in California, for the same earthquake magnitude, distance, and soil conditions. This is a strong assumption that is difficult to justify. Fundamental differences in ground-motion amplitude, frequency content, contribution of different wave types, and duration of shaking could apply between ENA and California, and still lead to the same MMI. Thus intensity-based methods should generally be given a low priority for the quantitative estimation of ground motion in ENA.

CONCLUSIONS

Estimates of strong earthquake ground motion in ENA are uncertain, because of the lack of instrumental data documenting this phenomenon and the perception that certain characteristics of earthquakes in ENA are different from those in

California. The uncertainties can be characterized by the physical phenomena causing them, and can be modeled as alternative hypotheses that cannot be resolved given the scatter of available data. The important uncertainties are:

☐ The stresses released during ENA earthquakes. Are the dynamic stress drops different from those of California shocks, and if so, how are they different (in particular, do they increase with earthquake size)?
☐ The anelastic attenuation in ENA. Does it vary within the eastern part of the continent? What is the average anelastic attenuation for ENA?
☐ The effects of soils on earthquake ground motions. How have soils modified recorded ground motions, and how might they affect motion during future shocks? Particular uncertainty is attached to the response of shallow soils overlying competent bedrock, and whether geologic conditions of this type can be categorized sufficiently accurately to allow approximate amplification factors to be used without site-specific investigations.

Of available methods to estimate ground motion, the calibrated theoretical model is best able to represent these uncertainties in an explicit way for engineering estimates of ground motion. Semitheoretical methods based on California observations at near-source distances cannot reflect uncertainties in source characteristics, and intensity-based methods as usually applied are not theoretically sound.

It should be the goal of ground-motion studies, and of seismic hazard and risk studies, to represent honestly the uncertainties in predicted ground-motion amplitudes for earthquakes in ENA, and to determine the resulting uncertainty in hazard and risk. Only in this way will optimal solutions be found to mitigate earthquake risks.

SUMMARY

Proper characterization of uncertainties in ground-motion estimation in eastern North America is important for any assessment of seismic hazard. The scientific issues surrounding these uncertainties are reviewed, as are the earthquake data available in eastern North America. A method of estimating ground motion that is based on a theoretical source spectrum attenuated to the site and modified by local soil conditions is preferred for estimating ground motion, as it can explicitly incorporate the scientific uncertainties that cause uncertainty in ground motion.

ACKNOWLEDGMENTS

This research is part of a larger program being conducted by the Electric Power Research Institute under the direction of J. Carl Stepp to develop seismic wave attenuation curves for engineering use in the central and eastern United States. This source of support is gratefully acknowledged.

REFERENCES

1. BRUNE, J. N. 1971. Tectonic stress and the spectra of seismic shear waves from earthquakes. J. Geophys. Res. 75: 4997–5002.
2. BRUNE, J. N. 1972. Correction. J. Geophys. Res. 76: 5002.

3. PAPAGEORGIOU, A. S. 1987. Earthquake source spectra—A comparison between western and eastern U.S. earthquakes. *In* Proceedings of the Workshop on Estimation of Ground Motion in the Eastern United States. EPRI Project RP2556-16 Rep. NP-5875.
4. HANKS, T. C. & R. K. MCGUIRE. 1981. Character of high-frequency strong ground motion. Bull. Seismol. Am. **71:** 2071–2095.
5. BOORE, D. M. 1983. Stochastic simulation of high-frequency ground motions based on seismological models of the radiated spectra. Bull. Seismol. Soc. Am. **73:** 1865–1984.
6. MCGUIRE, R. K., A. M. BECKER & N. C. DONOVAN. 1984. Spectral estimates of seismic shear waves. Bull. Seismol. Soc. Am. **74:** 1427–1440.
7. BOORE, D. M. & G. M. ATKINSON. 1987. Prediction of ground motion and spectral response parameters in eastern North America. Bull. Seismol. Soc. Am. **77:** 440–467.
8. TORO, G. R. & R. K. MCGUIRE. 1987. An investigation into earthquake ground motion characteristics in eastern North America. Bull. Seismol. Soc. Am. **77:** 468–489.
9. TORO, G. R. 1988. Prediction of earthquake ground motions in eastern North America using simple models of source spectrum and wave propagation. *In* Proceedings of the Workshop on Earthquake Ground-Motion Estimation in Eastern North America. EPRI Rep. NP-5875.
10. BOORE, D. M. 1986. Short period P- and S-wave radiation from large earthquakes: Implications for spectral scaling relations. Bull. Seismol. Soc. Am. **76:** 43.
11. LIU, H. L. & H. KANAMORI. 1980. Determination of source parameters of mid-plate earthquakes from the waveforms of body waves. Bull. Seismol. Soc. Am. **70:** 1989.
12. KANAMORI, H. & C. R. ALLEN. 1986. Earthquake repeat time and average stress drop. *In* Earthquake Source Mechanics, Maurice Ewing Symposium, **6,** 227. American Geophysical Union.
13. SCHOLZ, C., C. A. AVILES & S. G. WESNOUSKI. 1986. Scaling differences between large interplate and intraplate earthquakes. Bull. Seismol. Soc. Am. **76:** 65.
14. NUTTLI, O. W. 1983. Average seismic source parameter relations for mid-plate earthquakes. Bull. Seismol. Soc. Am. **73:** 519–535.
15. NUTTLI, O. W. 1983. Empirical magnitude and spectral scaling relations for mid-plate and plate-margin earthquakes. Tectonophysics **93:** 207.
16. SOMERVILLE, P. G., J. P. MCLAREN, L. V. LEFEVRE, R. W. BURGER & D. V. HELMBERGER. 1987. Comparison of source relations of eastern and western North American earthquakes. Bull. Seismol. Soc. Am. **77:** 322–346.
17. NUTTLI, O. W., Personal communication with G. R. Toro, March 1987.
18. BURGER, R. W., P. G. SOMERVILLE, J. S. BARKER, R. B. HERRMANN & D. V. HELMBERGER. The effect of crustal structure on strong ground motion attenuation relations in eastern North America. Bull. Seismol. Soc. Am. **77:**
19. BARKER, J. S., P. G. SOMERVILLE & J. P. MACLAREN. 1988. Wave propagation modeling of ground motion attenuation in the northeast United States and adjacent Canada. *In* Proceedings of the Workshop on Estimation of Ground Motion in the Eastern United States. EPRI Rep. NP-5875.
20. SHIN, T. C. & R. B. HERRMANN. 1987. Lg attenuation and source studies using the 1982 Miramichi data. Bull. Seismol. Soc. Am. **77:**384–397.
21. DWYER, J. J., R. B. HERRMANN & O. W. NUTTLI. 1983. Spectral attenuation of the Lg in the central United States. Bull. Seismol. Soc. Am. **73:** p. 781.
22. H. S. HASEGAWA. 1983. Lg spectra of local earthquakes recorded by the Eastern Canada Telemetered Network and spectral scaling. Bull. Seismol. Soc. Am. **73:** 1041–1061.
23. I. N. GUPTA & K. L. MCLAUGHLIN. 1988. Regional crustal frequency dependent Q and site effects estimated from LRSM and RSTN data. *In* Proceedings of the Workshop on Estimation of Ground Motion in Eastern North America. EPRI Rep. NP-5875.
24. ATKINSON, G. & L. SINCLAIR. 1986. Ground Motion Studies for Eastern North America, Report to Ontario Hydro, Acres International Ltd., Toronto.

25. BERNREUTER, D. L., J. B. SAVY, R. W. MENSING & D. H. CHUNG. Development of eastern United States ground motion models. Appendix C *in* Seismic Hazard Characterization of the Eastern United States: Methodology and Interim Results for Ten Sites. U.S. Nuclear Regulatory Commission Report NUREG/CR-3756.
26. CAMPBELL, K. W. 1981. A ground motion model for the central United States based on near-source acceleration data. *In* Proceedings of Earthquakes and Earthquake Engineering: Eastern United States. 213–232: Knoxville, TN.
27. CAMPBELL, K. W. 1983. The effects of site characteristics on near-source recordings of strong ground motion. *In* Proceedings of Workshop on Site-Specific Effects of Soil and Rock on Ground Motion and the Implications for Earthquake-Resistant Design, Santa Fe, July. U.S. Geol. Surv. Open File Rep. **83-845**: 280–309.
28. JOYNER, W. B. & T. E. FUMAL. 1985. Predictive mapping of earthquake ground motion. *In* Evaluating Earthquake Hazards in the Los Angeles Region—An Earth Science Perspective. J. I. Ziony, Ed. U.S. Geol. Surv. Prof. Pap. **1360**: 203–220.
29. ROGERS, A. M., J. C. TINSLEY & R. D. BORCHERDT. 1985. Predicting relative ground response. *In* Evaluating Earthquake Hazards in the Los Angeles Region—An Earth Science Perspective. J. I. Ziony, Ed. U.S. Geol. Survey Prof. Paper 1360: 221–247.
30. R. B. HERRMANN & O. W. NUTTLI. 1980. Strong motion investigations in the central United States. *In* Proceedings of the 7th World Conference on Earthquake Engineering, Istanbul. **II**: 533–536.
31. ATKINSON, G. M., 1984. Attenuation of strong ground motion in Canada form a random vibrations approach. Bull. Seismol. Soc. Am. **74**: 2629–2653.
32. NUTTLI, O. W. 1979. The relation of sustained maximum ground acceleration and velocity to earthquake intensity and magnitude. Misc. Paper S-73-1, Report 16, U.S. Army Waterways Exp. Sta., Vicksburg.
33. ALGERMISSEN, S. T. & D. M. PERKINS. 1976. A probabilistic estimate of maximum acceleration in rock in the contiguous United States. U.S. Geol. Survey Open File Rep. 76-416.
34. ALGERMISSEN, S. T. *et al.* 1982. Probabilistic estimates of maximum acceleration and velocity in rock in the Contiguous U.S. U.S. Geol. Survey Open File Rep. 82-1033.
35. CORNELL, C. A., H. BANON & A. F. SHAKAL. 1979. Seismic motion and response prediction alternatives. Earthquake Eng. Struct. Dyn. **9**: 295–315.
36. VENEZIANO, D. 1988. The use of intensity data in ground motion estimation. *In* Proceedings of the Workshop on Earthquake Ground-Motion Estimation in Eastern North America. EPRI Rep. NP-5875.

Combination of Seismic-Source and Historic Estimates of Earthquake Hazard[a]

DANIELE VENEZIANO AND LUC CHOUINARD

Department of Civil Engineering
Massachusetts Institute of Technology
Cambridge, Massachusetts 02139

INTRODUCTION

A frequently used method for earthquake hazard estimation[1] partitions the geographical region around the site of interest into provinces ("sources") and assumes that, within each source, earthquakes occur according to a stationary and homogeneous Poisson process. Another frequent assumption, one that is not essential to the method, however, is that earthquake magnitude m has truncated exponential distribution, and hence that earthquakes inside source i have a recurrence law of the type

$$\nu_i(m) = \begin{cases} 10^{a_i - b_i m} - 10^{a_i - b_i M_i}, & m \leq M_i \\ 0, & m \geq M_i \end{cases} \tag{1}$$

where $\nu_i(m)$ is the expected number of events per unit time and unit area with magnitude larger than m and M_i, a_i, and b_i are source-specific parameters.

For the calculation of earthquake hazard, one needs also know the attenuation law, i.e., the probability distribution of earthquake intensity at the site, Y, as a function of the magnitude m and the location X_0 of the earthquake. We write this attenuation law as

$$Y = g(m, X_0, \varepsilon) \tag{2}$$

where ε is a random variable.

From the recurrence model in Eq. 1 and the attenuation law in Eq. 2, one can calculate the exceedance rate function (seismic hazard function) at the site,

$$\lambda(y) = \text{rate of events with intensity higher than } y \tag{3}$$

We call the above method of seismic hazard analysis the seismic-source (SS) method and we denote by $\lambda_{SS}(y)$ the associated estimator of $\lambda(y)$.

As an alternative to the SS method, one may use historic (H) procedures.

[a] Financial support for this work has been provided by the National Science Foundation through the National Center for Earthquake Engineering Research, Project No. 871306.

These procedures estimate $\lambda(y)$ directly from a catalog of historic events, for example as

$$\lambda_H(y) = \frac{1}{T} \sum_{\text{historic events, } i} \frac{1}{P_D(m_i, X_{0_i})} P[Y_i > y | m_i, X_{0_i}] \tag{4}$$

where T is the time period covered by the catalog and $P_D(m, X_0)$ the probability that a generic event in T with characteristics (m, X_0) is recorded in the catalog. Hence, the product $TP_D(m, X_0)$ is the equivalent period of complete recording at location X_0, for events with magnitude m.

The relative accuracy of the estimators $\lambda_{SS}(y)$ and $\lambda_H(y)$ depends on the value of y. One should notice in particular that $\lambda_H(y)$ is a nonparametric estimator and is unbiased, irrespective of the spatial variation of earthquake activity and of the probability distribution of magnitude. The variance of $\lambda_H(y)$ is small at low intensities, but it becomes large at high intensities, especially for values of y such that $\lambda(y) < 1/T$. The estimator $\lambda_{SS}(y)$ has a smaller variance. However, if the geometry of the earthquake sources or the type of magnitude distribution are incorrectly specified, $\lambda_{SS}(y)$ is biased. The net result is that, in typical applications, the mean squared error MSE (variance plus squared bias) of λ_H is smaller than that of λ_{SS} for small y, whereas the reverse is true for large y.

We propose to use $\lambda_{SS}(y)$ and $\lambda_H(y)$ in combination to form estimators $\lambda_{SS-H}(y) = C \cdot \lambda_{SS}(y)$ that are more accurate than either λ_{SS} or λ_H over the (high) intensities of interest for earthquake risk assessment. The basic idea is to choose the constant C such that, at some low intensity, the combined estimate coincides with the historical estimate. Different definitions of C produce different combined estimators. The estimators $\lambda_{SS-H}(y)$ are suggested as practical alternatives to more sophisticated "local" estimators $\lambda_L(y)$ that result from allowing the parameters a and b in Eq. 1 to vary in space within each earthquake source.[2,3] In order to evaluate the combined estimators, we compare λ_{SS} and λ_{SS-H} with λ_L at many sites in the northeastern U.S. The main conclusions from the comparison are that: (1) the estimators λ_{SS-H} perform best if they are calibrated to λ_H at a site intensity that is exceeded about 15 to 20 times in the historical record; (2) for this choice of the calibration intensity, the estimator λ_L is closer to λ_{SS-H} than to λ_{SS}; and (3) the estimators λ_{SS-H} are considerably more robust than λ_{SS} with respect to the configuration of the earthquake sources. Therefore, the new estimators are especially recommended when the source geometry is poorly constrained.

COMBINED ESTIMATORS

The estimators $\lambda_{SS}(y)$ and $\lambda_H(y)$ give the rate at which any specified intensity y is exceeded at the site. Both estimators make corrections for catalog incompleteness and account for attenuation uncertainty. For one of the combined estimators introduced below, the calibration factor C is defined in terms of the functions λ_{SS} and λ_H, but other combined estimators studied here require calculation of the hazard for the incomplete earthquake sequence and for the case when the random attenuation law $g(m, X_0, \varepsilon)$ is replaced with the median law $\bar{g}(m, X_0)$. The function $\bar{g}$ is such that earthquakes with characteristics (m, X_0) produce site intensities above and below $\bar{Y} = \bar{g}(M, X_0)$ with equal probability. We denote the hazard estimators for the case of incomplete catalog and median attentuation by $\lambda'_L(y)$, $\lambda'_{SS}(y)$, and $\lambda'_H(y)$, depending on the method of estimation. The first two such estimators are given by

$$\begin{bmatrix} \lambda_L'(y) \\ \lambda_{SS}'(y) \end{bmatrix} = - \int_{X_0} \int_m P_D(m, X_0) \frac{\partial v(m, X_0)}{\partial m} K(y, m, X_0) \, dm \, dX_0 \qquad (5)$$

where $v(m, X_0)$ is the function in Eq. 1 at the geographical point X_0 and $K(y, m, X_0)$ is an indicator function with value 1 if $\bar{g}(m, X_0) > y$ and value 0 otherwise. The function $v(m, X_0)$ is estimated locally for λ_L' and is found under the assumption of homogeneous seismic sources for λ_{SS}'.

Consistently with Eq. 4, the estimator $\lambda_H'(y)$ might take the form

$$\lambda_H'(y) = \frac{n(y)}{T} \qquad (6)$$

where $n(y)$ is the number of historic events with median attenuated intensity $\bar{y}_i = \bar{g}(m_i, X_{0i})$ in excess of y. Another possibility is to use

$$\lambda_H''(\bar{y}_i) = \lambda_{H_i}'' = \frac{i}{T} \frac{n}{n+1} \qquad (7)$$

This last estimator is defined only at the median historic intensities ($\bar{y}_i$), which are ordered such that $\bar{y}_1 > \bar{y}_2 > \ldots > \bar{y}_n$.

We use λ_{SS}, λ_H, λ_{SS}', λ_H', and λ_H'' to form three combined estimators of $\lambda(y)$:

$$\lambda_{SS\text{-}H}^{(1)}(y) = \frac{\lambda_H(y^*)}{\lambda_{SS}(y^*)} \lambda_{SS}(y)$$

$$\lambda_{SS\text{-}H}^{(2)}(y) = \frac{\lambda_H'(y^*)}{\lambda_{SS}'(y^*)} \lambda_{SS}(y) \qquad (8)$$

$$\lambda_{SS\text{-}H}^{(3)}(y) = \frac{\lambda_H''(y^*)}{\lambda_{SS}'(y^*)} \lambda_{SS}(y)$$

In all cases, y^* is a calibration intensity, which we chose as described next.

CHOICE OF THE CALIBRATION INTENSITY AND EVALUATION OF THE COMBINED ESTIMATORS

It is convenient not to specify y^* externally and rather set y^* equal to one of the order statistics $\bar{y}_i$. [This is a necessity for the estimator $\lambda_{SS\text{-}H}^{(3)}$, which is defined only at the points $\bar{y}_i$.] The criterion we use to select the calibration intensity $\bar{y}_i$ is to minimize the mean squared error of the log exceedance rate, which for the kth combined estimator ($k = 1, 2, 3$) is

$$\text{MSE}^{(k)}(y; i) = \text{E}\{[\text{Log } \lambda_{SS\text{-}H}^{(k)}(y; y^* = \bar{y}_i) - \text{Log } \lambda_L(y)]^2\} \qquad (9)$$

In practice, it is impossible to calculate the mean squared error in Eq. 9, because only one earthquake catalog is available for a given region. One might resort to Monte Carlo simulation and, for example, assume that the earthquake process is Poisson with the recurrence law used in the calculation of λ_L. A drawback of the simulation method is that the geometry of the sources usually reflects the spatial distribution of historical seismicity. Therefore, one should redefine the sources for each simulation.

We find it preferable to replace the expectation in Eq. 9 with the average of the squared log error over a grid of sites. We further account for regional variations of seismicity by setting y to the intensity that is exceeded at each site with a given frequency; that is, we fix $\lambda_L(y) = \lambda$ and minimize with respect to i the quantity

$$\text{ASE}^{(k)}(\lambda; i) = \text{spatial average of } [\text{Log } \lambda^{(k)}_{\text{SS-H}}(y; y^* = \bar{y}_i) - \text{Log } \lambda]^2 \quad \textbf{(10)}$$

where y is an intensity that varies from site to site and satisfies $\lambda_L(y) = \lambda$. FIGURE 1 illustrates the calculation of $\text{ASE}^{(1)}$. Similar procedures apply to $\text{ASE}^{(2)}$ and $\text{ASE}^{(3)}$.

Numerical results are obtained using the Chiburis[4] catalog for the Northeast-

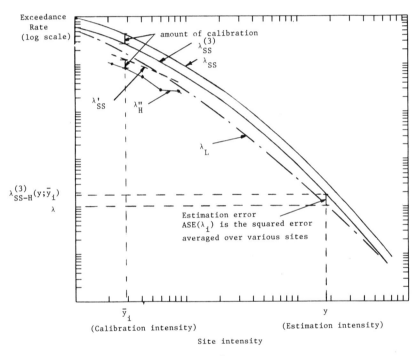

FIGURE 1. The combined estimator $\lambda^{(3)}_{\text{SS-H}}$ and its error with respect to λ_L.

ern U.S. A plot of main events, in the region 39–46°N, 66–77°W and for the period 1627–1981, is shown in FIGURE 2. Because for most of the large earthquakes the only available size measure is MM epicentral intensity I_0, in all calculations we use I_0 in place of m and convert magnitudes when needed using the formula proposed by Chiburis, $I_0 = (m - 1)/0.6$. The maximum possible intensity, which is the equivalent for I_0 of M_i in Eq. 1, is taken everywhere to be IX-X.

The seismicity model for $\lambda_L(y)$ and $\lambda'_L(y)$ has spatially varying a and b coefficients, as shown in FIGURE 3. These coefficients have been obtained through the "local neighborhood" method described in Veneziano and Chouinard[3] and refer to a recurrence relationship of the type

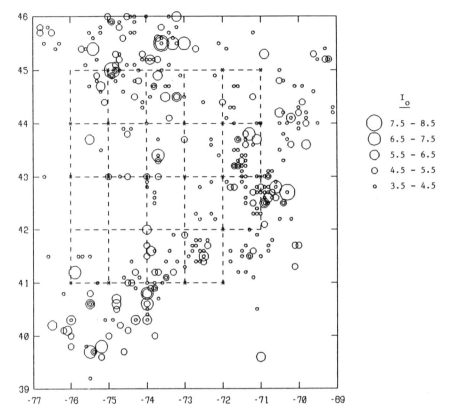

FIGURE 2. Earthquakes with MM intensity greater than 3.5 from 1627 to 1981.[4] The starred points on the grid are used to evaluate and rank different hazard estimators.

$$\nu(m, x) = 10^{a(x)-b(x)(I_0-3.5)} - 10^{a(x)-b(x)(9.5-3.5)} \tag{11}$$

The unit area in the definition of a is that of a square equatorial degree $(111.11 \text{ km})^2$.

For $\lambda_{SS}(y)$, we consider two alternative source configurations: In one case we partition the region into 11 sources, which closely reflect the spatial variation of historical seismicity. The sources are shown in FIGURE 4, which is an adaptation from FIGURE 1 of Ref. 5. In the other case we use a single homogeneous source for the entire region. The latter assumption is unrealistic, but is useful to generate an upper bound to $ASE^{(k)}$ over all reasonable choices of the seismic sources and to compare the robustness of the estimators λ_{SS} and $\lambda_{SS-H}^{(k)}$ with respect to source geometry.

In all the calculations, y is taken to be peak ground acceleration (cm/sec^2) and the attenuation law is that proposed by Heidari[6] for peak horizontal acceleration on rock in the eastern and central U.S., that is,

$$y = \exp\{2.00 + 1.14 \, m_{Lg} - 1.03 \, lnR - 0.003 \, R + \varepsilon\} \tag{12}$$

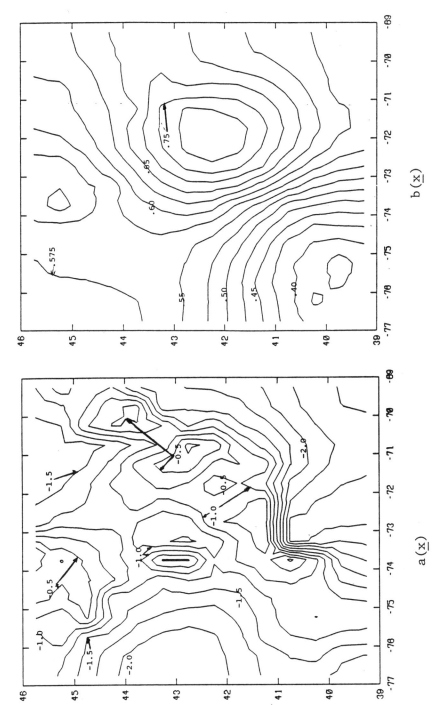

FIGURE 3. Contour plots of the seismicity parameters a and b in Eq. 1 used for the local estimator λ_L.

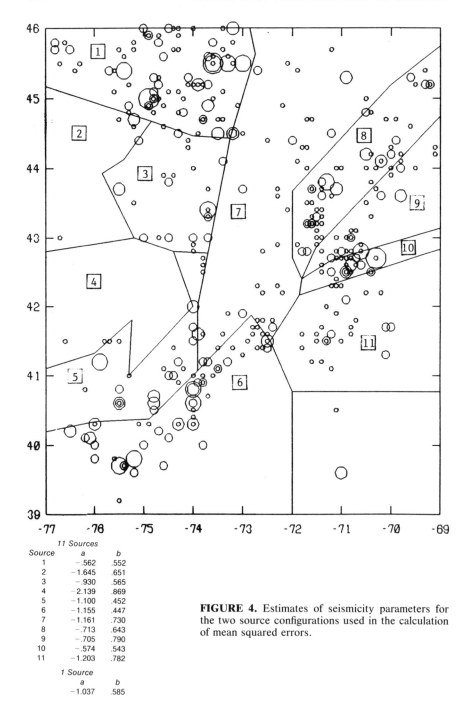

11 Sources		
Source	a	b
1	−.562	.552
2	−1.645	.651
3	−.930	.565
4	−2.139	.869
5	−1.100	.452
6	−1.155	.447
7	−1.161	.730
8	−.713	.643
9	−.705	.790
10	−.574	.543
11	−1.203	.782

1 Source	
a	b
−1.037	.585

FIGURE 4. Estimates of seismicity parameters for the two source configurations used in the calculation of mean squared errors.

where R is hypocentral distance in kilometers for a focal depth of 10 kilometers and ε is a normal random variable with zero mean, standard deviation 0.6, and symmetrical truncation at ± 1.8. The Lg magnitude is obtained from I_0 using $m_{Lg} = 1 + 0.6I_0$ and median attenuated values are generated by setting $\varepsilon = 0$.

The squared error is averaged over the 19 sites shown as stars on the grid of FIGURE 2 and calculations are repeated for three prediction rates: $\lambda = 10^{-2}$, 10^{-3}, and 10^{-4} events/year. Other sites on the grid are excluded from averaging, because the historical seismicity at these sites does not conform to the assumptions of the model. Lack of fit of the model has been detected by applying the Kolmogorov-Smirnov test at significance at level 10% to the median historic intensities $\bar{y}_i$, regarded as a random sample from the Poisson process with exceedance rate $\lambda'_{SS}(y)$ in Eq. 5 (11-source solution).

Results are presented in FIGURE 5a for the 11-source configuration and in FIGURE 5b for the single-source case. For each combination of seismic source geometry and exceedance rate, the average squared errors in Eq. 10 are plotted against i (and against λ''_{H_i} in Eq. 7, where for the present catalog $T = 354$ years and $n = 423$) and are compared with the average squared error of the seismic-source estimator,

$$\text{ASE}_{SS}(\lambda) = \text{spatial average of } [\text{Log } \lambda_{SS}(y) - \text{Log } \lambda]^2 \qquad (13)$$

In analogy with Eq. 10, the intensity y in Eq. 13 varies from site to site to satisfy $\lambda_L(y) = \lambda$. Notice that ASE_{SS} in Eq. 13 does not depend on the calibration intensity $\bar{y}_i$ and therefore plots in FIGURE 5 as a horizontal line.

FIGURE 5 indicates that the combined estimators $\lambda^{(2)}_{SS-H}$ and $\lambda^{(3)}_{SS-H}$ have similar average squared errors with respect to λ_L and are better than λ_{SS} if the calibration intensity $\bar{y}_i$ is chosen appropriately. The optimum value of i, i^*, decreases slightly (the calibration intensity increases slightly) as the rate λ at which hazard is estimated decreases. Also, i^* is slightly smaller (the calibration intensity is slightly higher) for a poorer choice of the earthquake sources. These variations, as well as the variation of i^* with the type of combined estimator, are, however, small, and one may in all cases use a value of i around 15, which corresponds for the present catalog to a historical exceedance rate λ''_{H_i} of about one event in 25 years. Over different seismicity conditions, the optimum value of i is expected to remain stable and the optimum calibration rate is expected to vary as $15/T$, where T is the period covered by the catalog.

The estimator $\lambda^{(1)}_{SS-H}$ is slightly less accurate than either $\lambda^{(2)}_{SS-H}$ or $\lambda^{(3)}_{SS-H}$, but is still superior to λ_{SS}, especially for low prediction rates and poor source configurations. The best value of i for $\lambda^{(1)}_{SS-H}$ is somewhat smaller than for the other combined estimators, but the choice $i^* = 15$ is still nearly optimal.

TABLE 1 gives estimates of the "error factors"

$$\text{EF}^{(k)} = 10^{\sqrt{\text{ASE}^{(k)}}} \qquad (14)$$

which express the degree of dissimilarity between $\lambda^{(k)}_{SS-H}$ and λ_L and the analogous error factor for λ_{SS}. Different values are given for "accurate" and "poor" source configurations, by which we mean source geometries that respectively do and do not reflect the spatial distribution of historic seismicity. The values for poor configurations are intermediate between those of FIGURES 5a and 5b, in consideration of the very crude assumption of complete homogeneity in FIGURE 5b.

An important conclusion from TABLE 1 and FIGURE 5 is that the combined estimators $\lambda^{(k)}_{SS-H}$ are more robust than λ_{SS} with respect to the specification of the earthquake sources. Therefore, combined estimators reduce the consequences of errors in the source configuration and are particularly recommended when the interpretation of historical seismicity is controversial or when homogeneous earthquake sources do not exist.

The previous analysis is based on the comparison of various hazard estimators with the local estimator λ_L. In the APPENDIX, a semi-theoretical analysis is made of the error of $\lambda_{SS-H}^{(3)}(y)$ with respect to the *true* hazard $\lambda(y)$. This

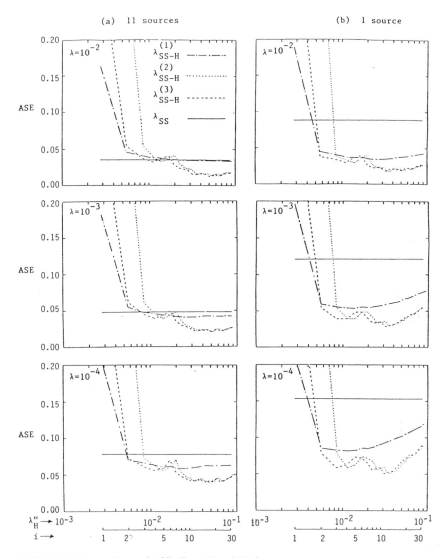

FIGURE 5. Comparisons of ASE (Eqs. 10 and 12) for two source configurations and three exceedance rates.

TABLE 1. Estimated Error Factors with Respect to the Local Estimator $\lambda_L{}^a$

	Estimator	Prediction Rate, λ (events/year)		
		10^{-2}	10^{-3}	10^{-4}
Accurate source configuration	$\lambda_{SS-H}^{(1)}$	1.55	1.60	1.75
	$\lambda_{SS-H}^{(2)}, \lambda_{SS-H}^{(3)}$	1.35	1.40	1.60
	λ_{SS}	1.55	1.65	1.90
Poor source configuration	$\lambda_{SS-H}^{(1)}$	1.55	1.65	1.85
	$\lambda_{SS-H}^{(2)}, \lambda_{SS-H}^{(3)}$	1.40	1.50	1.70
	λ_{SS}	2.00	2.20	2.50

^{a} See Eq. 14.

analysis indicates that the optimum calibration rate $\bar{y}_i$ is probably closer to $\bar{y}_{(20)}$ than to $\bar{y}_{(15)}$ and that the error factors of the combined estimator $\lambda_{SS-H}^{(3)}$ with respect to the true hazards are about 10% higher than the values reported in TABLE 1.

CONCLUSIONS

Historic estimators of earthquake hazard, $\lambda_H(y)$, have the desirable properties of being unbiased, of requiring little external information, and of being accurate at low intensities. However, for the high intensities of interest in earthquake risk mitigation, estimators $\lambda_{SS}(y)$ based on homogeneous earthquake sources and on parametric magnitude distributions are in most cases preferable. A problem with the latter estimators is that they are biased if the earthquake sources or the distribution or earthquake size are chosen incorrectly.

The bias of $\lambda_{SS}(y)$ can be reduced by scaling this estimator so that it coincides with $\lambda_H(y)$ at a specified site intensity y^* (a few variants of this idea are considered in the paper). We find that the resulting combined estimators perform best if y^* is an intensity that has been exceeded at the site about 20 times, according to the historical catalog. The optimum calibration intensity depends somewhat on the exceedance rate of interest (it is higher if one wants to estimate lower exceedance rates) and on the accuracy of the earthquake sources (it is smaller for source configurations that closely reflect the pattern of historical seismicity). These variations are not large, however.

Optimally calibrated combined estimators are superior to uncalibrated seismic-source estimators, in the sense of being closer to the exceedance rates obtained from detailed local models of seismicity. Another important property of the combined estimators is that they are robust with respect to misspecifications of the earthquake sources. Therefore, these estimators are useful when the source boundaries cannot be estimated accurately, and are even more useful when the very existence of homogeneous earthquake sources is in doubt.

ACKNOWLEDGMENTS

The basic idea of combining seismic-source with historic estimators through calibration was generated in discussions of the senior author with Professor G. Grandori of the Polytechnic of Milan, Italy.

REFERENCES

1. CORNELL, C. A. 1968. Engineering seismic risk analysis. Bull. Seismol. Soc. Am. **54:** 1583–1606.
2. VENEZIANO, D. & L. VAN DYCK. 1987. Statistical analysis of earthquake catalogs for seismic hazard. *In* Stochastic Approaches in Earthquake Engineering, Y. K. Lin & R. Minai, Eds. Lecture Notes in Engineering No. 32. Springer-Verlag. New York.
3. VENEZIANO, D. & L. CHOUINARD. 1987. Local models of seismicity and their estimation. *In* Proceedings of the Symposium on Seismic Hazards, Ground Motions, Soil-Liquefaction and Engineering Practice in Eastern North America, Sterling Forest, NY. Technical Report NCEER-87-0025. Klaus H. Jacob, Ed. National Center For Earthquake Engineering Research. Buffalo, NY.
4. CHIBURIS, E. F. 1981. Seismicity, Recurrence Rates, and Regionalization of the Northeastern U.S. and Adjacent Southeastern Canada. NUREG/CR-2309. Washington, DC.
5. WESTON GEOPHYSICAL CORPORATION. 1980. Site Dependent Response Spectra: Yankee Rowe. WGC. Report to Yankee Atomic Electric Company. Westboro, MA.
6. HEIDARI, M. 1987. Statistical Methods of Earthquake Attenuation. Ph.D. thesis, Department of Civil Engineering, MIT, Cambridge, MA.
7. JOHNSON, N.L. & S. KOTZ. 1970. Distributions in Statistics: Continuous Univariate Distributions-2: 38. Houghton Mifflin. Boston, MA.

APPENDIX
MEAN SQUARED ERROR OF $\lambda_{SS-H}^{(3)}$ WITH RESPECT TO THE TRUE RATE λ

In the body of the paper, we have compared different hazard estimators $\hat{\lambda}(y)$ based on the difference between Log $\hat{\lambda}(y)$ and the logarithm of the local hazard estimator, $\lambda_L(y)$. The justification for this criterion is that $\lambda_L(y)$ is an accurate estimator of the true hazard function $\lambda(y)$. In reality, λ_L is itself random and is positively correlated with all the other estimators $\hat{\lambda}(y)$, because all estimators use the same earthquake data.

Here we show some results on the mean squared error of $\lambda_{SS-H}^{(3)}$, when the error is defined as the difference between Log $\lambda_{SS-H}^{(3)}(y)$ and the logarithm of the true hazard, $\lambda(y)$. Hence we are interested in

$$MSE_T^{(3)}(\lambda; i) = E\{[\text{Log } \lambda_{SS-H}^{(3)}(y; y^* = \bar{y}_i) - \text{Log } \lambda]^2\} \qquad \textbf{(A-1)}$$

where y is such that $\lambda(y) = \lambda$ and the subscript T denotes "true." Analogous quantities for $\lambda_{SS-H}^{(1)}$ and $\lambda_{SS-H}^{(2)}$ in Eq. 8 are tedious to calculate, but they should be close to $MSE_T^{(3)}$.

First we write the logarithmic difference in Eq. A-1 as the sum of two terms,

$$\Delta_a(\bar{y}_i) = \text{Log } \lambda_H''(\bar{y}_i) - \text{Log } \lambda'(\bar{y}_i) \qquad \textbf{(a)}$$

and $\qquad\qquad\qquad\qquad\qquad\qquad\qquad\qquad\qquad\qquad\qquad\qquad\qquad\qquad$ **(A-2)**

$$\Delta_b(y, \bar{y}_i) = [\text{Log } \lambda_{SS}(y) - \text{Log } \lambda] - [\text{Log } \lambda'_{SS}(\bar{y}_i) - \text{Log } \lambda'(\bar{y}_i)] \quad \textbf{(b)}$$

so that Eq. A-1 becomes

$$MSE_T^{(3)}(\lambda; i) = E\{[\Delta_a(\bar{y}_i) + \Delta_b(y, \bar{y}_i)]^2\}$$

$$= (m_a + m_b)^2 + \sigma_a^2 + \sigma_b^2 + \rho\sigma_a\sigma_b \qquad \textbf{(A-3)}$$

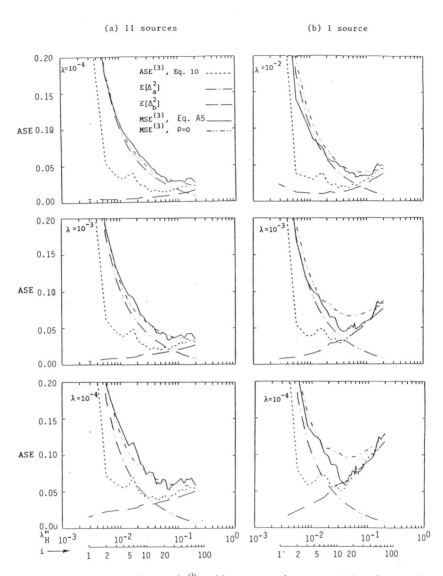

FIGURE 6. Mean squared error of $\lambda_{SS\text{-}H}^{(3)}$ with respect to the true rate λ (see APPENDIX).

where m_a and σ_a^2 are the mean value and variance of $\Delta_a(\bar{y}_i)$, m_b and σ_b^2 are the mean value and variance of $\Delta_b(y, \bar{y}_i)$, and ρ is the correlation coefficient between $\Delta_a(\bar{y}_i)$ and $\Delta_b(y, \bar{y}_i)$. The term $\Delta_a(\bar{y}_i)$ is the error of the historic estimator λ_H'' in Eq. 7 at the calibration intensity $\bar{y}_i$ and the term $\Delta_b(y, \bar{y}_i)$ is the error of prediction of the seismic source estimator if λ_{SS} is calibrated to the exact incomplete rate λ'. The first two moments of these errors, which are needed for the calculation of $MSE_T^{(3)}$, are obtained in a semiempirical way, as follows:

The mean value m_a and the variance σ_a^2 can be calculated theoretically: The error $\Delta_a(\bar{y}_i)$ is random because $\lambda'(\bar{y}_i)$ is random. This incomplete rate can be written as

$$\lambda'(\bar{y}_i) = \lambda_0'[1 - F'(\bar{y}_i)] \qquad \text{(A-4)}$$

where λ_0' is the total rate of events for the incomplete catalog and F' is the cumulative distribution function of site intensity for the generic event of the same catalog. The total rate λ_0' may be considered known with value N/T, where N is

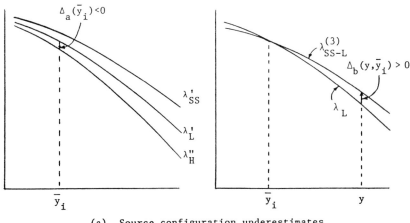

(a)   Source configuration underestimates
b($\underline{x}$) locally

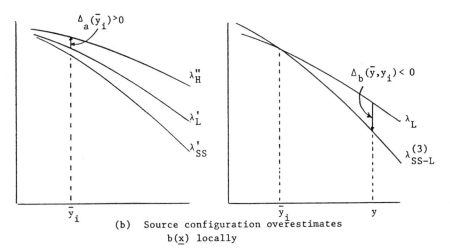

(b)   Source configuration overestimates
b($\underline{x}$) locally

FIGURE 7. Typical situations resulting in negative correlation between Δ_a and Δ_b. Notice that the estimators λ_H'', λ_L' and λ_{SS}' are similar for site intensities smaller than the calibration intensity $\bar{y}$ (plots on the left). The symbol $\lambda_{SS\text{-}L}^{(3)}$ is used for the SS curve calibrated to the local estimator.

the total number of events in the catalog and T is the period of recording. Therefore, the term $[1 - F'(\bar{y}_i)]$ is the only important source of randomness for $\Delta_a(\bar{y}_i)$. The distribution of $[1 - F'(\bar{y}_i)]$ is known to be Beta, with parameters $(i, N - i + 1)$.[7] This result can be used to calculate m_a and σ_a^2 for given N, T, and i.

Theoretical calculation of the other terms in Eq. A-3 is much more complicated. For them, we have resorted to numerical estimation under the assumption that Δ_b should not vary much if one replaces the true rates $\lambda(y) = \lambda$ and $\lambda'(\bar{y}_i)$ in Eq. A-2b with the corresponding local estimates $\lambda_L(y)$ and $\lambda_L'(\bar{y}_i)$. With this replacement, we have obtained m_b, σ_b^2, and ρ as sample values from the 19 sites used earlier to rank various estimators (see FIGURE 2).

The quantities

$$E[\Delta_a^2(\bar{y}_i)] = m_a^2 + \sigma_a^2$$

$$E[\Delta_b^2(y, \bar{y}_i)] = m_b^2 + \sigma_b^2 \qquad \text{(A-5)}$$

$$\text{MSE}_T^{(3)}(\lambda; i) = E\{[\Delta_a(\bar{y}_i) + \Delta_b(y, \bar{y}_i)]^2\}$$

are plotted in FIGURE 6. For comparison, the last quantity in Eq. A-5 for the case when $\rho = 0$ and the averaged squared error $\text{ASE}_{\text{SS-H}}^{(3)}$ in Eq. 10 are also shown.

As one would expect considering the correlation between $\lambda_{\text{SS-H}}^{(3)}$ and λ_L, the mean squared error in Eq. A-1 is larger than the average squared error $\text{ASE}_{\text{SS-H}}^{(3)}$. The difference between the two quantities increases with increasing calibration rate (with increasing i). As a consequence, the value of i that minimizes $\text{MSE}_T^{(3)}$ in Eq. A-1 increases from about 15 to about 30 for the 11-source case. For the case of a single source, the optimum value of i remains around 15–20. The increase in the optimal calibration rate for the 11-source case is probably exaggerated by the fact that the replacement of $\lambda(y)$ and $\lambda'(y)$ with $\lambda_L(y)$ and $\lambda_L'(y)$ reduces the value of $E[\Delta_b(y, \bar{y}_i)]$ and hence increases the optimum value of i. In consideration of this fact, we recommend a calibration value of i around 20.

The correlation ρ is small in the case of 11 sources, but is non-negligible and negative in the case of one source, adding to the robustness of the estimator.

The negative correlation is caused by the difference in the slope parameter b among various seismicity models. The parameter b has a direct influence on the slope of the hazard curves at the site. Because of the various degrees of spatial smoothing of b, one typically observes that the slope of the local hazard estimator λ_L is intermediate between the slope of λ_{SS} for a single source and the slope of the historical hazard estimator λ_H. FIGURE 7 illustrates typical situations, which explain the negative correlation between Δ_a and Δ_b.

General Discussion: Part III

ROBERT WHITMAN (*Massachusetts Institute of Technology, Cambridge, Mass.*): I have not heard duration mentioned much in these discussions, and I ask Gail Atkinson in particular: How might your curves comparing eastern and western North America change if some measure of duration were to be convoluted together with the maximum amplitude of some ground-motion quantity? And I ask all three speakers to give us any insight into bringing duration into probabilistic seismic hazard assessments.

PETER BASHAM, CHAIRMAN (*Geological Survey of Canada, Ottawa, Ontario, Canada*): In my line, when a response spectrum ground-motion parameter is shown, the duration is, in effect, accommodated and included in the derivation of that parameter. So, it's implicit in the computation of a response spectral parameter. I'll let Gail or anyone else answer that question as well.

GAIL ATKINSON: I'd like something clarified: Were you referring to the constant probability plots or the actual response spectral plots for a given magnitude and distance?

BASHAM: Take your pick.

ATKINSON: That's not very clarifying.

BASHAM: All right; constant probability.

ATKINSON: For a constant probability. Well, the duration is built into the curves to a certain extent. At low frequencies the constant probability curve is going to be made up largely of contributions from the larger-magnitude earthquakes, and at the higher frequencies the contributions are going to be rather more slanted to the lower-magnitude earthquakes. The larger-magnitude earthquakes last longer and that's part of the reason why the response is larger— because of the longer durations. So, in fact, there is duration built in there.

ROBIN MCGUIRE (*Risk Engineering, Inc., Golden, Colorado*): A corollary question: When Gail Atkinson showed magnitude-5 time histories generated by the random process method, the eastern duration was shorter than the western duration. Is that because a higher stress drop is being used in the East than in the West? Or is there another reason for that?

BASHAM: I don't remember them being different lengths.

MCGUIRE: They did have different durations.

BASHAM: Did they?

ATKINSON: The simulated ones?

BASHAM: The simulated magnitude 5, where you showed the higher frequencies and higher peaks in the East; I thought they were the same length.

ATKINSON: The simulated record was slightly longer in the East than it was in the West, but the same stress drop was used. I'm having trouble thinking right now of exactly why that is.

MCGUIRE: I've considered this duration question on and off over the last decade. Let me point out a couple of problems with using duration in seismic design in the way that people ordinarily want to use it.

A useful calculation is a plot of peak acceleration versus duration of strong shaking, using, for example, the so-called Trifunic-Brady method to estimate duration). For strong motion data from California you see a certain trend—at least for this definition you only get long durations of shaking at very low amplitudes of ground motion.

If you produce a seismic hazard curve, you obtain some level of PGA associated with some probability. Then you do a separate analysis on the duration of ground motion and you produce a seismic hazard curve for a certain duration of shaking associated with a certain probability.

What a designer might want to do then is to take those two and put them together and design a structure for an earthquake ground motion with that acceleration and that duration. He will get a motion that has never been observed, and he'll have a hard time justifying a design for that earthquake ground motion. So the problem is not a simple one.

The second complication I see is that if we give designers an earthquake response spectrum as an expected maximum ground motion appropriate for design over some period of time, they can handle that. They can translate that into appropriate forces, member stresses, and so on. If you then additionally give them duration, they don't know what to do with it; they don't know how to increase or decrease that seismic design to account for that duration of ten seconds. So, before we make a decision on producing seismic hazards incorporating the duration of ground motion explicitly, we should find out how the designer plans to use that quantity in his design.

Perhaps a more appropriate and easier way to incorporate duration into the design process is to use implicit measures of duration. That is: account for the fact that at certain levels of ground-shaking large magnitudes produce shaking over long durations and produce more damageability or more damage in structures and represent a more damageable ground motion than do small magnitudes with the same energy content at that frequency.

We have done some work in nonlinear analysis of structures, and others have done similar work in producing so-called F factors, which weight the ground motions according to their damageability. That is how we might be able either to discount small magnitudes or to take account of large magnitudes to put them on an equivalent basis, let's say for the energy content, at a certain frequency level. That is a step to be made in the future, but also an easy way to account for different durations of motions in the seismic design process.

ATKINSON: I have one more comment on duration. Because the response spectrum is most often used in a linear analysis procedure, the duration doesn't really count all that much in that you're looking strictly at the elastic range. The duration can be more explicitly taken into account in a nonlinear analysis in which you're using an actual time history. In this case the duration will enter because, the larger the magnitude of earthquake, the longer the time history. And that is where perhaps Dr. McGuire's point about the cumulative damage could come into play.

I'd also like to now answer the earlier question concerning the reason why the simulated eastern time histories had longer durations than the western ones. Duration is longer in the East because in our model we used the increasing duration term found by Bob Herrmann of .05 times R, and we used that at all distances. So, even at the short 10-kilometer distance that I showed, .05 times 10 is a half a second, and that's what you were seeing.

ED KAVAZANJIAN (*Parsons, Brinckerhoff, Quade & Douglas, New York, N.Y.*): The question of duration is perhaps more important to a soils engineer than to a structural engineer. In liquefaction analysis it becomes very important. In the hazard analyses that we've done, we've found it necessary to not put all the magnitudes into one basket, so to speak, but to do hazard analysis separately for different magnitude ranges. Conventionally, you assign a certain number of significant cycles to certain magnitude ranges, at least for West Coast earthquakes. So, one question would be: Would those same number of significant cycles be applicable to a liquefaction analysis for East Coast earthquakes?

Also, there are other parameters besides duration that we sometimes use in our analyses, such as the root mean square of acceleration or the sigma ratio, the ratio of the root mean square to the peak acceleration. I wonder whether there is any difference in the sigma ratio between East Coast earthquakes and West Coast earthquakes.

BASHAM: Is there a difference in the ratio between peak and RMS accelerations East and West? Dave Boore. Yes or no?

DAVID BOORE (*United States Geological Survey, Menlo Park, Calif.*): I can't speak from any data that I've actually seen. But theoretically if the stress drops were the same, then you'd expect exactly the same ratio between the RMS and the peak. It's possible that the stress drops in the West are somewhat less than in the East. In that case you would see some small difference, but I don't think it would be very large.

JENNIFER O'CONNOR (*Johns Hopkins University, Baltimore, Md.*): If longer durations are associated with larger-magnitude earthquakes, which have more damage potential for structures, yet this does not correlate well with peak acceleration, doesn't that imply that peak acceleration isn't necessarily the most meaningful parameter in terms of structural damage potential for ground motion? Is there some way to account for that in this kind of analysis?

BASHAM: I would ask one of the speakers who have so strongly promoted computation of response spectra to respond to that. Why don't you try, Dr. McGuire?

McGUIRE: If you use, instead of the peak acceleration, the response spectrum to represent the level of shaking or the response of the structure, and if you have a longer, larger magnitude occurring at a longer distance that produces an average RMS level of ground motion, you'll have a larger structural response from that motion than you will from a similar average level of response caused by a small magnitude closer to you that occurs over a short duration. That's a short answer to your question. But your hesitancy about PGA is well taken.

UNIDENTIFIED SPEAKER: Both Gail Atkinson and David Boore have assumed or used data to justify the idea that high-frequency ground-motion amplitudes are greater in the East than in the West. I'd like to know whether they have more observations or illustrations of that. I know that Gail Atkinson compared one eastern time history with one western time history, but I wonder whether there is some way that she can readily show the difference using more data.

ATKINSON: The quick answer to that is simply to look at the other eastern records. There aren't very many. And it's true, that it is a problem. But on the near field records of the Miramichi earthquakes, and now the Nahanni earthquakes to some extent, you can also see this same thing—the higher-frequency content as compared to that in the western records.

JACOBO BIELAK (*Carnegie-Mellon University, Pittsburgh, Pa.*): Earlier, reference was made to two earthquakes—one in Ohio and the other in California—that were of almost the same magnitude, but the records were quite different.

The question of duration is extremely important. In trying to answer the question, "Is peak acceleration very important?", I think the answer is yes for high frequencies. For structures with high frequencies the peak acceleration will determine the response. For other structures, not necessarily so.

If we are talking about nonlinear response, which may or may not be as important in the East as in the West, then the actual details of the time history, not just peak values, become important. It is important to realize differences between earthquakes in the East and the West, and then to look at details, and not just the peak values as we have.

Summary Panel Discussion I

ROBIN MCGUIRE, CHAIRMAN (*Risk Engineering, Inc., Golden, Colorado*): Let me start by posing a question and having the panel members spend a few minutes giving an off-the-cuff response. We have just heard some good presentations on issues of tectonics, seismic sources, and ground motions in eastern North America. What did these presentations suggest to you is the next step in reducing earthquake risks? How should we proceed to mitigate earthquake risks? Is the current knowledge consistent with the current seismic building codes, which, for instance, may combine modes of prevention for two different types of earthquakes? How can we modify or update building codes to incorporate the latest information on earthquake hazards? Do we do it by additional scientific studies or by political solutions?

What are your current thoughts, then, on how the latest scientific information should be used to make decisions to reduce earthquake risks?

ROBERT HAMILTON (*United States Geological Survey, Reston, Virginia*): One lesson to be gotten is that a lot can be learned from better definitions of the seismic source zones, as witness the case of New Madrid, where over a ten- to fifteen-year period we went from not knowing the tectonic basis for the earthquakes to having a fairly good model of why they are there. The problem is that for any given seismic source zone, say on the order of 100 by 100 kilometers, it probably would take somewhere between 5 and 50 million dollars and about 10 years to really achieve a good understanding, assuming that the techniques used would be successful.

Five to 50 million dollars sounds like a lot of money, and it is certainly far more than is currently available, but compared with the consequences that hinge on the knowledge of the source zones, the amount is fairly small, particularly with regard to the cost associated with nuclear reactor or waste isolation site delays. So, from my point of view I would say that a lot can be gained through better understanding of the basic tectonics of the seismic source zones.

MCGUIRE: Where would you go for a study of that type? What part of eastern North America appears to be an appropriate place for that kind of study?

HAMILTON: A lot of effort in that direction has been expended in the Charleston area without a lot of success; perhaps some more deep seismic work there might be fruitful. The work being done in the Canadian area of the St. Lawrence Valley area seems to be meeting with a lot of success, and this area is a very important seismic source zone to learn about. A number of the Mesozoic basins along the Atlantic margin, the rifted margin that was identified as one of the source areas, might be a good one to work in as would some of the other reactivated rift zones of the eastern United States.

I. M. IDRISS (*Woodward-Clyde Consultants, Santa Ana, Calif.*): We are heading in the right direction in evaluating earthquake ground motions. But until we have the recordings in the East that we have had in the West, we won't be able to know absolutely. Nevertheless, we can't afford to wait until we have absolutely correct indications because there are far more critical issues to address at the moment.

We have to think very seriously about how to translate these parameters into design parameters. There are two totally different sets of numbers. But how do we take these numbers and apply them in practice? How does the structural engineer tomorrow design a building in location X or location Y? And how do we handle an earth dam? Much of the material to be discussed at this conference deals with how to do things once a number is given, and not how to derive that number or what we might record and estimate based on some theoretical approach. This is one issue worthy of further discussion.

We learned a lot from the West in how to take recorded ground motions and translate them into design parameters. The experience in San Fernando was extremely illuminating—to see that buildings subjected to about three times the level they were designed for actually suffered very little structural damage, although there was a lot of architectural damage. That kind of information would be extremely important to start thinking about as well as how to translate the information coming from the East into design parameters.

KLAUS JACOB (*Lamont-Doherty Geological Observatory, Palisades, N.Y.*): I would like to comment on the nagging issue of what engineers should now practice versus what researchers can now provide. It is interesting to compare a *design spectrum* for a given building site in a given region, say at a New York City site, and compare it with an actually observed *response spectrum*. The observed response spectrum is obviously for an earthquake that occurred somewhere, with a given magnitude, and at some distance from a recording site. Let us assume such an event to be a "typical" or "design earthquake," that is, an earthquake that will affect the given site during the lifetime of a building with a reasonable likelihood. The proposed comparison is instructive because a code-related design spectrum (such as one given by ATC 3-06 or ATC-14) may differ significantly from the response spectra obtained for any particular earthquake. And the degree of difference may tell us whether a model code is likely to provide some acceptable protection during expected events. Remember: a design spectrum, as given by ATC-14, for example, is an average spectrum falling between the occurrences of nearby smaller earthquakes and distant larger earthquakes. Therefore, a code-related design spectrum attempts to reflect, on average, a variety of seismic events. By necessity, it cannot be expected to fit well the response spectral shape of any individual earthquake at any specific distance.

Let us use ATC-14 as the model code applied to a New York City site. The shape of the ATC-14 design spectrum for New York City is determined by the formulae given in ATC-14; the spectral shape is based on three input parameters, A_a and A_v and the soil-type coefficient S. The quantities A_a and A_v, the acceleration- and velocity-related effective horizontal accelerations, can be gleaned from Algermissen's map attached to ATC-14, and both happen to be equal and are given as 0.1 g for a New York City location. Choice of the particular soil type, given by the soil coefficient S_1, S_2, or S_3 applicable to the particular construction site modifies the shape of the design spectrum for periods ≥ 0.3 seconds. We have made a preliminary soil-type map for New York City showing the distributions of S_1, S_2 and S_3. Let us assume a stiff soil site, characterized by a soil coefficient S_2.

As a plausible typical earthquake for the New York City site, we select the San Salvador earthquake of 1986, magnitude 5.1 (5.4) on the Richter scale m_b (or M_s, respectively). The event had a magnitude only slightly larger than that of the historical New York City earthquake of 1884 (M = 5); the San Salvador event had a shallow depth, ≤ 10 km, which is likely to apply also to New York City area earthquakes. We chose the recording of the Hotel Camino Real site, at a 10-km epicentral distance from the event. We use 10 km only as a reference distance. We can scale the observed spectrum up or down by assuming as a first-order approach a simple $1/r$ distance-amplitude relationship. We realize that 10 km is a possible but unlikely short distance to be experienced by a New York City site in less than 500 years of exposure, for a magnitude 5 earthquake. But ≥ 40 km is a likely distance for this region, span, and magnitude.

When we compare the ATC-14 design spectrum to the measured response spectrum for San Salvador for a 10-km distance, but applied to New York City for the reasons given above, we note the following: At low periods, <0.2 seconds, the observed response spectrum for a 10-km distance exceeds the ATC-14 design spectrum by factors of 2 to 4, while at distances of about 40 km the down-scaled response spectrum approximately coincides with the ATC-14 design spectrum. At intermediate periods, between 0.2 and 3 seconds, the observed response spectrum for 10 km substantially exceeds the design spectrum by factors reaching almost 10 at 0.5 seconds (2 Hz), that is, at periods where the peak in the observed response spectrum occurs for this particular event and recording. To make the peak of the scaled response spectrum coincide with the ATC-14 design spectrum level, one would have to be at least 40 km away from the epicenter. At the longest periods (3 to 10 seconds), the observed spectrum for the 10-km distance nicely coincides in level with the ATC-14 design spectrum, while at a distance of 40 km it is smaller by a factor of $\sim$5.

Summarizing this comparison, we conclude that for short distances ($\leq$20 km) and at short periods (<3 seconds) the ATC-14-recommended design spectrum for average soil conditions (S_2) substantially underpredicts the spectral levels observed for an $M = 5 +$ event, while for distances $\geq$40 km ATC-14 provides reasonable estimates. At longer periods ($\geq$3 seconds), ATC-14 provides perhaps conservative, safe design levels for all distances $\geq$10 km for a $M = 5+$ earthquake, but probably not for much larger ($M \geq 6$) earthquakes. The reason for this apparent conservatism at long periods for a moderate $M = 5$ earthquake is that the ATC-14 design spectrum is controlled for these longer periods by contributions from larger, more distant events than the Salvador recordings represent.

Why make such a comparison? It may help to answer the question repeatedly raised by the practicing community: Should engineers adopt and use the so-far only recommended, yet existing codes such as ATC 3-06, ATC-14, UBC-88? Or should the engineering community wait until improved ground-motion estimates for eastern North America can be adopted into updated building codes?

A balanced conclusion from the above comparison may be: for the next ten years it seems prudent to use *existing* code recommendations. Using them for new constructions and renovations clearly provides *some* needed level of protection. Not using any seismic provisions at all is not only unwise and economically shortsighted, but, given the apparent hazards, appears socially irresponsible. The Salvador example showed that there will be cases and period ranges where the code-recommended spectra may well underpredict the ground motion levels that can be reasonably expected in the eastern regions. However, the fact that the presently existing codes and ground-motion estimates need improvements is no valid excuse to continue construction projects without earthquake-mitigating provisions.

Therefore, my proposal is: at latest by 1990 adopt and enforce for New York City and other eastern localities one of the existing codes, whether ATC 3-06, ATC-14, NEHRP, UBC-88, or any similar seismic code. It is relatively unimportant which code is chosen, but it is exceedingly important that a presently recognized code apply. Seismologists must diligently work during the next 10 years to collect badly needed ground-motion data for eastern North America and analyze them to prepare improved code recommendations that are to be integrated with updated engineering information. I strongly feel that existing codes can indeed provide urgently needed protection during the next 10 years for new and renovated structures if and when the codes were legally binding and were rigorously enforced. Beyond that, experts must prepare improved code provisions to be targeted for adoption not later than the year 2000.

DAVID M. BOORE (*United States Geological Survey, Menlo Park, Calif.*): I will use this opportunity to clarify something. When I make a prediction of ground motion using the stochastic method or empirical methods, the predictions are not for the ground motion from a specific earthquake. The predictions are for the average motion we would expect for a suite of earthquakes of a given magnitude. And so, when you make a comparison of the ground motions predicted by either the ATC spectra or by our equations with the records from a specific actual earthquake, it wouldn't be surprising if the comparison was not that good.

The ground motions from one specific earthquake might be low in some instances and high in other instances, relative to the average for all earthquakes of that magnitude.

There is a real danger in looking at the ground motions of one specific earthquake and comparing it to something that is supposed to represent a median value for a strong motion.

That brings up an issue I haven't heard discussed very much: the uncertainty in ground-motion predictions. I did not raise that issue in my presentation. But the uncertainties in our predictions of ground motions, whether they be from empirical analyses or theoretical analyses, are certainly going to be quite large, and we have to take that into account in estimates of seismic hazard.

PETER BASHAM (*Geological Survey of Canada, Ottawa, Ontario, Canada*): I can just repeat what Dave Boore just said. The thing that struck me during the day was the large degree of uncertainty in our seismic hazard estimates. I know that in Canada when we deal with the engineering community with the zoning maps referenced by the National Building Code, we get frustrated at times when the engineers start computing probabilities to four decimal places. You can grab them by the lapels and try to convince them that there's at least an uncertainty of a factor of 2 in any ground-motion parameter that might come off our zoning maps. I haven't yet had the heart to tell them that I think the uncertainty is much greater than that.

Robin McGuire showed us seismic source zone models by two different groups of experts that produce real differences in ground-motion parameters at a fixed probability. There is evidence in some of Robin's data that these uncertainties can be a factor of 5 in the ground-motion parameter.

I'm torn on what to do with that because if we keep emphasizing our uncertainties, we give the policy-makers the impression that we can't do anything. But on the other hand, it is very important to recognize that these seismic hazard estimates are very uncertain.

It's difficult to treat those uncertainties if you are designing a common building, for example, using moderate-probability ground-motion parameters. And perhaps you don't need to. I would suggest you take the median value and run with it. But you can make use of those uncertainties if your facility is a high-hazard facility that requires a more conservative type of design, and you could very easily argue that if you have a critical facility, you must accommodate those uncertainties by intentionally being more conservative.

So, the message that I would like to convey is: don't forget how little we know about seismic hazards in eastern North America.

DANIELE VENEZIANO (*Massachusetts Institute of Technology, Cambridge, Mass.*): I would like to comment upon nonstationarity of seismicity. What one observes over periods of a few decades, and certainly of two or three centuries, are migrations of seismicity and the activation and deactivation of sources. This raises several questions: Which sources will be active in the next 50 or 100 years? At what levels? How much can we trust historic seismicity to predict future seismicity?

One exercise we went through in order to investigate these problems was to fit seismicity models to different parts of the catalog for the northeastern region. In one case we used the entire data set, which covers a period of about 300 years. In another case we used about only the most recent 50 years of data. The models estimated in the two cases are substantially different. Then we asked: which of the two models should we use to predict future seismicity.

Using statistical (cross-validation) techniques, we found that the accuracy of prediction of the two models is comparable. The reason for this is the following: If there are some nonstationarities in seismicity and one fits the model to the more recent data, and one reduces the bias, but because the data are limited, statistical uncertainty on the parameters is large. If one includes more data, the bias increases, but the parameters of the model are better constrained.

We are now in the process of modeling variations of seismicity. The objective is to integrate them in probabilistic earthquake predictions. This area ought to receive more attention in the future.

IDRISS: I want to emphasize Dr. Jacob's point that we need to separate the process from its results. ATC-14, which is built upon many years of work in the West, has led to numbers that are highly applicable on the basis of our understanding of ground motions recorded in the West. Ground motions in the East could be quite different, however, so we need to back off and ask how we should modify a document like ATC-14, the principles of which may be quite applicable, to reflect actual numbers that may be different. That's what I was referring to earlier when I spoke about taking the numbers we're getting from recordings and theoretical calculations and seeing how they would be translated into design parameters.

The other issue that we have not touched upon is that of local site conditions and what effect they may have.

UGO MORELLI: I have a comment about ATC-14. I am starting with two strikes against me. As the project officer from FEMA I'm sort of a "paper tiger," and I'm neither a geologist nor a seismologist, nor an engineer, nor an architect, for that matter. I'm a social scientist. And I am here because it is through associating with experts in various scientific fields that I learn a little bit each time.

Now, to my comment: Let me urge you not to adopt ATC-14 yet. ATC-14 is going through a review, which I hope will be very thorough. The code is being expanded to cover different types of buildings so that it will be more applicable to the eastern part of the United States. The code already has another number— ATC-22—and it's being done by ATC under contract to FEMA.

After it's been developed, it will be subjected to a consensus-backed review, similar to the one that was given to the NEHRP-recommended provisions so you will have ample opportunity to participate. I urge you to take the opportunity to expand ATC-14 and make it more applicable to the East.

JIM MALLEY (*Degenkolb Associates, San Francisco, Calif.*): My firm was a subcontractor for this famous ATC-14 project. I would like to note that ATC-14 was developed to evaluate existing buildings. During the development of the project, the decision was made to use a lower spectrum than that used for new building design. The ATC-3 spectrum was developed with one sigma amplification factor and a 10 percent chance of exceedance in 50 years, while the ATC-14 spectrum was developed with median amplification factors and a 10 percent chance of exceedance in 50 years. I mention this because I thought Dr. Jacob's remarks may have been a little misleading if you were thinking he was presenting a design earthquake for a new building.

JACOB: No. I applied it to New York simply because the New York City situation involves mostly existing buildings.

MALLEY: I just wanted to make the point that it wasn't a new building spectrum. It was for an existing building. . . .

JACOB: I realize that. That's why you have a factor of 2.12 instead of 2.5. I'm fully aware of that. But that still doesn't take away from the problem that it's a lumped-together spectrum, very much like Gail Atkinson's constant probability spectrum, which is more clearly defined versus an individual spectrum. What the community has to address here is that the seismologists start with individual spectra that have a particular shape for a certain magnitude, stress and distance. And we're talking lumped spectra here. I wonder how much the practicing community is going to be confused by these new seismological contributions, which are important and correct. That's all I'm trying to say.

OVE GUDMESTAD: I am from Norway and I'm here because the North Sea area and the eastern U.S. have many geological features and earthquake source mechanisms in common. In the North Seas we have a very huge data base of geophysical profiles. The oil companies have collected these data in the search for oil and gas. Similar large data bases from the Grand Banks must also exist, and I would suggest that these data bases could be used to look for the faults and to determine how they go up towards the surface. I would also suggest looking for a correlation with recent earthquakes and with offshore slides. This could help in understanding the sources and the potential for earthquakes in these areas.

JOHN ADAMS (*Geological Survey of Canada, Ottawa, Ontario, Canada*): We have looked in a very general way at the offshore profiles, and some of the offshore well-site surveys are very interesting. I lack the expertise to go through those data systematically. I think that you would learn from them that there is a belt of fairly deep seated rift faults, maybe 100 kilometers wide, which are the ones that Peter Basham and I suggest are active. We see some of these faults in some places cutting sediments less than forty million years old, and in a few places there are suggestions that they come right up to the surface.

What is not quite clear when you look at these faults in the very soft sediments at the surface is whether you're seeing tectonics or some sort of growth faulting in response to slow subsidence. But we do not have the data or the people to get into this analysis in great detail.

In May 1988 a NATO symposium, of which Peter Basham is the co-convener, will be held in Copenhagen, which will look at seismicity and seismic hazard on passive margins on both sides of the Atlantic. I'm certainly hoping to learn from our Norwegian colleagues what more we could be doing and what the experience there has been. We should not look at the passive margin of North America and say that this is the only one; there are a lot of other passive margins out there that have been studied and a lot more earthquakes along them that should be looked at.

MIKE UZUMERI (*University of Toronto, Toronto, Ontario, Canada*): As a structural engineer, and as the chairman of a code-writing committee which is making final the Canadian Code, let me try to answer the question of what I have learned at this meeting.

I've learned a lot, but nothing that I'm going to use immediately in developing the building code. This conference is intended to be for "constructed facilities," which includes complex structures such as nuclear power plants and dams. Building codes, on the other hand, deal with structures that are pretty close to being routine. These routine buildings constitute essentially 95 percent of what is being put up. So, the obligation of the building code writer to the profession is to deemphasize the importance of numbers, emphasize the selection of structural systems, and emphasize good detailing. We must not become a slave to the numbers. In Canada we have followed this philosophy. Drs. Basham, Adams, and Weichert, who are seismologists, provide the answers from the seismological end to our committee, that is, they tell us what is likely to occur with the probability of 10 percent in fifty years.

The political decision to take 10 percent probability in fifty years was made by the Committee. Having decided that, the Committee decided what the zonal boundaries should be. Seismologists gave us the zones. From that point on, the structural requirements of the code are written.

GUY NORDENSON (*Ove Arup and Partners, Int., Ltd., New York, N.Y.*): I'd like to address Klaus Jacob's remarks. The response spectrum shape he used was originally derived from the ATC-3 in shape; and that came out in 1978 and was in turn derived from work that was done about five years before that, after the San Fernando earthquake. Nationally we have seen at various conferences a lot of response spectra for different earthquakes—in Chile, in Mexico City, in San Salvador—that have shapes that look to me, and I'm an engineer and don't entirely understand all the seismological details, quite different. In addition, we have the problem of coming up with a shape for code work on the East Coast.

It seems to me that the time is ripe for an effort on the scale of ATC-3 to bring a lot of this information together and come up with a consensus based on what we've learned in the last fifteen years about spectral shapes for the nation overall. Professor Bertero, for instance, has pointed out several times that if the shapes we've obtained from the earthquake in Mexico City or in Chile are transferrable to earthquakes we might get here, we'll be very unconservative out West and we may be very unconservative here in the East. We have a lot to digest and to get into our codes.

JACOB: We actually agree. Any other impression I may have inadvertently created by comparing the San Salvador response spectra with the ATC design spectra for New York City, and by pointing out some mismatch between them does not change the fact that we fundamentally agree on the most essential point: ATC (or similar) spectra will provide a substantial protective measure to buildings in eastern cities which the present construction practice in the absence of any obligatory seismic code just cannot provide. It is true that much important research is presently in progress, especially regarding site effects on soft soils. But to wait until the fruits of this research can be incorporated into upgraded codes by a slow consensus may result in no codes' being adopted in eastern states and cities for another 10 years. This could have irreparable effects which we must avoid. I endorse the adoption of imperfect codes. We should insist, however, that firm provisions for periodic updating be built into the codes, and that the next update should be finalized not later than the year 2000. It should be updated earlier, provided a consensus based on important new findings demands it.

Some Basic Aspects of Soil Liquefaction during Earthquakes[a]

RICARDO DOBRY

Civil Engineering Department
Rensselaer Polytechnic Institute
Troy, New York 12180

INTRODUCTION

Seismic liquefaction of saturated sand deposits in level or almost level ground is an important cause of damage, both to the ground itself and to structures supported by (or buried in) the soil. Much research has been done on the subject, which has clarified significant aspects including the basic character of cyclic shear strain amplitude in determining the rate of pore water pressure buildup and the proven usefulness of empirical charts based on *in situ* Standard Penetration Tests (SPT).[1-3]

However, some issues of great relevance to engineers are still not clear and are under discussion. They include the relative importance of partially drained versus purely undrained failure mechanisms, and the associated problem of the engineering significance of sand boil observations.[3,4] This paper explores those issues in the light of available field, laboratory, and analytical evidence. Possible answers are proposed, and it is suggested that a main reason why the SPT charts work is because of the strong correlation between penetration measurements and the amount of water expelled by the liquefied sand.

PORE PRESSURE BUILDUP

FIGURE 1 sketches the typical situation in the field. A liquefiable saturated sand layer is located at a certain depth, overlain by a nonliquefiable stratum. This nonliquefiable shallow layer includes all soils above the groundwater level, as well as submerged soils that are not susceptible to liquefaction because of their high density, cohesion, and other favorable properties.[5] Excess pore pressures can develop in the liquefiable layer because of the cyclic shear straining induced by the earthquake shaking. As sketched in FIGURE 1, the ground seismic accelerations a are associated with shear stresses τ, with τ being proportional to a. A more useful parameter is the associated shear strain, $\gamma = \tau/G$, where G is the shear modulus of the sand. Although G varies with γ because of the nonlinear stress-strain behavior of soil, G can be computed once its small-strain value, G_{max}, is

---

[a] This paper was prepared as part of NCEER Research Project No. 86-2031 entitled "Investigation of Seismic Failure and Permanent Deformations of Soils during Earthquakes."

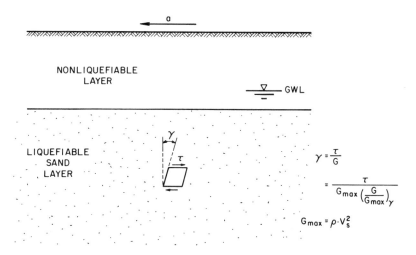

FIGURE 1. Typical soil and seismic loading conditions in liquefaction problems.

known, and G_{max} is in turn readily obtained from *in situ* geophysical measurements of shear wave velocity.[1,3,6,7]

FIGURE 2 presents the influence of cyclic strain amplitude, γ_c, on the pore pressure ratio after ten cycles; this correlation includes all available cyclic triaxial laboratory data and is valid for both loose and dense sands, as well as for clean and silty sands. The plot is very consistent, and it shows that any liquefiable sand deposit subjected to earthquake cyclic loading in one horizontal direction, having

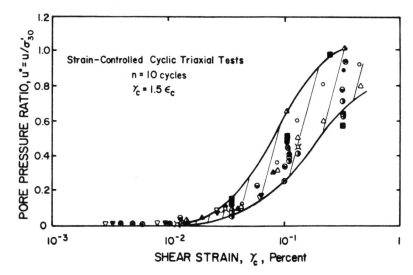

FIGURE 2. Pore pressure ratio in sands versus cyclic shear strain after ten cycles (correlation developed by me and included in the NRC report[3]).

an equivalent duration of shaking of 10 cycles, will develop a pore pressure ratio, $u^* \simeq 1$, if the induced γ_c is about 0.3% or larger. For the usual calculations of seismic strain using a measured G_{max} and nondegraded G/G_{max} a more relevant value to enter the plot is $u^* \simeq 0.5$, because De Alba et al.[8] have shown that once u^* reaches 0.5, the soil weakens significantly, and the seismic strains and excess pore pressure increase rapidly during further earthquake loading. A new $\gamma_c \simeq$ 0.1% is obtained for $u^* = 0.5$ and 10 cycles, which is further reduced to $\gamma_c \simeq$ 0.06% once the influence of the second horizontal acceleration component is considered.[9] The corresponding value for a longer earthquake duration of 30–50 cycles is $\gamma_c \simeq 0.04\%$

SAND BOIL GENERATION

Liu and Qiao[10] have reported shaking table experiments on stratified sand which clarify the mechanics of sand boil generation during earthquakes, both in the free field and around structures. One of these tests is represented in FIGURE. 3. Both in the free field and at great depths under the foundation, $u^* \simeq 1$ was reached during the shaking. Significant amounts of water expelled from the pores of the soil accumulated under the more impervious layers, forming there "water interlayers." Sand boiling occurred when one of these water interlayers broke up to the ground surface.

In Liu and Qiao's tests, the liquefiable sand deposit extends to the ground surface. As pointed out by Ishihara[5] a much more common situation that can also produce sand boils is that shown by FIGURE 4, where a liquefiable sand stratum of thickness H_2 is located under a nonliquefiable layer of thickness H_1. If the sand stratum reaches a pore pressure ratio u^* close to 1 because of shaking, it will reconsolidate, and the water expelled will tend to flow upwards to the ground surface. This can induce significant vertical gradients in the upper layer, as shown in the figure. These gradients, and the possibility that sand boils do appear, will be enhanced if the *upper* soil is thin (small H_1), relatively impervious and incompressible, and has vertical fissures or cracks. Features of the liquefiable, *lower* layer that enhance the process are a large compressibility and thickness H2 (which increase the total volume of water available for upward flow), and a large permeability (which makes the water available fast).[4,11-13] Castro[4] suggests that the rela-

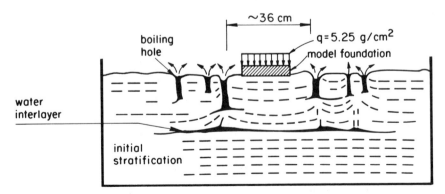

FIGURE 3. Shaking table test of stratified sand deposit. (After Liu and Qiao.[10])

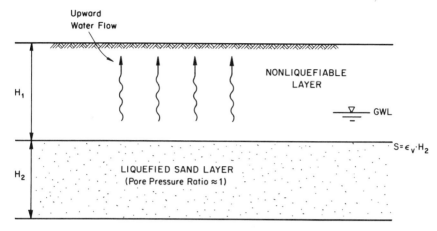

FIGURE 4. Upward water flow after liquefaction of lower layer.

tively infrequently observed occurrence of sand boils generated by the reconsolidation of silty sands is due to the low permeability of these soils.

Therefore, sand boils are mainly an indication of high upward gradients in the layer located between the ground surface and the stratum liquefied by the earthquake, caused by the water expelled by this lower liquefied stratum. In cases in which the upper layer is also made up of sand, high gradients have been observed to produce a delayed liquefaction of this layer near the ground surface ("quicksand phenomenon"), often after the end of the shaking, as described in the next section. On the other hand, the fact that the lower stratum develops large values of u^* ("it liquefies") does not automatically mean that high gradients and sand boils will appear in the upper layer; this may or may not occur, depending on the other factors previously mentioned.

SAND BOILS AND ENGINEERING FAILURE

As noted by Castro,[4] sand boils by themselves are of very little engineering significance. They are potentially more important as indicators of the existence of water interlayers, such as shown in FIGURE 3, and of high gradients in the upper layer overlying the liquefied deposit. The NRC report[3] defined several possible failure mechanisms which could produce a loss of stability of liquefying terrain due to a decrease of the soil shear strength in the presence of driving (static) shear stresses. Two of those mechanisms are directly relevant to this discussion. In mechanism B, a loose film of soil with a high water content and reduced shear strength develops at the top of the liquefied layer, similar to the "water interlayers" of FIGURE 3, and the failure surface passes by this film (see also Seed[14]). In mechanism C, the high upward gradients near the ground surface decrease the effective stresses acting on the soil, with the corresponding decrease in shear strength of the shallow layer, which may cause heavy structures supported by it to sink or tilt, and buried light structures to float up. Therefore, mechanisms B and C are directly related to the same phenomena that generate the sand boils; and the engineering significance of sand boils as indicators will depend on how important these two failure mechanisms are in explaining observed engineering failures. Another important aspect of the sand boil issue derives from the fact that the observation of (or lack of) sand boils after an earthquake is often taken as equivalent to the statement that the site did (or did not) liquefy. For example, the SPT charts by Seed et al.[2] are partially based on sand boil observations.

A cursory review of the literature reveals that, indeed, the appearance of sand boils and observations of upward water now are well correlated with most types of engineering damage and ground failure caused by liquefaction. Photos taken for about half an hour starting immediately after the end of shaking, in an area that liquefied in Niigata, Japan in the 1964 earthquake, reveal that strong water boiling started within 2–3 minutes after the end of shaking and lasted for about 20 minutes; and the Annex building moved horizontally 2 meters after boiling had started.[15] In fact, most of the damage to structures in this earthquake was caused by the upward water now near the ground surface. Summarizing a number of field observations for the same city and event, Ambraseys and Sarma[12] state that "fountains of water continued to play for nearly 20 minutes after the earthquake . . . [and] settlements and tilts [of buildings] occurred gradually during some minutes of time following the earthquake, and considerable time, possibly many minutes, elapsed whilst structures continued not only to settle and tilt, but also to rise." A lateral-spread ground failure occurred in part of the Heber Road site during the Imperial Valley, California earthquake of 1979, accompanied by many sand boils; another part of the site, located on denser sand, did not fail and also did not exhibit sand boils.[16] A number of examples for other earthquakes of correlations between sand boils and engineering manifestations of liquefaction are cited by Housner,[11] Ambraseys and Sarma[12] and Tohno and Yasuda.[17]

VOLUME OF WATER EXPELLED BY LIQUEFIED LAYER

A key parameter in the development of water interlayers, high upward gradients near the ground surface, and sand boils, as well as of engineering failures having mechanisms B or C, is the total volume of water expelled by the liquefied sand layer shown in FIGURE 4. If this water volume is very small, it will not be able to sustain the process, and no liquefaction effect of any kind may be observed in the upper layer, *even if $u^* \approx 1$ in the liquefied lower layer.* This water volume is measured by the settlement S of the liquefied layer, and it is $S = \varepsilon_v \cdot H_2$, where ε_v = permanent volumetric strain of the soil. Therefore, it should be expected that both H_2 and ε_v play a role in controlling the appearance of sand boils and engineering failure after the earthquake generates large pore pressures in the liquefied layer.

Ishihara[5] discussed in detail the thickness H_2 necessary for surface manifestations of liquefaction, and related it to the thickness H_1 of the nonliquefiable layer. For the 1964 Niigata earthquake, he concluded that surface manifestations took place only when $H_2 > 3$ m and $H_1 < 3$ m; this criterion is also consistent with observations from the 1977 Vrancea, Rumania earthquake and the 1979 Montenegro, Yugoslavia earthquake. FIGURE 5 presents a similar correlation for the 1983 Nihonkai–Chubu, Japan event. Finally, Ishihara also reproduced a qualitatively similar correlation between H_1 and H_2 obtained by Chinese researchers after the 1976 Tangshan earthquake. In summary, there is overwhelming evidence that, for a given earthquake and type of soil profile, it is not enough that high pore pressures develop at depth for ground failure and other surface manifestations of

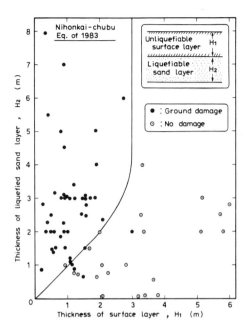

FIGURE 5. Liquefaction, ground damage and layer thicknesses in 1983 Nihonkai–chubu earthquake. (After Ishihara.[5])

liquefaction to happen; in addition, H_2 must be large enough and H_1 must be small enough. An illustrative example from Niigata is reproduced in FIGURE 6, which includes profiles from two sites: one which exhibited surface evidence of liquefaction (River Site) and another where this did not happen (South Bank Site). Both sites were subjected to essentially the same shaking; both have about 1 m of

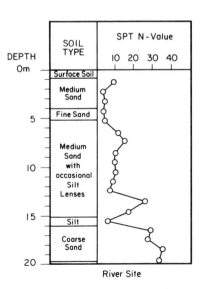

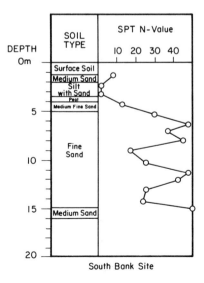

FIGURE 6. Soil profiles that exhibited surface evidence of liquefaction (River Site) and that did not (South Bank Site) during the 1964 Niigata earthquake. (Modified from Ishihara.[5])

nonliquefiable surface soil; the groundwater level is at the same depth; and both contain very loose sandy soils below the surface layer. The difference is that the very loose layers, which must have developed $u^* \approx 1$ during the shaking, have a total thickness of more than 10 m at the River Site and only about 2 m at South Bank Site. Another factor that may have also contributed to this different performance is the fact that the very loose layer between 2 and 4 m in the South Bank site profile is predominantly silt, and thus has a low permeability.

The second factor determining the value of S is the volumetric strain ε_v, which is mostly related to the degree of compaction or relative density of the sand. There are three main sources of information to estimate ε_v. The first source in laboratory data on reconsolidation of sands after developing $u^* \approx 1$ by undrained cyclic loading.[13,18,19] The second source is laboratory data on drained compaction of sand by cyclic straining, such as provided by Silver and Seed[20] and illustrated in FIGURE 7; this would correspond to the assumption that the excess water is completely expelled from the sand in the field while the shaking is still taking place. The third source of information is measurements of surface settlement S and calculated $\varepsilon_v = S/H_2$, from the field after actual earthquakes[18] and from model centrifuge tests after simulated earthquakes.[21] From all these measurements, a range of $\varepsilon_v \approx 1.5$ to 5% is obtained for loose sand, while a much smaller value, of the order of $\varepsilon_v \approx 0.2\%$ or less, is estimated for very dense sand. The highest values of ε_v for loose sand ($\sim$3–5%) were obtained from the field observations and centrifuge tests, probably because of the much higher cyclic strains developed in the loose sand layer once $u^* \approx 0.5$ to 1; this phenomenon occurs to a much lesser extent in dense sand. Therefore, it is reasonable to estimate that, typically, ε_v is 0.2% or less in a very dense sand and 3–5% in a loose sand. *That is, other things being equal, the amount of water expelled by a very loose sand can be as much as fifteen or twenty times larger than if the sand is dense!*

An example of the effect of this difference in ε_v values for loose and dense sand is provided by the liquefaction that occurred at Heber Road Site in the 1979 Imperial Valley, California earthquake. This case history, already mentioned,

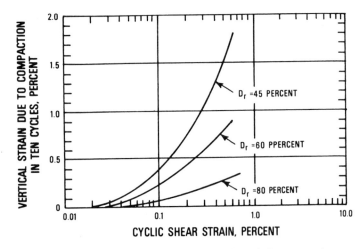

FIGURE 7. Volumetric strain ε_v due to compaction of sand after ten cycles versus cyclic shear strain γ_c and relative density of soil. (After Silver and Seed.[20])

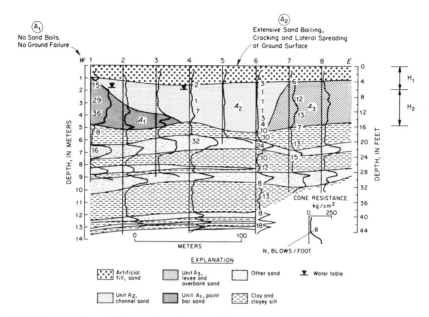

FIGURE 8. Soil conditions and surface manifestations of liquefaction, Heber Road Site, 1979 Imperial Valley earthquake. (After Youd and Bennett.[16])

was reported by Youd and Bennett[16] and is also discussed by Vucetic et al.[9] The profile of the site is reproduced in FIGURE 8. The stratigraphy is essentially identical throughout the site: about one meter of artificial sand fill (nonliquefiable layer) above the water table, underlain by about 4 m of saturated silty sand (liquefiable layer) followed by clay. The site was within 2 km of the fault rupture for this magnitude 6.6 event, and strong motions at comparable distances from the fault indicate that the site was subjected to very large accelerations, of the order of 0.5 g or greater. Shear-wave velocity measurements performed at the site by Sykora and Stokoe[7] allowed Vucetic et al.[9] to perform a detailed study of the seismic strains and pore pressures induced in the liquefied layer by the earthquake. In this last publication it is concluded that high values of $u^* \approx 1$ developed in the liquefiable layer *throughout* the whole site. However, although extensive sand boiling, cracking, and lateral spreading did in fact occur at the part of the site labeled A_2 in FIGURE 8, corresponding to very low SPT values between 1 and 7 blows/foot, none of these effects was observed in another part, labeled A_1, which had high SPT values between 29 and 36 blows/foot. After the previous discussion, the explanation for this dramatic difference in behavior was clearly the much larger value of ε_v in the loose sand A_2 compared with the dense sand A_1.

WHY DO THE SPT LIQUEFACTION CORRELATIONS WORK?

Until about ten years ago, the usual belief was that the SPT empirical correlations proposed by several authors to predict liquefaction were successful because the SPT, after correcting for the effect of overburden pressure, measures essentially the degree of compaction or relative density of the sand. The reasoning went like this: liquefaction in the field, including surface manifestations and engineering ground failure, occurs when a saturated sand layer at a certain depth reaches $u^* \approx$ 1. For a given intensity of earthquake shaking (as measured by the cyclic stress ratio), undrained cyclic stress–controlled tests show that the number of cycles of stress needed to reach the $u^* \approx 1$ condition is more or less uniquely related to the relative density of the sand. Therefore, the field SPT measures the key parameter (relative density) controlling rate of undrained pore pressure buildup and thus also liquefaction failure.

Extensive laboratory investigations showed later that this picture was too simplistic, and that a number of factors jointly control the rate of pore pressure buildup in stress-controlled tests, with relative density being only one of them. These other factors include the method of sample preparation or fabric effect, preshaking of the soil, lateral earth pressure coefficient and overconsolidation, and increased time under pressure (aging effect).[22] In that regard, Seed,[22] in his 1979 paper, suggested that the effect of all these factors on SPT is similar to their effect on pore pressure buildup, and this explains why the SPT charts work. In 1982 Dobry et al.[1] proposed an alternative explanation: the main reason for the success of the SPT is that penetration values are correlated to shear wave velocity and G_{max}, which controls pore pressure buildup through its effect on the amplitude of the seismic shear strain. Both of these explanations by Seed and Dobry et al. share a common feature: they focus on the undrained pore pressure buildup, and identify the reaching of high values of u^* with the appearance of surface manifestations of liquefaction and associated ground failure.

However, the discussion above—including the evidence from the Heber Road Site-suggests that in addition to its relation with G_{max}, another key to the success of the SPT lies in its correlation to relative density and hence to ε_v. That is, both the penetration measurements and ε_v are very sensitive to changes in the density of the sand, and as a result the SPT correlates well with ε_v.

For example, FIGURE 9 shows the analysis of the liquefaction at Heber Road Site performed by Youd and Bennett[16] for the 1979 event, using Seed's SPT charts

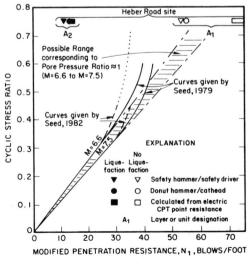

FIGURE 9. Application of liquefaction SPT chart to the Heber Road Site case history, and speculative relation proposed by me between chart and pore pressure buildup. (Modified from Youd and Bennett.[16])

for earthquake magnitudes M = 6.6 and 7.5. The chart is successful in predicting the surface liquefaction effects and lateral-spread failure on the loose sand A_2, as the corresponding points plot above Seed's curves; and it also successfully predicts the absence of any manifestation of liquefaction on the dense sand A_1, as the points plot below the curves. The chart is completely successful despite the fact that the pore pressure ratio u^* was close to 1 in both sands A_1 and A_2, as previously discussed.

Seed et al.[2] have suggested that the curve of cyclic stress ratio versus N_1 for M = 7.5 in FIGURE 9 corresponds to a constant cyclic shear strain, $\gamma_c \approx 0.03\%$ for $N_1 < 30$ blows/foot. The author agrees with this value, which is of the same order of magnitude of $\gamma_c \approx 0.04$ to 0.06%, estimated by the author at the beginning of this paper as the earthquake strain needed to reach $u^* \approx 1$ after 10–50 cycles. Therefore, up to $N_1 \approx 30$, Seed's curves in FIGURE 9 correspond to an approximately constant shear strain and to $u^* \approx 1$. The author speculates that the continuation of these curves of $u^* \approx 1$ beyond $N_1 = 30$ are either straight lines or downward curves, instead of the upward bending curves of the chart. One possible location for these $u^* \approx 1$ lines is indicated by the hatched band on the figure. Unlike Seed's curves this new band would have correctly predicted $u^* \approx 1$ for sand A_1 at Heber Road Site in 1979. The difference between this $u^* \approx 1$ band and Seed's curves, caused by the sharp turn upward of the latter for $N_1 > 30$, would correspond to dense sand layers that experience the $u^* \approx 1$ condition, but with this having no engineering consequence because of the low amount of water expelled by the soil during reconsolidation.

CONCLUSIONS

Some basic but still unclear aspects of soil liquefaction at level sites during earthquakes are discussed. These aspects, relevant both to an imroved understanding of the phenomenon and to its engineering consequences, includer the mechanism of formation and the significance of sand boils, the relative importance of partially drained versus purely undrained failure mechanisms, and the relation between the widely used Standard Penetration charts and the behavior *in situ* during earthquakes. On the basis of available evidence, I reached the following preliminary conclusions:

(1) the amount of water expelled by the liquefied layer plays an important role both in the development of partially drained failure mechanisms and in the appearance of sand boils;

(2) the good correlation observed between liquefaction-induced failures and sand boiling strongly suggests that failure mechanisms in partially drained soil are very significant;

(3) the amount of water expelled by the liquefied layer depends on both the degree of compaction of the sand and the thickness of the layer, and this explains the importance of both parameters in determining the effects of liquefaction; and

(4) a main reason why the Standard Penetration charts work is the good correlation between penetration measurements and the amount of water expelled by the sand after liquefaction. Specifically, in a dense sand it is possible to generate a pore water pressure ratio close to unity without much water being expelled, and with insignificant settlement and no sand boils or engineering failure taking place. In this case the penetration charts correctly predict the absence of liquefaction manifestations rather than the level of pore pressure buildup in the soil, as is commonly believed.

REFERENCES

1. DOBRY R., R. S. LADD, F. Y. YOKEL, R. M. CHUNG & D. POWELL. 1982. Prediction of pore water pressure buildup and liquefaction of sands during earthquakes by the cyclic strain method. NBS Building Science Series 138. National Bureau of Standards. Washington, DC.
2. SEED, H. B., I. M. IDRISS & I. ARANGO. 1983. Evaluation of liquefaction potential using field performance data. J. Geotech. Eng. ASCE 109: 458–482.
3. NATIONAL RESEARCH COUNCIL. 1985. Liquefaction of soils during earthquakes. Report by the Committee on Earthquake Engineering, National Research Council, National Academy Press. Washington, DC.
4. CASTRO, G. 1987. On the behavior of soils during earthquakes—liquefaction. Proceedings of the 3rd International Conference on Soil Dynamics and Earthquake Engineering, Soil Dynamics and Liquefaction. 42: 169–204. Princeton University, Princeton, NJ.
5. ISHIHARA, K. 1985. Stability of natural deposits during earthquakes. Proceedings of the 11th International Conference on Soil Mechanics and Foundation Engineering. 1: 321–376.
6. DOBRY, R., K. H. STOKOE, II, R. S. LADD & T. L. YOUD. 1981. Liquefaction susceptibility from S-wave velocity. In situ Testing to Evaluate Liquefaction Susceptibility, ASCE National Convention, St. Louis, MO.
7. SYKORA, D. W. & K. H. STOKOE, II. 1982. Seismic investigation of three Heber Road sites after October 15, 1979 Imperial Valley earthquake. Geotechnical Engineering Report GR82–24, Civil Engineering Department, University of Texas at Austin.
8. DE ALBA, P., C. K. CHAN & H. B. SEED. 1975. Determination of soil liquefaction characteristics by large-scale laboratory tests. Report No. EERC 75–14, College of Engineering, University of California, Berkeley, CA.
9. VUCETIC, M. et al. 1988. Analytical study of liquefaction observed at the Heber Road site during the October 15, 1979 Imperial Valley earthquake. Research report, Dept. of Civil Engineering, Rensselaer Polytechnic Institute, Troy, N.Y. In preparation.
10. LIU, H. & T. QIAO. 1984. Liquefaction potential of saturated sand deposits underlying foundation of structure. Proceedings of the 8th World Conference on Earthquake Engineering III: 199–206. San Francisco.
11. HOUSNER G. 1958. The mechanism of sandblows. 1958. BSSA 48: 155–161.
12. AMBRASEYS, N. & S. SARMA. 1969. Liquefaction of soils induced by earthquakes. BSSA 59(2): 651–664.
13. YOSHIMI, Y. & F. KUWABARA. 1973. Effect of subsurface liquefaction on the strength of surface soil. Soils and Foundations 13(2): 67–81.
14. SEED, H. B. 1986. Design problems in soil liquefaction. Report No. UCB/EERC–86/02, College of Engineering, University of California, Berkeley, CA.
15. ———. 1966. Pictures taken at the moment of the Niigata earthquake. Soil and Foundation VI(1): i–vi.
16. YOUD, T. L. & M. J. BENNETT. 1983. Liquefaction sites, Imperial Valley, California. J. Geotechnical Eng. 109: 440–457.
17. TOHNO, I. & S. YASUDA. 1981. Liquefaction of the ground during the 1978 Miyagiken–Oki earthquake. Soils and Foundations 21(3): 18–34.
18. LEE K. L. & A. ALBAISA. 1974. Earthquake-induced settlement in saturated sands. J. Soil Mechanics Found. Div., ASCE 100: 387–406.
19. BHATIA, S. K. 1980. The verification of relationships for effective stress method to evaluate liquefaction potential of saturated sands. Ph.D. thesis, Department of Civil Engineering, University of British Columbia.
20. SILVER, M. L. & H. B. SEED. 1971. Volume changes in sands during cyclic load. J. Soil Mechanics Found. Div., ASCE 97: 1171–1182.
21. WHITMAN, R. V., P. C. LAMBE & B. L. KUTTER. 1981. Initial results from a stacked ring apparatus for simulation of a soil profile. Proceedings of International Conference on Recent Advances in Geotechnical Engineering and Soil Dynamics.: 1105–1110. St. Louis, MO.
22. SEED, H. B. 1979. Soil liquefaction and cyclic mobility evaluation for level ground during earthquakes. J. Geotech. Eng. Div., ASCE 105: 201–255.

Liquefaction Evidence for Repeated Holocene Earthquakes in the Coastal Region of South Carolina[a]

S. F. OBERMEIER, R. E. WEEMS, R. B. JACOBSON, AND
G. S. GOHN

United States Geological Survey
Reston, Virginia 22092

Features thought to have originated from earthquake-induced liquefaction have been discovered throughout much of the coastal region in South Carolina and in extreme southeastern North Carolina. Nearly all these liquefaction features are sandblows presently manifested as filled craters. Prehistoric craters near Charleston formed in long-separated episodes at least three times within the past 7200 years. Ages of dated craters far from Charleston, beyond the farthest 1886 earthquake sandblows, differ from ages of craters near Charleston. Insufficient data have been collected to determine whether ages of all craters far from Charleston differ from ages of craters near Charleston. Both the size and relative abundance of pre-1886 craters are greater in the vicinity of Charleston (particularly in the 1886 meizoseismal zone) than elsewhere, even though the susceptibility to earthquake-induced liquefaction is approximately the same at many places throughout this coastal region. These data indicate that, in this coastal region, the strongest earthquake shaking during Holocene time has taken place repeatedly near Charleston.

INTRODUCTION

The strongest historic earthquake in the southeastern United States took place in 1886 near Charleston, South Carolina. The meizoseismal zone (encompassing Modified Mercalli intensity-X effects) was about 35 km wide and 50 km long,[1] and the estimated body-wave magnitude (m_b) was between 6.6 and 7.1.[2] The potential for a future earthquake as strong as that of 1886 is a major concern in engineering design in the Southeast. The concern is reinforced by a 300-year historical record of continuing weak seismic activity near Charleston. The source of the earthquakes in the Charleston area remains unidentified, and seismotectonic hypotheses are widely disparate despite many geologic, geophysical, and seismic studies during the past decade. No faults or fault systems have been identified that fully explain the large 1886 Charleston earthquake or the smaller, historic earthquakes that have occurred throughout much of South Carolina.[3-5] Therefore, because direct evidence of seismotectonic conditions was lacking and because the historic

[a] This research was supported by the U.S. Nuclear Regulatory Commission under Agreement No. RES-82-001.

earthquake record was too limited to estimate the frequency of moderate to strong earthquakes, we undertook a search for pre-1886 sandblows. Liquefaction-induced features, particularly sandblows, were a commonplace effect of the 1886 earthquake within the meizoseismal zone and caused us to first search that area. The initial discovery of pre-1886 sandblows was reported by Gohn et al.[6] and Obermeier et al.[7] The study was next extended throughout much of coastal South Carolina[8] and more recently into southeastern North Carolina.

FIGURE 1 shows the current results of ongoing research: (1) the approximate boundary of the 1886 Charleston earthquake meizoseismal zone; (2) areas conspicuous in 1886 for development of sandblows (described as "craterlets" by Dutton[9]); (3) the outer limit (most distant) site of reported 1886 sand blows[10]; and (4) the sites of pre-1886 sandblows that we have discovered. The unpatterned onshore part seen in FIGURE 1 encompasses our principal search area, which includes predominantly marine sediments. Fluvial deposits were searched only locally along the Edisto River.

None of the pre-1886 sand blows has any expression on the ground surface that is discernible by on-site examination or on airphotos. The sandblows are visible only where exposed in walls of excavations at least 1.5 m deep; generally, sandblow exposures have been found in drainage ditches and borrow pits. At most sites shown in FIGURE 1, at least three or four sandblows are exposed within a few hundred meters of one another.

The following sections focus on (1) the geologic setting in which these sandblows are found and on criteria for interpreting their earthquake origin; (2) the ages of the craters; (3) our technique for estimating relative shaking severity throughout the coastal region of South Carolina; and (4) our interpretation of relative shaking severity during Holocene time.

EARTHQUAKE-INDUCED LIQUEFACTION FEATURES

Earthquake Criteria

Identification of an earthquake origin for the features we have observed depends primarily on eliminating alternative mechanisms, the most likely of which is artesian springs (nonearthquake). Other mechanisms that must be eliminated include liquefaction induced by ocean-wave pounding, ground disruption by trees and landslides, compaction-induced de-watering, and physical and chemical weathering.

Geologic criteria that we have developed for interpreting that near-surface features are earthquake-induced sandblows generally consist of the following elements:

1. The features must have sedimentary characteristics consistent with an earthquake-induced liquefaction origin; that is, there is evidence of an upward-directed, strong hydraulic force that was suddenly applied and was of short duration.

2. Characteristics such as shape, width, and depth are consistent with historical observations of liquefaction features formed during the 1886 earthquake.

3. The features are in groundwater settings where a suddenly applied, strong hydraulic force of short duration could not be reasonably expected except from earthquake-induced liquefaction. In particular, these settings are extremely unlikely sites for artesian springs.

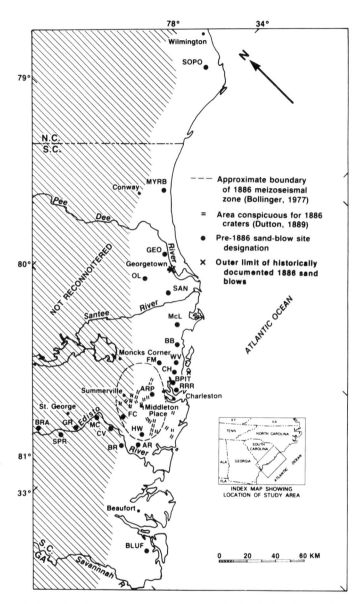

FIGURE 1. Map showing pre-1886 sand-blow sites. *Nonhatched* onshore region is made up predominantly of marine sediments younger than about 240,000 years. *Hatched area* denotes region of older marine sediments that was not reconnoitered. Younger fluvial sediments occur locally. All sand-blow sites in the *nonhatched* region are in marine-related deposits, and all sites in the *hatched* region are in fluvial deposits. Numerous sites discovered in 1886 meizoseismal zone are not shown because of lack of space. Abbreviations adjacent to sand-blow sites are specific site designations.

4. Similar features occur at multiple sites (within a few kilometers of one another) in similar geologic and groundwater settings. Where evidence of age is present, it should support the interpretation that the features formed in one or more discrete, short episodes.

The more of these criteria that are met, the greater the probability that a feature is earthquake-induced.

Geologic Setting

In South Carolina, the coastal region is known locally as the "low country" because it has low local relief (1–3 m) and low elevation (0–30 m) and because vast expanses of swamp and marshland are under water much of the year. Most of the Carolina low country is covered by a 5- to 10-m-thick blanket of unconsolidated Quaternary marine and fluvial deposits; this blanket lies on semilithified Tertiary sediments.[11] The Quaternary sediments occur primarily as a series of six well-defined, temporally discrete, interglacial beaches and associated back-barrier and shelf deposits that form belts subparallel to the present shoreline. The oldest beach deposits are farthest inland and at the highest altitudes; younger beach deposits are progressively closer to the ocean and at lower altitudes.

FIGURE 1 shows the approximate inland limits of the marine-related deposits (beach, shelf, open-sound back barrier) designated as units Q1, Q2, and Q3 by McCartan et al.[11] Q3 deposits are about 200,000 to 240,000 years old,[12] and are present about as far as 20 to 40 km inland from the modern coast; Q1 and Q2 deposits are younger and closer to the ocean. The search for sandblows was generally restricted to units Q1, Q2, and Q3 because older deposits have such a low susceptibility to liquefaction (due to weathering and deep groundwater) that the likelihood of sandblows forming during the Holocene and late Pleistocene has been extremely low.

The geologic setting most frequently associated with recognizable earthquake-induced liquefaction features in our study is the crest or flank of Pleistocene beach ridges (FIG. 2), where a thin cover of clay-bearing sand or humate-rich soil overlies clean, uniformly sized sand (i.e., containing no silt or clay). According to first-hand observations of effects of the 1886 earthquake by Earl Sloan, "These craterlets are found in greatest abundance in belts parallel with [beach] ridges and along their anticlines."[13] To a much lesser extent, sandblows have been found in fluvial and back-barrier sediments.

Types of Features

We have interpreted two types of pre-1886 earthquake-induced liquefaction features: sandblows caused by venting of sand and water to the surface, and sheared zones caused by sliding of a solid slab above a liquefied stratum.

At the great majority of liquefaction sites seen in FIGURE 1, we observed the "craterlet" (or crater) type of sandblow, which was the most abundant type produced by the 1886 earthquake. A few sites show evidence of earthquake-induced sandblows that formed as deposits vented to the surface, commonly through a tabular fissure, leaving relict sand mounds (see FIGURE 5 in Obermeier et al.[8]). Evidence of earthquake-induced oscillating ground movement on a liquefied stratum has been found at one site. A few sites show evidence of earthquake-induced lateral spreads (landslides that formed on nearly level ground far from any scarps downslope or upslope).

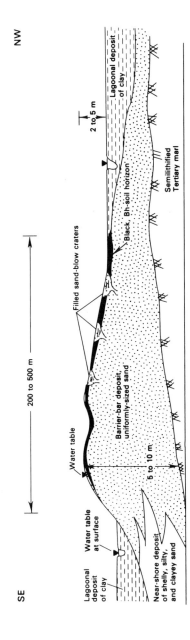

FIGURE 2. Schematic cross section of representative Pleistocene barrier showing sediment types, groundwater table location, filled sand-blow craters, and Bh-soil horizons. Modern shoreline is located southeastward. Lagoonal clay deposit at *left* is younger and lower in elevation than the barrier bar deposit.

Sand Blows

Only the crater type of sand blow is discussed herein because of the relative abundance of this type and because of the difficulty generally attendant in establishing an earthquake origin to the vented-sand-mound type of feature. Almost all the pre-1886 craters had an original morphology and size comparable to those of the 1886 craters described by Dutton[9]; however, the craters now are filled with sediment. The filled sand-blow craters share many common sedimentary structures and sequences in fill sedimentation, which are illustrated in FIGURE 3. The figure shows a soil horizon cut by an irregular crater, which is filled with stratified to massive (nonstratified) and graded sediments; five layers (identified in the figure) are characteristically present. Materials within the craters are sand and clasts from the Bh (humate-rich B horizon), B-C, and C soil horizons, and sand from depths much below the exposed C horizon. In a pre-1886 filled crater, the Bh horizon that developed on the in-filled sediment is generally thinner than that of the laterally adjacent, undisturbed soil. With increasing age, the Bh horizon on the filled crater becomes thicker, has more clay, and has better developed soil structure.

Interpreted phases in the formation of the filled sand-blow craters include the following[6,7]: (1) after earthquake-induced liquefaction at depth, excavation of a large hole at the surface was caused by the violent upward discharge of the liquefied mixture of sand and water; (2) a sand rim was accumulated around the hole by continued expulsion of liquefied sand and water after the violent discharge; (3) sand, soil clasts, and water churned in the lower part of the bowl, followed by settling of the larger clasts and formation of the graded fill sequence; and (4) the crater was filled from adjacent surface materials to form the thin stratified fill sequence during the weeks to years after the eruption. In the craters predating the 1886 earthquake, the sand blanket ejected from the crater is indistinguishable in the field from the surface and near-surface (A, E, and Bh) soil horizons because the blanket has been incorporated into these soil horizons.

We infer that the craters were formed by a short-lived process because of the preservation of very friable clasts of Bh- and C-horizon soil in the graded zone, and because of the very sharp boundary between the graded and layered zones (i.e., the contact between layers 3 and 4). We interpret the force to have been strong and upward-directed because many of the large clasts in layer 1 clearly have been rounded by tumbling in a fluidized bed. Where multiple craters occur near the topographically highest part of a beach ridge, and where multiple craters appear (on the basis of similar soil profiles or similar radiocarbon dates) to have formed at the same time, all four geologic criteria are satisfied for interpreting an earthquake origin. This interpretation of origin is strongly supported by historical accounts of the general morphology and geologic setting where the craters formed and by our discovery of craters very near where they were reported to have occurred in 1886[13] (at sites HW and ARP in FIGURE 1).

Reverse Shears

Along the flanks of some Pleistocene beach ridges, reverse shears associated with liquefaction features appear to have formed during the same earthquakes. The shears generally occur on ground that slopes less than 1 percent. These slopes are so gentle and the possibility of high artesian pressures is so remote that gravity-induced slumping is virtually impossible. Shear displacements commonly range from 1 to 4 cm. Reverse shears of earthquake origin also have been found on level ground. At one site (site RRR in Figure 1), for example, reverse shears dipping in opposite directions (toward one another) formed about 10 m apart; the shears occur above and in the stratum that liquefied during shaking, and sand blows having vents traceable to this liquefied stratum formed between the shears. It appears that the only possible mechanism that could have formed the opposite dipping shears was alternating directions of ground motion on the liquefied stratum. Such a mechanism is illustrated in Youd.[14]

Reverse shears also can occur as isolated features, but generally are associated with sandblows. Many filled craters have a reverse shear near the edge of the crater. The shear is invariably located on the downslope side of the crater and cuts otherwise undisturbed soil horizons and underlying sediments. The shear did not form after venting because the vent is not cut by the shear. At a few sites, the shears along crater edges could have formed only in response to earthquake-induced lateral-spread movement because the shears are traceable into and along the bedding of the stratum that liquefied. At these sites, gravity-induced (non-earthquake) slumping is precluded by the low slope angles and the high frictional strength of the sand materials. Only rarely is an exposure sufficiently deep to show that the shear goes into a stratum that liquefied, and thus an earthquake origin cannot be assigned definitely at all sites. However, we judge that these reverse shears along crater edges indicate an earthquake origin, even where they formed on gently sloping ground (less than 2 percent) as much as 5 m below the beach crest.

Reverse shears that formed at the toe of slopes of lateral spreads are much more difficult to locate than sand blows. Lateral spreads are relatively scarce, and, in addition, they generally formed in wet areas, near streams, where the regional water table is so high as to preclude drainage and expose outcrops, even with ditching.

Confidence in Interpretation of Origin

Features at all pre-1886 sites shown in Figure 1 are interpreted to be of earthquake origin, although the quantity and quality of evidence differs from site to site. The best evidence of earthquake origin is found at sites that have at least one of the following features: (1) craters near topographically high, beach-ridge crests; (2) numerous, especially large craters near where craters were reported to have been abundant during the 1886 earthquake; (3) ground-oscillation shears formed in opposite directions; (4) lateral spreads that could not be gravity-induced, and have shears traceable into a liquefied stratum; or (5) shattered ground (indicating forceful injection) cut by numerous sand-filled dikes in settings where high artesian pressures could not have been involved. Sites yielding the best evidence include those designated BR, AR, HW, ARP, RRR, CH, FM, WV, McL, SAN, OL, and SOPO in Figure 1.

All other sites in Figure 1 represent filled craters that are more than several meters below the crests of beach ridges or are on fluvial terraces (where groundwater conditions are not well known and periodically might be artesian). Evidence for an earthquake origin is less certain for these sites. However, some craters at site MYRB (Fig. 1) have reverse shears, which makes an earthquake origin probable.

All five of the types of sites and features listed above lie approximately within the boundary of the 1886 meizoseismal zone shown in Figure 1; this diversity of mutually reinforcing data is strong evidence for prehistoric, Holocene earthquake activity in that area. At the site farthest from Charleston (SOPO), filled craters are present on topographically high beach deposits, which again is relatively strong evidence for earthquake triggering.

Even if sites were diverse, reinforcing lines of evidence are neglected, sufficient sites would remain to support our interpretation of Holocene seismic activity (discussed below).

AGES OF CRATERS

Craters are generally the only features for which radiocarbon dating can be used to determine earthquake ages because other liquefaction-related features have not been found to contain organic matter. The following methods have been used to bracket the times of crater formation[15]: (1) radiocarbon dating showing ages of woody material (tree limbs or pine bark) that fell into the open crater soon after crater formation; (2) radiocarbon dating showing ages of roots that were sheared off at the edge of the crater (roots predating crater formation) and dating of roots that grew into the stratified fill portion of the crater (roots postdating crater formation); and (3) dating of clasts of Bh material that fell into the graded-fill zone of the crater. The first method yields a relatively accurate time of earthquake occurrence, whereas the other two yield only a broad time range within which the true age lies.

Sufficient data have been collected at site HW (near Charleston; FIG. 1) to show that at least three pre-1886 earthquakes have produced sandblows within the past 7200 years. The radiocarbon-determined age of pine bark in a crater at site ARP (also near Charleston; FIG. 1) independently supports the second of these three events. The only definitive statement about earthquake recurrence that can be made currently is that, near Charleston, at least four sandblow-producing earthquakes (m_b probably >5.5) have occurred within the past 7200 years (including the 1886 event).

A relatively accurate age of crater formation has been obtained at the site farthest from Charleston (site SOPO; FIG. 1), but that age differs from those between Charleston and site SOPO, thereby suggesting that the craters at site SOPO originated from an epicentral region far from Charleston. The radiocarbon ages that have been determined for liquefaction features throughout the Carolina coastal region are insufficient to define separate epicentral regions for separate earthquakes. At many sites far from Charleston, at least two generations of craters are present that are long separated in time of formation.

HOLOCENE EARTHQUAKE SHAKING

Methodology

Measurement of the size and number of craters at each site shown in FIGURE 1 provides a way to estimate the relative severity of shaking that affected the coastal region during the Holocene. The methodology for estimating shaking intensity is based on the premise that the number and the size of liquefaction features are greatest where earthquake shaking is strongest for a fixed geologic setting, liquefaction susceptibility, and depth to water table.

The condition of a fixed geologic setting is met almost ideally. Most sites shown in the unshaded area of FIGURE 1 are in Pleistocene beach deposits (units Q2 and Q3) that have a narrow range of thicknesses, which lie on Tertiary marl that is rock-like with respect to transmission of seismic energy. Because of the areal extent of this setting, shaking of the potentially liquefiable material probably has been comparable in the near-surface sediments at many places throughout the coastal region.

The condition of high liquefaction susceptibility is also widespread throughout that part of the coastal region in FIGURE 1 shown with no pattern. Source-stratum sands typically are loose (as judged by limited Standard Penetration data and numerous observations of ease of augering) and have about the same thickness. Moreover, the thickness and properties of nonliquefiable sediments overlying the source stratum lie within a narrow range. Also, recurrence of liquefaction at a site does not greatly diminish the ability to produce numerous large craters in the loose sands that are typical of the Carolina coastal region. This is verified by the observation that at site HW (FIG. 1) large numbers of large craters formed in each of at least three generations of Holocene earthquakes, each generation being widely spaced in time. Thus, at sites in beach deposits shown in FIGURE 1, liquefaction susceptibility is generally high and has not been reduced greatly by a previous occurrence of liquefaction.

Humate-rich (Bh) soil horizons provide evidence that depth to the water table, the other major variable, presently is generally very shallow and has been throughout the Holocene. Bh horizons are subsoil horizons that have accumulated significant amounts of organic matter; they tend to be dark gray to black, and they commonly extend 0.6 to 1.0 m below the land surface (FIG. 3). Bh soil horizons form throughout the coastal region in areas of high water tables and sandy parent materials. Generally, they are considered to form in acid soils in the zone of water-table fluctuation where seasonal lows in the water table allow vertical infiltration of organic material, and the generally high water table prevents oxidation of the organics. Radiocarbon-determined ages from bases of Bh horizons of soils in the meizoseismal area of the 1886 earthquake range from 5000 to 10,000 years,[15] which indicates that water tables have been high at least during most of the Holocene. In fact, because these radiocarbon ages almost certainly are mean residence times of organic matter, the presence of even older organic carbon is implied. Furthermore, field and laboratory data indicate that below the distinct black Bh horizons, these soils lack any evidence of previous periods of oxidation and vertical infiltration. Therefore, no evidence exists that the water table has been substantially below present-day levels during the Holocene.

Shaking Intensity

FIGURE 4 shows the relative number of pre-1886 craters and the ranges of crater diameters for selected areas along the coast. Four classes of crater diameters are shown: small, medium, large, and huge. These diameters are the maximum widths exposed in the ditch walls, at a depth of about 0.5 m below the present ground surface. The relative number is the measure of the number of craters per unit area found in the setting most susceptible to earthquake-induced liquefaction. The relative numbers constitute a semiquantitative index of crater density made on the basis of our exploration of numerous drainage-ditch networks throughout the region. An index value of 1000 has been assigned arbitrarily to the area encompassed by the 1886 meizoseismal zone. On this basis, a value of about

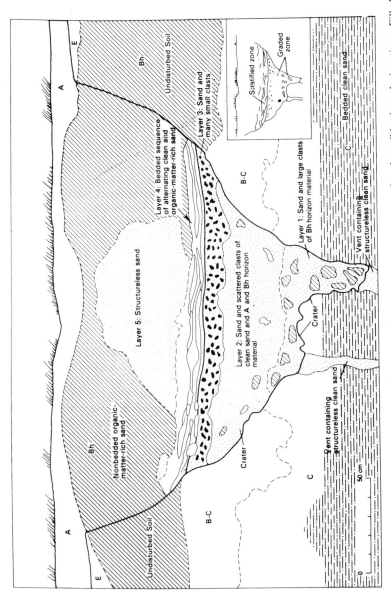

FIGURE 3. Schematic cross section of representative filled sand-blow crater. Letters correspond to the soil horizon designations. Filled crater in this figure significantly predates 1886 earthquake as determined by the thickness of the Bh horizon in the filled crater.

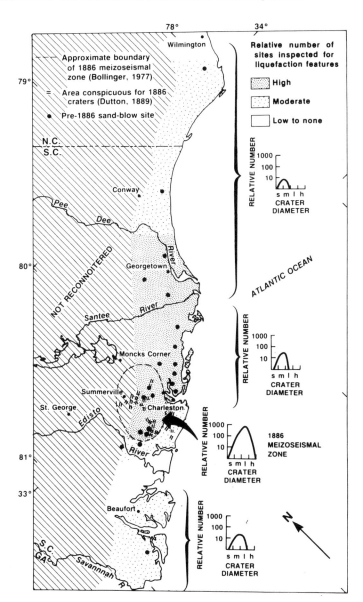

FIGURE 4. Relative number of filled craters and crater diameters for pre-1886 sand blows at sites on marine-related sediments. The relative number is a scaling based on comparison with abundance of craters in the 1886 meizoseismal zone, which has an arbitrary value of 1000. Crater diameters are small (s, less than 1 m), medium (m, 1–2 m), large (l, 2–3 m), and huge (h, greater than 3 m). Some sites in the 1886 meizoseismal zone are not shown in FIGURE 1; all other sites in this figure are also shown in FIGURE 1.

10 is appropriate for the area north of the Santee River; this indicates that that area has approximately 1 percent as many craters as are in the 1886 meizoseismal zone.

Both the relative number and the crater diameter are greatest within the 1886 meizoseismal zone. Both decrease with distance from the zone, although the shape of their curves remains about the same. The relation indicates that the distribution represents a variation in exposure to strong earthquake shaking. Therefore, we conclude that pre-1886 Holocene shaking has been strongest in the general area of the 1886 meizoseismal zone. North of the Santee River, shaking has been much weaker. Intermediate shaking has taken place between Charleston and the Santee River and also between Beaufort and the Savannah River.

Evidence for this interpretation is fairly substantial for the area between Charleston and Wilmington because of the hundreds of kilometers of ditches we searched in that region (see FIGURE 4). Evidence is very substantial for the 1886 meizoseismal zone and fairly substantial for the area southeast of the 1886 meizoseismal zone and for the area between Beaufort and the Savannah River. Data are too limited to make an interpretation for the area between Beaufort and the Edisto River because of the very limited number of ditches and pits available for inspection in that region.

Whether or not the pre-1886 Holocene shaking in the 1886 meizoseismal zone was stronger than the 1886 event can be determined definitely only by further radiocarbon-dating of craters at sites far beyond the 1886 meizoseismal zone.

PRINCIPAL CONCLUSIONS

1. At least three prehistoric liquefaction-inducing earthquakes have taken place within the past 7200 years near Charleston. Different ages of craters have been obtained far from Charleston, which suggests another epicenter(s) exists far from Charleston.

2. Preliminary data indicate that Holocene earthquake shaking has been stronger near Charleston than elsewhere along the coast of South Carolina and the coast of southeastern North Carolina.

ACKNOWLEDGEMENTS

This study was accomplished only with the generous assistance of many people and organizations. The field-intensive nature of the study required permission to work at sites scattered throughout South Carolina. We especially thank Don McConaughy and Jack Lacy, Westvaco Corporation; James Ralston, International Paper Company; and J. G. McGavin II and M. B. Walden, Union Camp Corporation. We also thank Professor Joyce Bagwell and the Baptist College of Charleston for generously permitting us to use their facilities.

REFERENCES

1. BOLLINGER, G. A. 1977. Reinterpretation of the intensity data for the 1886 Charleston, South Carolina, earthquake. *In* Studies related to the Charleston, South Carolina, earthquake of 1886—A preliminary report. D. W. Rankin, Ed. U.S. Geol. Surv. Prof. Pap. **1028:** 17–32.

2. NUTTLI, O. W. 1983. 1886 Charleston, South Carolina, earthquake revisited. *In* Proceedings of Conference XX, a workshop on "The 1886 Charleston, South Carolina, earthquake and its implications for today." W. W. Hays & P. L. Gori, Eds. U.S. Geol. Surv. Open File Rep. **83-843:** 44–50.

3. HAYES, W. W. & P. L. GORI, EDS. 1983. Proceedings of Conference XX, a workshop on "The 1886 Charleston, South Carolina, earthquake and its implications for today." U.S. Geol. Surv. Open File Rep. **83-843:** 1–502.

4. DEWEY, J. W. 1985. A review of recent research on the seismotectonics of the Southeastern seaboard and an evaluation of hypotheses on the source of the 1886 Charleston, South Carolina, earthquake. NUREG/CR-4439: 1–44. U.S. Nuclear Regulatory Commission. Washington, DC.

5. ———. 1986. A century after the Charleston earthquake. Science News **129**(17): 263.

6. GOHN, G. S., R. E. WEEMS, S. F. OBERMEIER & R. L. GELINAS. 1984. Field studies of earthquake-induced liquefaction-flowage features in the Charleston, South Carolina area: Preliminary report. U.S. Geol. Surv. Open File Rep. **84-670:** 1–26.

7. OBERMEIER, S. F., G. S. GOHN, R. E. WEEMS, R. L. GELINAS & M. RUBIN. 1985. Geologic evidence for recurrent moderate to large earthquakes near Charleston, South Carolina. Science **227:** 408–411.

8. OBERMEIER, S. F., R. B. JACOBSON, D. S. POWARS, R. E. WEEMS, D. C. HALLBICK, G. S. GOHN & H. W. MARKEWICH. 1986. Holocene and later Pleistocene(?) earthquake-induced sandblows in coastal South Carolina. *In* Proceedings of the Third U.S. National Conference on Earthquake Engineering: **1:** 197–208. Earthquake Engineering Research Institute. El Cerrito, CA.

9. DUTTON, C. E. 1889. The Charleston earthquake of August 31, 1886. U.S. Geological Survey Ninth Annual Report 1887–88: 203–528.

10. SEEBER, L. & T. G. ARMBRUSTER. 1981. The 1886 Charleston, South Carolina earthquake and the Appalachian detachment. J. Geophys. Res. **86**(B9): 7874–7894.

11. McCARTAN, L., E. M. LEMON, JR. & R. E. WEEMS. 1984. Geologic map of the area between Charleston and Orangeburg, South Carolina. U.S. Geological Survey Miscellaneous Investigations Map I-1472: scale 1 : 250,000.

12. SZABO, B. J. 1985. Uranium-series dating of fossil corals from marine sediments of southeastern United States Coastal Plain. Geol. Soc. Am. Bull. **96:** 398–406.

13. PETERS, K. E. & R. B. HERRMANN, Compilers and Eds. 1986. First-hand observations of the Charleston earthquake of August 31, 1886, and other earthquake materials. South Carolina Geol. Surv. Bull. **41:** 116.

14. YOUD, T. L. 1984. Geologic effects—liquefaction and associated ground failure. *In* Proceedings of the Geologic and Hydrologic Hazards Training Program. U.S. Geol. Surv. Open File Rep. **84-760:** 210–232.

15. WEEMS, R. E., S. F. OBERMEIER, M. J. PAVICH, G. S. GOHN, M. RUBIN, R. L. PHIPPS & R. B. JACOBSON. 1986. Evidence for three moderate to large prehistoric Holocene earthquakes near Charleston, South Carolina. *In* Proceedings of the Third U.S. National Conference on Earthquake Engineering. **1:** 197–208. Earthquake Engineering Research Institute. El Cerrito, CA.

Earthquake-Induced Liquefaction in the Northeastern United States: Historical Effects and Geological Constraints

MARTITIA P. TUTTLE AND LEONARDO SEEBER

Lamont-Doherty Geological Observatory
Palisades, New York 10964

INTRODUCTION

Liquefaction causing permanent ground deformation in unconsolidated sediments has been reported for several earthquakes in the eastern United States. Features that formed as a result of liquefaction during the 1886 Charleston, South Carolina earthquake have been studied and characterized by various teams of scientists.[1-5] Several conclusions can be drawn from studies of this and other earthquakes that have triggered liquefaction. In any one event, several different kinds of liquefaction features can be formed. These features tend to be concentrated in different parts of the meizoseismal area. The localization of liquefaction features is strongly controlled by characteristics of the unconsolidated surficial sediments, such as grain-size distribution and thickness and sequence of stratigraphic units. Furthermore, because liquefaction features can be recognized, the timing of large prehistoric earthquakes may be retrieved by studying deformation structures in sediments that are susceptible to liquefaction. In the eastern U.S., this has been achieved in the meizoseismal areas of the 1886 Charleston, South Carolina[1-5] and the 1811–1812 New Madrid, Missouri earthquakes,[6] both large historic events known to have caused liquefaction.

Liquefaction of subsurface materials can lead to different types of ground failure, including development of quick conditions, formation of sand boils or ejection of water, settlement of the ground surface, and mass movements on low to moderate slopes. These types of ground failures can have serious impacts on structures. As the result of liquefaction, buildings may tilt or sink into the ground, buried structures may float to the surface, and building foundations may be displaced laterally.[7] In the investigation of historic earthquakes, reports of these types of ground failure serve as liquefaction indicators.

During the past three centuries, several moderate to large earthquakes may have triggered liquefaction and ground failure in the Northeast (FIG. 1). The 1727 Newbury earthquake is one of the better documented cases and provides the opportunity to study an historic case of earthquake-induced liquefaction in this glaciated region. The purpose of the study in the Newbury area is to locate and characterize 1727 and possibly earlier earthquake-induced deformation structures and to identify the geologic settings of ground failure during the relatively small 1727 event. This study lays the groundwork for recognizing earthquake-induced deformation structures in glacial sediments in the Northeast and for identifying

196

the geologic conditions that may control liquefaction and ground failure during future earthquakes.

EARTHQUAKES AND LIQUEFACTION IN THE NORTHEAST

As mentioned above, several moderate to large earthquakes may have induced liquefaction in the Northeast. On the basis of numerous reports of ground failure indicative of liquefaction, strong cases of earthquake-induced liquefaction can be made for the 1727 Mfa = 5 Newbury earthquake and the 1755 Mfa = 5.5 Cape Ann

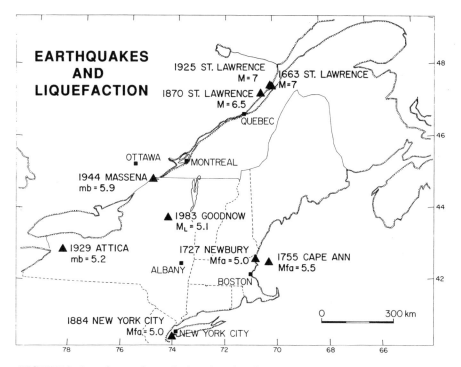

FIGURE 1. Locations and magnitudes of earthquakes that may have triggered liquefaction in the northeastern United States and southeastern Canada. On the basis of accounts of ground failure during these events, strong cases of earthquake-induced liquefaction can be made for the 1727 Newbury, MA, the 1755 Cape Ann, MA and the 1944 Massena, NY earthquakes. More tenuous cases include the 1884 New York City, the 1929 Attica, NY and the 1983 Goodnow, NY earthquakes. See text for references.

earthquake, both in Massachusettss,[8,9] and for the 1944 mb = 5.9 Massena earthquake, in New York.[10] The evidence of liquefaction is less strong for the 1884 Mfa = 5.0 New York City earthquake, the 1929 mb = 5.2 Attica, New York earthquake, and the 1983 M_L = 5.1 Goodnow earthquake, located in the Adirondacks of upper New York State.[9,11]

Effects of the 1727 Newbury Earthquake

The Newbury area was colonized by the English in the early 1600s. The original landing and settlement, known today as Newbury Old Town, was on high ground along the Parker River (FIG. 2). From there, settlement spread north toward the Merrimack River. Homes were built on the uplands, cattle was grazed in the lowlands, and the marshes were harvested for hay. The mouth of the Merrimack River provided an ideal harbor and became the prominent port of Newburyport. Between 1635 and 1845, numerous earthquakes were felt in the Newbury area. The most strongly felt events were the 1638, 1727, and 1755

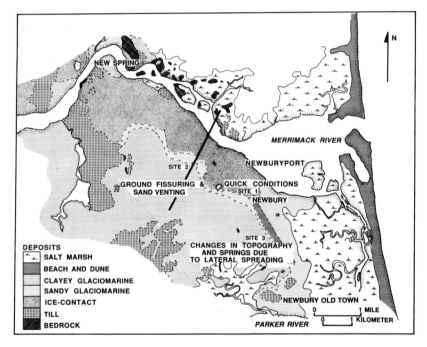

FIGURE 2. Surficial geology of the Newbury-Newburyport area in northeastern Massachusetts (modified from Sammel[25] and Edwards and Oldale[26]) and locations of ground effects reported during the 1727 earthquake and of site investigations (FIGS. 3–5).

earthquakes. By the time of the 1727 earthquake, the Newbury-Newburyport area was sparsely settled.

According to historical reports, the 1727 earthquake caused the following ground effects in the Newbury area: (1) upland was locally changed to "quagmire"; (2) places in the marsh lowland were elevated, becoming too dry to support native grasses; (3) in at least ten places in the clay lowlands, water and sand vented from 0.3- to 0.6-meter-wide "chasms," depositing, in one case, 16 to 20 "cartloads" of fine sand on the ground surface; and (4) also in the clay lowlands, new springs opened and others went dry[8,12–15] (FIG. 2). These reports indicate that liquefaction of sediments caused quick conditions in the uplands and ground-fissuring and sand-venting in the lowlands. There were similar reports of ground-fissuring and sand-venting in Hampton Falls, New Hampshire, about 20 km north of Newbury.[16] Thus, the localization of liquefaction probably is not reflecting solely the distribution of shaking. A very shallow source below Newbury can be ruled out because of the occurrence of liquefaction in New Hampshire.

Effects of the 1755 Cape Ann and the 1944 Massena Earthquakes

In addition to toppling walls and chimneys in Boston, the 1755 Cape Ann earthquake triggered ground-fissuring and sand-venting in Scituate, about 25 km southeast of Boston.[8,17] There were reports of ground-fissuring and dewatering also during the 1944 Massena earthquake.[18,19] This event caused a great deal of damage, amounting to about $8M by 1984 standards; the areas of greatest damage are underlain by the sensitive Leda clay, suggesting that damage was related to a strength loss of or amplification by this material during the event.[20]

Tenuous Cases of Earthquake-Induced Liquefaction

During the 1884 New York City earthquake, beach houses on Long Island tilted and subsided, possibly due to the liquefaction of beach deposits.[21] Water spouts immediately followed the 1929 Attica earthquake.[9] In the Adirondacks of upper New York State, an earthen dam, within about 5 km of the epicenter, failed during the 1983 Goodnow earthquake. Preliminary results suggest that very fine sand beneath the portion of the dam that failed may have liquefied during shaking, causing that failure.

St. Lawrence Valley Earthquakes

There are several St. Lawrence Valley earthquakes that may have liquefied subsurface materials. They include the 1663, M = 7, the 1870, M = 6.5, and the 1925, M = 7 events.[22] The 1663 earthquake triggered earth flows along the banks of the St. Maurice, Batiscan, and St. Lawrence Rivers. The 1870 earthquake caused ground-fissuring and sand-venting near Baie-St. Paul, Quebec, where water spouts reached 8 to possibly 15 ft in the air following that event. During the 1925 earthquake, water and sand oozed from cracks that formed in the frozen ground near St. Urbain, Quebec.[23]

SURFICIAL GEOLOGY OF THE NEWBURY-NEWBURYPORT AREA

In the Newbury-Newburyport area, bedrock (ranging in elevation from 50 m above to at least 32 m below mean sea level) is overlain by till, ice-contact, and glaciomarine deposits.[24-26] The town of Newburyport is built on a linear, southeast-trending ice-contact delta (FIG. 2). The delta was probably deposited about 13,000 years B.P. by glacial meltwater in a subaqueous marine environment within a recess of the late Wisconsin glacier. The delta now forms a ridge (upland) 20 to 30 m above present sea level. The delta is comprised primarily of sand and gravel and serves as the recharge area for local aquifers.[27] Distal and lateral facies of the delta are characteristically finer-grained. At Newbury, the trend of the delta becomes more southerly. The southern portion of the delta is thought to have been reworked by nearshore processes and elongated to the south by longshore currents.[26] Glaciomarine deposits predominate in the lowlands (10 m elevation) to either side of the delta.

A cross-section based on borehole data[25,27] and drawn perpendicular to the long dimension of the delta illustrates the local stratigraphic relationships between the ice-contact and glaciomarine deposits (FIG. 3). Along both flanks of the delta, clay overlies fine sand. This sand is confined to bedrock lows adjacent to the delta and may represent either the lateral-most facies of the ice-contact deposits or reworked materials derived from the delta as its lateral support was removed during retreat of the ice margin.

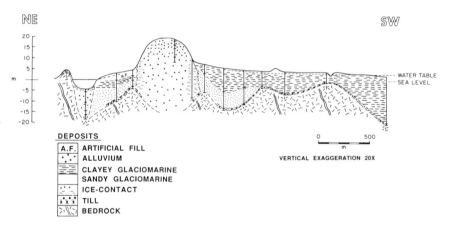

FIGURE 3. A cross-section, located on FIGURE 2, of surficial deposits in Newburyport based on pre-existing borehole data[25,27] illustrates the geologic control of ground failure. Great lateral and vertical variations within glacial sediments can lead to a variety of ground effects over a small area during earthquake shaking.

The Newbury clay, possibly correlative with the Presumscott Formation of Maine and the "blue clay" of Boston,[28,29] is ubiquitous in the lowlands and onlaps bedrock highs and portions of the delta up to approximately 15 m elevation. The top 10 feet of the Newbury clay deposit is similar in appearance to "outwash III" described in the central Boston area. Like this and other glaciomarine clays mapped in the region, the Newbury clay was derived from a glacier as rock flour and deposited in coastal marine waters some distance from the ice-margin. The fining-upward sequence in the Newbury lowlands is similar to glaciomarine sequences recognized elsewhere along the New England Coast which have been interpreted as transgressive sequences deposited during retreat of the marine-based glacier.[28-32] The absence of till above the Newbury clay suggests that the deposit was never overridden by the glacier. West and north of Newburyport, the clay is overlain, in ascending order, by sand, silt and peat, reflecting the subsequent marine regression caused by crustal rebound.[26]

EARTHQUAKE-INDUCED DEFORMATION FEATURES IN GLACIAL SEDIMENTS

The identification of earthquake-induced deformation features is complicated by the occurrence of deformation structures in glacial sediments that form as the result of other processes. These include depositional processes, ice-tectonics, periglacial processes, compaction and dewatering, piping, windfall of trees and spring action.[33–37] In order to conclude that a feature is earthquake-induced, one must be able to rule out these nonseismic origins.[38–40] Previous studies that have dealt with earthquake-induced deformation in the Northeast include one in the St. Lawrence Lowlands of New York,[41] and another in central Connecticut.[37,42] In both studies, an earthquake origin was hypothesized for soft-sediment deformation structures in the New York case, and for clastic dikes in the Connecticut case. Although both areas are seismically active, none of the features in either area has been linked to any particular earthquake. The origins of the features are still in question. Our findings in Newbury, where we know that an earthquake caused liquefaction and ground failure, may help to resolve the origins of these features and to identify earthquake-induced structures elsewhere in the Northeast.

A reconnaissance of sand and gravel pits, excavations, and stream-channel exposures within a 20-km radius centered on the 1727 meizoseismal area has revealed a high concentration of deformation structures in ice-contact as well as fine- and coarse-grained glaciomarine deposits in the Newburyport area. Features most likely to be earthquake-induced are postdepositional deformation structures that are consistent in style with descriptions of ground effects during the 1727 earthquake. These structures include sand diapirs in ice-contact deposits in the uplands and clastic dikes in glaciomarine deposits in the lowlands. Descriptions of several examples follow.

Ice-contact and overlying loess deposits are exposed in a sand pit in the distal and finer-grained portion of the delta in Newburyport (Site 1; FIG. 2). The boundary between these two deposits is deformed in association with a complex pattern of soft-sediment deformation structures, which suggests multiple liquidization events (FIG. 4). In addition, the ice-contact and loess deposits are mixed together above a highly deformed zone within the ice-contact deposits. Although we have no absolute age of deformation, the involvement of the younger loess indicates that at least some of the deformation postdates deposition and is late-glacial or younger in age. The style of deformation indicates that the sediments liquefied and is consistent with descriptions of quick conditions in the uplands during the 1727 earthquake.

In the general vicinity where reportedly the ground broke open and 16 to 20 cartloads of fine sand were vented to the ground surface during the 1727 earthquake, suspicious subsurface features have been investigated by a variety of techniques including profiling with ground-penetrating radar, augering, coring, and trenching. The site is located on the fringe of the clay lowlands and adjacent to the southwestern flank of the Newburyport delta (Site 2; FIG. 2). At this site, the otherwise subhorizontal boundary between glaciomarine sand and clay is found to be abruptly terminated near a bedrock high, where sand is exposed at the present ground surface (FIG. 5). The concentration of clastic dikes and the fragmentation of the clay is the greatest near this termination. Clastic dikes of very fine, white sand cut fine- to medium-grained reddish sand become sills along the base of the overlying clay, and also extend into the clay. Furthermore, discontinuous domains of the coarser-grained sand in the clay may have been mobilized

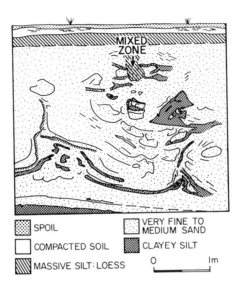

FIGURE 4. Soft-sediment deformation structures including sand diapirs in ice-contact deposits at Site 1 (Hillers Sand Pit). Cross-cutting relationships indicate at least two liquidization events. Deformation of the boundary between ice-contact and overlying loess deposits and mixing of the two deposits indicate that some of the deformation is postdepositional and late-glacial or younger in age. The latest stage of deformation may have occurred during the 1727 earthquake and would be consistent with reports of quick conditions in the uplands during that event.

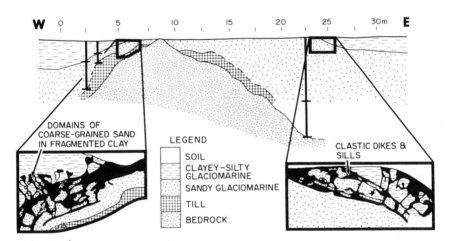

FIGURE 5. Compilation of subsurface data (ground-penetrating radar profiles and logs of auger holes and trenches) at Site 2 (Hale Street), located in an area where reported ground-fissuring and sand-venting occurred during the 1727 earthquake. The concentration of sand dikes and fragmentation of clay along the abrupt termination of the boundary between sandy and clay-rich glaciomarine deposits suggests that glaciomarine sand was forcefully injected into the overlying clay.

during an earlier phase or event. These characteristics suggest that this feature may be a remnant of one or more superimposed sand blows and that bedrock highs may have played an important role in the localization of ground-rupturing and sand-venting. If the glaciomarine sand liquefied during shaking it no longer would be capable of supporting the overlying clay. The clay would founder, rupturing over the bedrock high and providing an escape route for the liquefied sand. Unfortunately, the removal of the topsoil in preparation for building at this site limits our opportunities to fully characterize this feature and to date liquefaction and injection events.

A few kilometers to the southeast of Site 2 and also in the lowlands, there is an area of hummocky topography and dome-shaped structures. This peculiar topography may be due to sand-venting during the 1727 earthquake. Trenching of one of the domes in this area (Site 3; FIG. 2) revealed clastic dikes and sills of very fine, gray sand cutting the C soil horizon composed of interbedded glaciomarine sand and clay (FIG. 6). Clasts of the organic-rich A horizon and of the C horizon were

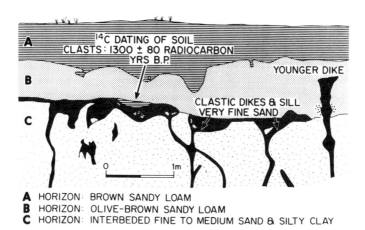

A HORIZON: BROWN SANDY LOAM
B HORIZON: OLIVE-BROWN SANDY LOAM
C HORIZON: INTERBEDED FINE TO MEDIUM SAND & SILTY CLAY

FIGURE 6. Clastic dikes and sills exposed in a trench of a low-relief dome at Site 3 (Wilkinson's Field Trench #1). The incorporation of A- and C-horizon material in the sill indicates that sand was forcefully injected from below. A radiocarbon age of ~1300 years on clasts of the A horizon in dike sand and considerable soil development since the injection event suggest that these features predate the 1727 earthquake.

found in a clastic sill of very fine, gray sand above the dikes and below an irregularity in the boundary between the A and B soil horizons. The association of the irregular soil boundary and the soil clasts out of stratigraphic sequence with the clastic dikes is reminiscent of earthquake-induced liquefaction features identified in the Charleston, South Carolina area.

Radiocarbon dating of displaced clasts of organic-rich soil by the Illinois State Survey yielded an age of 1300 ± 80 radiocarbon years before 1950. Two samples taken from similar positions within the profile were used in the analysis because individual samples contained insufficient quantities of carbon for the test. The result suggests that the clasts of organic-rich soil may have been disturbed and isolated from the A horizon approximately half a millenium before the 1727 event, possibly as a result of liquefaction induced by a prehistoric earthquake. In addition, residual carbon in the soil and possible contamination of the sample by various processes may seriously skew the age of the soil, complicating the conversion of the radiocarbon age of the soil to a calender year. Although we are unlikely to resolve all the questions that arise when using this dating technique, we hope with additional dating to determine whether prehistoric events have occurred in this area, and if so, the relative ages of these events.

STRATIGRAPHIC SETTINGS OF GROUND FAILURE

Liquefaction of a subsurface sand layer may be manifested at the ground surface by different types of ground failure, including the development of quick conditions, sand boils or sand blows, ground-fissuring, land subsidence, lateral spreads and flow failures. Ground failures can also result from strength loss of cohesive sediments during ground-shaking. On the basis of stratigraphic settings in which different liquefaction features were observed, certain hypotheses can be drawn about geologic settings where similar types of ground failure may result in the future from the liquefaction of subsurface materials.

The presence of postdepositional soft-sediment deformation structures at Site 1 suggests that quick conditions developed in ice-contact deposits where near-surface, fairly well-sorted, fine- and medium-grained sands are underlain by less-permeable, laminated silt and clay. Large- and small-scale sand dikes at Site 2 suggest that ground-fissuring and sand-venting possibly related to lateral spreading occurred above bedrock highs where stiff, glaciomarine clay caps very fine- to medium-grained glaciomarine sand. Groundwater entering local aquifers through the ice-contact deposits of the uplands exhibits high rates of flow when tapped in the lowlands. A high-pressure head in the glaciomarine sand may have contributed to their susceptibility to liquefaction. At Site 3, sand dikes and sills were observed in interbedded glaciomarine sands and silty clays. Although in this case liquefaction and fluidization did not appear to result in ground failure, disturbance of the organic-rich A horizon presents the opportunity to date liquefaction events.

Ice-contact and glaciomarine deposits do occur elsewhere in the Northeast and are especially abundant along the New England coast and the St. Lawrence River Valley[45,46] (FIG. 7). Moderate and large earthquakes have caused liquefaction in these types of deposits. Other types of deposits that are known to be susceptible to liquefaction, such as fluvial and beach deposits, also occur in the Northeast. A great deal of information is already available on surficial geology and the liquefaction susceptibility of materials in the region. The compilation of these data is needed to assess the liquefaction potential in the Northeast. This is currently being done by various groups affiliated with the National Center for Earthquake Engineerng Research for both New York City and the Boston area.

CONCLUSIONS

Accounts of specific types of ground failure during historic earthquakes indicate that several moderate earthquakes within the region, as well as three St. Lawrence Valley events, triggered liquefaction. One of the better documented cases is for the moderate 1727 Newbury earthquake. An ongoing study in the meizoseismal area of that event is yielding information about the character of earthquake-induced deformation features in glacial sediments and about the geologic conditions that controlled ground failure during the moderate 1727 earthquake. The combination of geologic and historical data could be used to assess the liquefaction potential of the region and to predict the probability of liquefaction at a site. Glacial deposits similar to those found in Newbury are widespread in the region and are especially abundant along the New England coast and the St. Lawrence River Valley. If these deposits are as susceptible as those in Newbury, then they may liquefy even during moderate earthquakes, which are considered possible in most of the region.

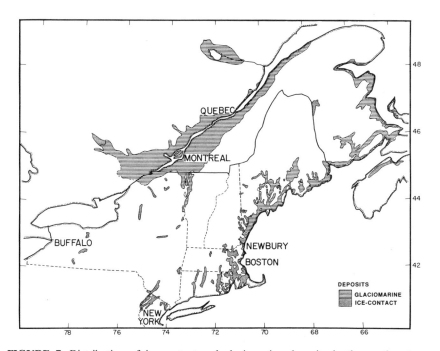

FIGURE 7. Distribution of ice-contact and glaciomarine deposits in the northeastern United States and in southeastern Canada. If similar to the equivalent deposits in Newbury that liquified during the 1727 earthquake, ice-contact and glaciomarine deposits elsewhere in the region may also be susceptible to liquefaction during moderate earthquakes. (Modified from Flint et al.[43] and Prest et al.[44])

1. Moderate, (M = 5 to 6) earthquakes have caused liquefaction in the northeastern United States.

2. Soft-sediment deformation structures most likely to be earthquake-induced are postdepositional structures that are consistent in style with descriptions of ground failure during the 1727 earthquake, such as sand dikes and diapirs that disturb soil horizons.

3. Different types of ground failure can be related to specific stratigraphic settings.

4. Deposits similar to those that liquefied in Newbury during the 1727 earth-quake are widely distributed in the northeastern United States.

5. Therefore, the potential for earthquake-induced liquefaction and ground failure needs to be considered in areas of the northeast that are mantled by Pleistocene sediments.

REFERENCES

1. Cox, J. H. M. 1984. Paleoseismology studies in South Carolina. M.S. thesis. University of South Carolina, Columbia, SC.
2. Obermeier, S. F., G. S. Gohn, R. E. Weems, R. L. Gelinas & M. Rubin. 1985. Geologic evidence for recurrent moderate to large earthquakes near Charleston, South Carolina. Science 227: 408–411.
3. Ebasco Services, Inc. 1987. Paleoliquefaction features on the Atlantic Seaboard. Report to U.S. Nuclear Regulatory Commission. Contract NRC-04-86-117: 1–68.
4. Weems, R. E., S. F. Obermeier, M. J. Pavich, G. S. Gohn, M. Rubin, R. L. Phipps & R. B. Jacobson. 1986. Evidence for three moderate to large prehistoric Holocene earthquakes near Charleston, S.C. In Third U.S. National Conference on Earthquake Engineering. I: 3–13.
5. Obermeier, S. F., R. B. Jacobsen, D. S. Powers, R. E. Weems, D. C. Hallbick, G. S. Gohn & H. W. Markenwich. 1986. Holocene and Late Pleistocene (?) earthquake-induced sand blows in coastal South Carolina. In Third U.S. National Conference on Earthquake Engineering. I: 197–208.
6. Russ, D. P. 1979. Late holocene faulting and earthquake recurrence in the Reelfoot Lake area, northwestern Tennessee. Geol. Soc. Am. Bull. Part 1, 90: 1013–1018.
7. Seed, H. B. & I. M. Idriss. 1982. Ground Motions and Soil Liquefaction during Earthquakes. Earthquake Engineering Research Institute. Berkeley, CA.
8. Boston Edison Company. 1976. Historical Seismicity of New England. BE-S67601: 1–641.
9. Armbruster, J. 1987. Personal communication.
10. Barstow, N. L., K. G. Brill, Jr., O. W. Nuttli & P. W. Pomeroy. 1981. An approach to seismic zonation for siting nuclear electric power generating facilities in the eastern United States. NUREG/CR-1577: 1–143.
11. Seeber, L. & J. G. Armbruster. 1986. A Study of earthquake hazards in New York State and adjacent areas. NUREG/CR-4750: 1–98.
12. Coffin, J. 1845. A Sketch of the History of Newbury, Newburyport, and West Newbury, 1635–1845. Samuel Drake.
13. Perley, S. 1891. Historic Storms of New England. Salem Press. Salem, MA.
14. Currier, J. J. 1906. History of Newburyport, Massachusetts. Currier.
15. Knight, Betty. 1987. Personal communication.
16. Brown, W. 1900. History of the Town of Hampton Falls, New Hampshire. Manchester, NH.
17. Coffman, J. L. & C. A. Von Hake, Eds. 1973. Earthquake History of the United States. Publication 41-1. U.S. Department of Commerce. Washington, DC.
18. The Toronto Globe & Mail, September 6, 1944.
19. Berkey, C. P. 1944. Geological study of the Massena-Cornwall earthquake of September 5, 1944 and its bearing on the proposed St. Lawrence River Project. U.S. Engineer's Office Report. New York District.
20. Revetta, F. A., D. Wiegand & K. McGregor. 1985. Seismic hazard investigation of the Massena-Cornwall area, northern New York. Geological Society of America. Abstracts with program: 60.
21. The New York Tribune, August 11 & 12, 1884.
22. Basham, P. W., D. H. Weichert & M. J. Berry. 1979. Regional assessment of seismic risk in eastern Canada. Bull. Seismol. Soc. Am. 69: 1567–1602.
23. Smith, W. E. T. 1962. Earthquakes of eastern Canada and adjacent areas, 1534–1927. Pub. Dom. Obs. Ottawa 26: 1–301.

24. CHUTE, N. E. & R. L. NICHOLS. 1941. Geology of the coast of northeastern Massachusetts. Bulletin No. 7: 1–47. Commonwealth of Massachusetts, Boston, MA.
25. SAMMEL, E. A. 1967. Water Resources of the Parker and Rowley River Basins, Massachusetts. U.S. Geological Survey Hydrologic Investigations, Atlas HA-247.
26. EDWARDS, G. B. & R. N. OLDALE. 1987. Late-Quaternary paleodeltas of the Merrimack River, Western Gulf of Maine 22nd Annual Meeting of the Geological Society of America. Northeastern Section. Abstracts with Programs:12.
27. DELANEY, D. F. & F. B. GAY. 1981. Data of the Lower Merrimack River Basin, Massachusetts, from Concord River, Lowell to Plum Island, Newburyport. U.S. Geol. Surv. Open File Rep. **81-1185:** 1–34.
28. BLOOM, A. L. 1960. Late Pleistocene changes of sea level in southwestern Maine. Maine Geological Survey. Department of Economic Development: 1–143.
29. KAYE, C. A. 1976. Outline of Pleistocene geology of the Boston Basin. *In* Geology of Southeastern New England: A Guidebook for Field Trips. B. Cameron, Ed. 68th Annual Meeting NEIGC: 46–63. Science Press. Princeton, NJ.
30. BLOOM, A. L. 1963. Late-Pleistocene fluctuations of sea level and postglacial crustal rebound in coastal Maine. Am. J. Sci. **261:** 862–879.
31. SMITH, G. W. 1984. Glaciomarine sediments and facies associations, Southern York County, Maine. *In* Geology of the Coastal Lowlands, Boston to Kennebunk, Maine. L. S. Hanson, Ed. 76th Annual Meeting NEIGC: 352–369.
32. OLDALE, R. N. 1964. Surficial geology of the Salem Quadrangle, MA. U.S. Geological Survey. Geologic Quadrangle Map: GQ-271.
33. SCHAFER, J. P. 1968. Periglacial features and pre-Wisconsin weathered rock in the Oxford-Waterbury-Thomaston area, Western Connecticut. *In* NEIGC Guidebook No. 2 for Fieldtrips in Connecticut. B-2: 1–5.
34. LOWE, D. R. 1975. Water escape structures in coarse-grained sediments. Sedimentology **22:** 157–204.
35. ALLEN, J. R. L. 1982. Sedimentary Structures: Their Character and Physical Basis. Elsevier. Amsterdam.
36. BLACK, R. F. 1983. Pseudo-ice-wedge casts of Connecticut, northeastern United States. Quaternary Res. **20:** 74–89.
37. SHAFER, J. P., S. F. OBERMEIER & J. R. STONE. 1987. On the origin of wedge structures in southern New England. 22nd Annual Meeting of the Geological Society of America. Northeastern Section. Abstracts with Programs: 1–55.
38. SIMS, J. D. 1972. Earthquake-induced structures in sediments of Van Norman Lake, San Fernando, California. Science **182:** 161–163.
39. MORNER, N. 1985. Paleoseismicity and geodynamics in Sweden. Tectonophysics **117:** 139–153.
40. EL-ISA, Z. H. & H. MUSTAFA. 1986. Earthquake deformation in the Lisan deposits and seismotectonic implications. Geophys. J. R. Astr. Soc. **86:** 413–424.
41. COATES, D. R., ED. 1975. Quaternary Deformed Sediments of the St. Lawrence Lowland as an Index of Seismicity. Report to New York State Atomic and Space Development Authority. Binghamton, NY.
42. THORSON, R. M., W. S. CLAYTON & L. SEEBER. 1986. Geologic evidence for a large prehistoric earthquake in eastern Connecticut. Geology **14:** 463–467.
43. FLINT, R. F., R. B. COLTON, R. P. GOLDTHWAIT & H. B. WILMAN, EDS. 1959. Glacial Map of the United States East of the Rocky Mountains. Geological Society of America. New York.
44. PREST, V. K., D. R. GRANT & V. N. RAMPTON, EDS. 1968. Glacial map of Canada. Geological Survey of Canada. Ottawa, Ontario.

Extent and Character of Soil Liquefaction during the 1811–12 New Madrid Earthquakes

STEVEN G. WESNOUSKY, EUGENE S. SCHWEIG, AND
SILVIO K. PEZZOPANE

Center for Earthquake Research and Information
Memphis State University
Memphis, Tennessee 38152

INTRODUCTION

The great 1811–12 New Madrid earthquakes produced extensive liquefaction which is still very much in evidence today. Visible as a myriad of light-colored and often irregular shapes against the dark brown soils of the Mississippi embayment, sands liquefied and extruded during the 1811–12 earthquakes are still readily recognized both in the field and on aerial photographs. No systematic studies of the geological effects of the earthquake were undertaken immediately after the earthquakes. Consequently, the extent, magnitude, and style of liquefaction produced by the earthquakes can be assessed from compilations of graphic accounts of contemporaries who witnessed the events and from later studies of evidence registered in the geologic record. Our intent here is not to duplicate those efforts. Rather, we will limit ourselves to a brief synopsis of work bearing on liquefaction which took place during 1811–12 as a basis for discussing the potential that still exists to advance our understanding of liquefaction processes and seismic hazard in the Mississippi embayment through further study of the geologic record.

THE NEW MADRID SEISMIC ZONE

Seismological Characteristics

The New Madrid Seismic Zone strikes about 175 km in a northeasterly direction through the Mississippi embayment, from near Memphis, Tennessee in the south to Cairo, Illinois in the north[1] (FIG. 1). The seismic zone is not expressed on the ground surface by an active and mappable fault zone, though subtle evidence of tectonic warping and faulting of recent sediments have been reported in a limited region overlying the seismic zone.[2–6] Nonetheless, reviews of isoseismal data and secondary ground deformations resulting from the New Madrid earthquakes lend strong support to assertions that the New Madrid Seismic Zone was indeed the source of the displacements that produced the 1811–12 New Madrid earthquakes.[7] Isoseismal data show that the 1811–12 sequence of earthquakes was arguably the largest seismic disturbance in the conterminous United States during historical time (FIG. 2). The magnitude and extent of observed ground deformations are consistent with such an argument.

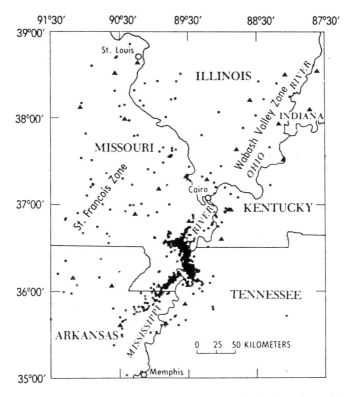

FIGURE 1. The location of the New Madrid Seismic Zone is clearly delineated by this plot of earthquake epicenters and extends from south of Cairo, Illinois to northwest of Memphis, Tennessee. (Adapted directly from Stauder.[1])

FIGURE 2. Isoseismals of Modified Mercalli (MM) VI and VII for four major United States earthquakes. Regions sustaining MM VII shaking or greater are *hachured*.

Contemporary Accounts of Liquefaction Phenomena during 1811–12

Initial reports of ground deformations and damage during the earthquakes are primarily the result of eyewitness accounts of local inhabitants. Several compilations of such material have been published since 1811–12.[8–11] Although personal accounts of the earthquakes are often and understandably biased toward the sensational, compilations of those accounts leave no doubt regarding the extensive nature of liquefaction during 1811–12. Perusal of the accounts provides evidence of major liquefaction phenomena, including extensive ground fissuring, the ejection of sand, water, and other debris through fissure systems, the settling of extensive tracts of land below the water table, and numerous landslides along the bluffs that border the Mississippi River. Evidence of the style, magnitude, and extent of liquefaction that took place in 1811–12 is also provided by geologic studies subsequent to the earthquakes.

Geologic Accounts of Liquefaction Phenomena

Sir Charles Lyell was among the first geologists to visit and provide a graphic description of liquefaction phenomena as recorded in the geologic record.[12] Lyell recorded the still relatively fresh evidence of fissuring, sand blows, landslides, and "sunken" lands. A number of other descriptions of geologic deformations that occurred during the 1811–12 earthquake were reported during the century after the earthquake. Among them, Usher[2] and McGee[3] cited evidence for doming and uplift of young alluvial sediments in the New Madrid region. The first work to systematically document the style, extent, and magnitude of deformations resulting from the New Madrid earthquakes is Fuller's[9] synthesis of prior accounts of the earthquakes, and report of his own geological traverses across the region nearly 100 years after the event. Fuller's work shows that liquefaction during 1811–12 was pervasive within a zone ranging from 20 to 50 km wide extending northeasterly for a distance of about 150 km from near Memphis, Tennessee in the south to New Madrid, Missouri in the north (FIG. 3). Fuller concluded that fissuring of the ground surface was the most common and widespread form of liquefaction phenomena within this zone. He cited contemporary accounts indicating fissures reaching to 5 miles in length and 600–700 feet in width. His study of landforms showed that the fissures commonly produced the down-faulting of narrow blocks to 5 or 6 feet or more, and were generally limited to the portion of the zone south of New Madrid, within the broad flat alluvial bottoms of the Mississippi and St. Francis drainage basins. The creation of fissures was often accompanied by the ejection of water, sand, mud, and gas. Ejecta commonly were produced through sandblows, leaving distinct patches of sand reaching diameters of 100 feet or more for circular varieties, or lengths of 200 feet and breadths of 25–50 feet for the linear varieties. In other cases, the amount of ejecta was sufficient to cover tracts of land many miles in extent by sand and water 3 to 4 feet in thickness and depth, respectively. Local settling or warping of alluvial deposits due to ground shaking also resulted in flooding of tracts of land miles in extent and, in turn, the widespread destruction of forest lands. Fuller[9] also corroborated eyewitness accounts of extensive landslide failures by observing the scars of landslides still well preserved and concentrated along the set of bluffs that border the eastern edge of the Mississippi River between New Madrid, Missouri and Cairo, Illinois. Contemporary and geologic accounts thus show that essentially every type of liquefaction failure that has been observed during recent earth-

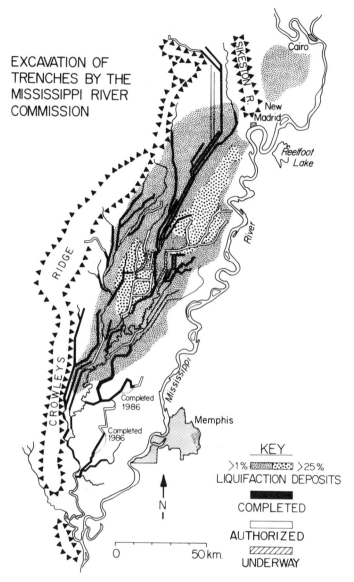

EXCAVATION OF
TRENCHES BY THE
MISSISSIPPI RIVER
COMMISSION

KEY

>1% >25%
LIQUIFACTION DEPOSITS

COMPLETED

AUTHORIZED

UNDERWAY

FIGURE 3. Aerial photography and field studies[9,13,14] show that liquefaction phenomena were pervasive during 1811–12 in the region extending from near Cairo, Illinois to northwest of Memphis, Tennessee. Areas in which liquefaction deposits still comprise ≥1% and ≥25% of the ground cover are *shaded* and *stippled,* respectively. The region of liquefaction south of New Madrid encompasses the St. Francis drainage basin, which empties into the Mississippi River south of Memphis. It is this region that Fuller[9] referred to as the "St. Francis Sunk Lands." For purposes of flood control and land reclamation, the U.S. Army Corps of Engineers has excavated an extensive network of drainage channels. The channels, which range from completed (*black*) through under way (*hachured*) to authorized (*open*) for excavation in the near future, total hundreds of kilometers in length, are generally several or more meters in depth, and are tens of meters wide. Excavation of these channels provides kilometers of new exposure each year, which is ideal for examining the geologic record of liquefaction within the St. Francis Sunk Lands.

quakes was pervasive over a region measured in thousands of square kilometers during 1811–12. Indeed, liquefaction phenomena reported for other large earthquakes within the conterminous United States during historical time pale in comparison to those registered during the New Madrid earthquakes.

During the last decade, investigators have used aerial photography to reexamine the extent of liquefaction during the 1811–12 earthquakes. Jibson and Keefer[15] examined landslide deposits along the bluffs that run along the eastern edge of the Mississippi River between about New Madrid, Missouri and Memphis, Tennessee. They concluded that the majority of landslides in evidence were produced by shaking during 1811–12, and that the entire extent of bluffs remains extremely susceptible to landsliding during earthquakes. Heyl and McKeown's[13] and Obermeier's[14] recent use of aerial photography generally confirmed Fuller's conclusions regarding the extent of liquefaction in 1811–12 (FIG. 3). Of considerable interest, the above zone of concentrated liquefaction deposits overlies the zone of microearthquakes which define the New Madrid Seismic Zone (FIGS. 1 and 3).

Potential for Further Study

It is thus evident from reports of the 1811–12 earthquakes that deposits of the Mississippi embayment are extremely prone to liquefaction. The New Madrid region is a vastly different place than it was in 1811–12. Characterized by a population measured in the thousands in 1811–12, the number of people living within the zone marked by strong ground motions in 1811–12 now measures in the millions. In that regard, we may be certain that the recurrence of earthquakes similar to those in 1811–12 would produce equally extensive liquefaction and, in turn, immense losses to property and life. However, there are few data that bear upon how often events similar to those of 1811–12 occur, or whether such earthquakes can be expected to occur elsewhere in the Central United States. For example, Johnston and Nava[16] recently estimated the average repeat time of New Madrid-type earthquakes to equal about 600 years, but their estimate was based on the extrapolation of instrumentally recorded data reaching back only 10 years and historical records for the period after 1811. A similar value of repeat time for large earthquakes in the region was also put forth by Russ[4] primarily on the basis of geomorphic study if displaced and deformed near-surface sediments exposed along Reelfoot Fault. That estimate, however, was limited because it is not certain that the discrete displacements registered in the trench were due to earthquakes equivalent in size and origin to the New Madrid earthquakes.[4] The widespread liquefaction phenomena recorded so well in the stratigraphy of the Mississippi embayment represent an excellent opportunity to further address the question of seismic potential in the Mississippi embayment.

Recent studies in Charleston, South Carolina have shown the potential value that geological study of liquefaction effects may play in understanding the prehistoric record of large earthquakes in a region.[17,18] As yet, a systematic study of liquefaction phenomena to identify deformation due to prehistoric earthquakes within the Mississippi embayment has not been done. Ongoing flood control efforts by the U.S. Army Corps of Engineers have resulted in an extensive system of major drainage channels throughout the New Madrid Seismic Zone (FIG. 3). The most recently excavated channels provide an excellent opportunity to search for liquefaction features that possibly predate the 1811–12 sequence and to examine the mechanics of liquefaction in cross-section as well. Thus far, we have examined several sites along these channels on a reconnaissance basis. Although evidence of earthquakes prior to 1811–12 has not been observed, the resulting logs of exposed sediments show the excellent exposure of liquefaction phenomena afforded by the Corps' channels.

Logs of two exposures examined near Big Lake, Arkansas are shown in FigURES 4 and 5. At site No. 1 (Fig. 4) the exposed strata consist of thick soil horizons underlain by fine- to medium-grained alluvial sands alternating with clay beds. The logs clearly show that impermeable clay layers play a controlling role in the liquefaction process by limiting the vertical flow of sands, as evidenced by sills

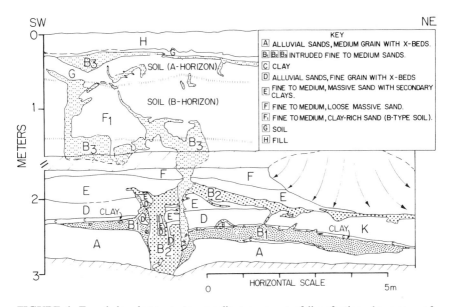

FIGURE 4. Trench log demonstrates excellent exposure of liquefaction phenomena afforded by channels recently excavated by the Army Corps of Engineers. At this site (No. 1), near Big Lake, Arkansas, cross-cutting relations show evidence of three episodes or phases of sand injection (*stippled units* B_1, B_2, and B_3). The competent clay (unit C) and clay-rich (unit E) layers inhibited the upward propagation of liquefied sand, as evidenced by the intruded sills of liquefied sand which underlie the respective units. The horizontal extent of these sills reaches to near 10 m in this exposure.

of sand (stippled) that extend from the central pipe to a distance of approximately 5 meters horizontally beneath impermeable clay layers. Cross-cutting relationships further indicate several phases of sand injection (Fig. 4). In this case, the liquefied sand was primarily limited to injection into dikes and sills in the subsurface. Site No. 2 (Fig. 5) shows the cross section of a sandblow or fissure that shows apparent extension of about 50 cm and from which extruded sands reached to about 1 meter in thickness. The exposures are limited to between 2–3 meters in depth by the water table at both sites Nos. 1 and 2, and the source of liquefied sands is below this level in each case. Although evidence of any liquefaction prior to the 1811–12 earthquakes has not been observed in this brief reconnaissance, our work shows that the preexisting and growing network of drainage channels provides perhaps the most viable and economic opportunity to systematically examine the mechanics of liquefaction and to search for evidence of large earthquakes prior to 1811–12 within the Mississippi embayment. This resource currently remains untapped.

SUMMARY

The great 1811–12 New Madrid earthquakes produced extensive liquefaction which is still very much in evidence today. Visible as a myriad of light-colored and irregular shapes against the dark brown soils of the Mississippi embayment, sands liquefied and extruded during the 1811–12 earthquakes are readily recognized

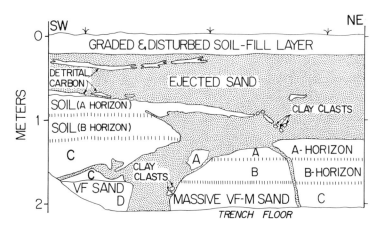

FIGURE 5. A second example of liquefaction exposed near Big Lake, Arkansas (site No. 2) shows the extrusion of sand through soil horizons, resulting in an overlying deposit of sand up to 1 m thick.

both in the field and on aerial photographs. The extent of surficial liquefaction deposits produced by the 1811–12 earthquakes has been well established by both field studies and airphoto analyses. Liquefaction deposits are most concentrated in a zone approximately 20–50 km wide that strikes southwestward about 150 km along the western edge of the Mississippi River from near New Madrid, Missouri to Marked Tree, Arkansas. Extruded sands account for more than 25% of the surface deposits in much of this area, and excavations show the sand deposits reaching to more than a meter in thickness. It is certain that the occurrence today of similar sized earthquakes in the New Madrid region would produce equally destructive liquefaction. However, few data exist to bear upon how often events similar to those of 1811–12 recur, or whether such earthquakes can be expected elsewhere in the Central United States. Statistical analysis of historical seismicity cannot confidently address these questions because of the relative brevity of recorded history. But clues to the past occurrence of large earthquakes may be recorded in the geology, and application of paleoseismologic techniques may be the key to determining the expected location and occurrence rate of large earthquakes in the area. Ongoing flood control efforts by the Army Corps of Engineers have resulted in an extensive system of major drainage ditches throughout the New Madrid Seismic Zone. The most recently excavated drainage ditches provide an opportunity to look for liquefaction features that possibly predate the 1811–12 sequence and to examine the mechanics of liquefaction in the vertical dimension. Several drainage areas have thus far been examined. Exposed strata are generally composed of fine- to medium-grained alluvial sands alternating with clay beds. Exposures are limited to a depth of about 3 meters by the water table, and the source of liquefied sands is below this level in each case. Exposures examined show that impermeable clay layers play a controlling role in the liquefaction process by limiting the vertical flow of sands, as evidenced by sills of sand extending more than 5 meters from the central pipe beneath impermeable clay layers. Cross-cutting relationships indicate at least several phases of sand injection, but evidence of liquefaction prior to the 1811–12 earthquakes has not been observed. This preliminary work shows that the preexisting and growing network of drainage ditches might provide a good opportunity to examine the mechanics of liquefaction and to search for paleoseismic evidence of large pre-1811 earthquakes within the Mississippi embayment.

REFERENCES

1. STAUDER, W. 1982. Present-day seismicity and identification of active faults in the New Madrid seismic zone. U.S. Geol. Surv. Prof. Pap. **1236:** 20–30.
2. USHER, F. C. 1837. On the elevation of the banks of the Mississippi in 1811. Silliman's J. (Am. J. Sci., first series) **31:** 291–294
3. McGEE, W. J. 1892. A fossil earthquake [abstract]. Geol. Soc. Am. Bull. **4:** 411–415.
4. RUSS, D. P. 1979. Late Holocene faulting and earthquake recurrence in the Reelfoot Lake area, northwestern Tennessee. Geol. Soc. Am. Bull. **90:** 1013–1018.
5. RUSS, D. P. 1982. Style and significance of surface deformation in the vicinity of New Madrid, Missouri. U.S. Geol. Surv. Prof. Pap. **1236-H:** 94–114.
6. STEARNS, R. G. 1979. Recent vertical movement of the land surface in the Lake County uplift and Reelfoot Lake basin areas, Tennessee, Missouri, and Kentucky. *NUREG/CR-0874.* U.S. Nuclear Regulatory Commission. Washington, DC.
7. NUTTLI, O. W. 1973. The Mississippi Valley earthquakes of 1811 and 1812 intensities, ground motion and magnitudes. Seismol. Soc. Am. Bull. **63:** 227–248.
8. MITCHELL, S. L. 1815. A detailed narrative of the earthquakes which occurred on the 16th day of December, 1811. Trans. Lit. Philos. Soc. New York **1:** 281–307.
9. FULLER, M. L. 1912. The New Madrid Earthquake. U.S. Geol. Surv. Bull. **494.**
10. PENICK, J. 1981. The New Madrid earthquakes of 1811–12. The University of Missouri Press. Columbia, MO.
11. RANKIN, D. W. 1977. Studies related to the Charleston, South Carolina, earthquake of 1886—introduction and discussion. U.S. Geol. Surv. Prof. Pap. **1028:** 1–16.
12. LYELL, C. 1849. A second visit to the United States, Vol. 2. Harper and Brothers. New York.
13. HEYL, A. & F. McKEOWN. 1978. Preliminary seismotectonic map of central Mississippi valley and environs. U.S. Geological Survey Miscellaneous Field Studies Map MF-1011.
14. OBERMEIER, S. 1984. Liquefaction potential for the central Mississippi Valley. U.S. Geol. Surv. Open File Rep. **84-770:** 391–446.

15. JIBSON, R. & D. KEEFER. 1984. Earthquake-induced landslide potential for the central Mississippi Valley, Tennessee and Kentucky. U.S. Geol. Surv. Open File Rep. **84-770:** 353–390.
16. JOHNSTON, A. C. & S. NAVA. 1985. Recurrence rates and probability estimates for the New Madrid seismic zone. J. Geophys. Res. **90**(B7).
17. OBERMEIER, S., G. GOHN, R. WEEMS, R. GELINAS & M. RUBIN. 1985. Geologic evidence for recurrent moderate to large earthquakes near Charleston, South Carolina. Science **227:** 408–411.
18. TALWANI, P. & J. COX. 1985. Paleoseismic evidence for recurrence of earthquakes near Charleston, South Carolina. Science **229:** 379–381.

General Discussion: Part IV

I. M. IDRISS, CHAIRMAN (*Woodward-Clyde Consultants, Santa Ana, California*): This session contains four excellent discussions on the topic of liquefaction, and the fact has been pointed out that the liquefaction phenomenon, if it occurs during an earthquake, can be very destructive. Liquefaction can be very extensive and can occur in the same place over and over again. The fact that it had occurred does not provide "immunity" in the future.

There was also discussion of how we can estimate whether liquefaction would or would not occur at a specific location. What makes one area more susceptible than another?

We also saw excellent examples of what liquefaction has done: It has resulted in loss of foundation support, tilting of apartments and breakage of buildings. It has resulted in increased lateral loads and deformations along seafronts. It has caused massive slides, such as that of the dam in San Fernando, and several landslides in Japan and Alaska. And liquefaction has also resulted in a lot of settlement.

Thus, liquefaction is a phenomenon we should take into account wherever earthquakes might occur, wherever soil conditions are susceptible to it.

Given all that, what options are available to us? What options have we exercised over the years in seismic areas both in the United States and elsewhere?

When we examine the options that have been used, we find that there have really been only two: One is to accept the risk, and there are many areas in the United States, Canada, and elsewhere where we have done exactly that. The second is to mitigate the problem. For example, we can incorporate into the design the possibility of liquefaction's occurring. Dr. Dobry showed us a building supported on piles, where the area around the building had tremendous settlement, with lateral spreading. The building, itself, was supported on deep piles that went well below the liquefied zone, and as a result suffered practically no damage whatsoever. So there are ways to design for the consequences of liquefaction. We might have designed the laterally loaded areas for the full load of the liquefied soils and therefore may not have seen the type of deformation shown by Dr. Dobry.

Another way is to contain the liquefied area, and this we can do by sloping the areas and such. And, of course, the third way we can do this is to improve site conditions by altering susceptible soils to nonsusceptible ones.

Thus the problem of liquefaction can be mitigated if we so choose. But, as I said earlier, in many parts of the world and in many parts of this country we basically accept the risk.

ED KAVAZANJIAN (*Parsons, Brinckerhoff, Quade & Douglas, Inc., New York, N.Y.*): I would like to make two observations. First, there are kinds of ground failure other than liquefaction that can be induced by earthquakes and that we should be concerned with. There is good documentation that the New Madrid earthquake induced slope failures along the Mississippi River bluffs for approximately 200 miles up and down the Mississippi River. So we should be concerned not only with liquefaction, but also with other forms of earthquake-induced ground failure.

Second, the conclusion that moderate earthquakes of magnitude 5 and 5.2 can induce liquefaction is counter to evidence from earthquakes in the West Coast and Japan. Work that Youd at the U.S.G.S. and Japanese investigators have done has indicated that magnitude 5.5 is just about the cut-off for earthquake-induced liquefaction. At least three or four years ago, the last time I looked at the data, there was no documented evidence for liquefaction in California, the West Coast, or Japan for earthquakes of magnitude less than 5.5. I'm not sure what magnitude measure was used in those studies, so perhaps a 5.2 body wave magnitude is similar to a 5.5 for surface wave and I do not know whether these investigators used surface wave magnitude measurements.

On the basis of the kind of accelerations obtained from a 5 or 5.2 earthquake using conventional attenuation relationships, if you go into the SPT correlation and use typical densities for beach sand deposits, you'd have a very hard time demonstrating that liquefaction could occur. So I find it very curious that we can attribute liquefaction to events with magnitudes of 5 to 5.2. That indicates to me that either there is something wrong with those magnitude measures, since they are simply based on historical evidence, or there is something fundamentally different about earthquakes on the East Coast in terms of duration or frequency content that would make liquefaction more likely from smaller events.

RICARDO DOBRY (*Rensselaer Polytechnic Institute, Troy, N.Y.*): These lower limits—the magnitude 5 and the intensity 7—are certainly debatable. In fact, somebody could say that Modified Mercalli Intensity VIII is really the lowest threshold for liquefaction. The definition of Modified Mercalli Intensity VIII includes a phrase about "ejection of sand and mud in small amounts" while that for MMI VII does not. On the other hand, Obermeier found that in the New Madrid 1811–1812 earthquakes there were manifestations of liquefaction in areas of MMI VII. It's not completely clear; there are no clear-cut boundaries in this area.

The problem is also related to the fact that as you get to a smaller magnitude, the area subjected to very strong shaking becomes smaller and smaller. Therefore the chances of having a susceptible deposit within that area and of finding evidence also become smaller and smaller.

A number of earthquakes in Japan and the western part of the United States having magnitudes between 5.2 and 5.5 have caused liquefaction within a few kilometers from the epicenter.

MARTITIA TUTTLE (*Lamont-Doherty Geological Observatory, Palisades, N.Y.*): One factor that may contribute to the susceptibility of these materials to liquefy is initially high pore-water pressures.

GAIL ATKINSON: I would like to ask a question regarding the frequency content and its possible effect on liquefaction.

As mentioned earlier, for the same magnitude and distance the peak ground accelerations are higher in the East than in the West. And if you look at an empirical chart of liquefaction potential, which is based on cyclic shear stress, you would infer then that that same magnitude would have a higher liquefaction potential in the East than it does in the West. My question is: Is the higher frequency content a mitigating factor?

Another way of asking the question, in a laboratory context, would be: Suppose you took a sample of soil and subjected it to twenty cycles of .1 g at 5 Hz (that is 4 seconds of motion) and that was just enough to bring it to liquefaction. If you then took an identical soil sample and gave it twenty cycles of .1 g, but at a frequency of 20 Hz (that would last only 1 second), would that also cause liquefaction?

DOBRY: All the evidence we have from the laboratory indicates that frequency content is not very important. What is important is the cyclic stress or strain and the number of cycles. Another more indirect evidence is provided by predictions of pore-water pressure build-up in centrifuge tests from laboratory tests where the frequency is much smaller. For example, the frequency in the laboratory test may be 1 Hz and in the centrifuge may be 100 Hz and still a reasonable agreement is obtained.

Therefore what is important either in the East or in the West is the level of peak ground acceleration and the number of cycles. Now, you have to multiply the dominant frequency of the record by its duration to get the equivalent number of cycles. The frequency alone or the duration in seconds alone is not a good predictor of pore-pressure build-up and liquefaction.

KLAUS JACOB (*Lamont-Doherty Geological Observatory, Palisades, N.Y.*): I am mostly concerned with how liquefaction may manifest itself in urban environments. There is one feature that is of concern to some laymen and I want to know whether it's justified or not.

Specifically, a great deal of the waterfront is being developed in urban environments on the East Coast such as that now occurring in Boston and New York. I would like to know how these man-made fill type of subfoundational materials do even in the presence of piling, particularly where the bedrock may have strong sloping and the piling may or may not be anchored in it. If soil liquefaction should occur, could this piling hold up under the conditions one could expect?

Let me give four specific examples. The first is in New York City. All three airports—Kennedy, LaGuardia, and Newark Airport— are on either marshland or man-made fill. On a soil map, these are on S-3 soft soil. In New Jersey, the Meadowlands are now under pressure to be developed.

Also in New York City, there is a strip known as Battery Park, which is filled with sand carted in from beaches. It's beautiful sand; it's like a beach. All new buildings stand on piling. I want to know, from the experts, whether soil liquefaction, even in the presence of pilings is possible or not with a magnitude 5 earthquake.

The last example is the airport in Boston, where old harbor and industrial facilities are being redeveloped. I would like to know what soil and geotechnical engineers have to say about such situations.

IDRISS: Since I talked about mitigating measures, maybe I should say a couple of words. The fact that there are piles will not necessarily prevent liquefaction from occurring unless the driving of the piles resulted in compaction of the sand. Sometimes piles are driven to compact sand in place. We don't know what the conditions are.

If liquefaction does occur, would the system stay in place if it had been designed for the full liquefied weight, which is about 125 pounds per cubic foot? I doubt very much that it would fail.

Typically these kinds of design numbers are not used. More like a third of that weight with a factor of safety around 2 is used in the design. So, I would say that if liquefaction did occur, we would see the type of damage we have seen in Japan, where key walls and port-type facilities were subjected to massive lateral movements because they had not been designed for the full force that was exerted by the liquefied material.

Now, whether the sand as it exists is or is not susceptible to liquefaction is something that needs to be looked into because some of that might have been placed a lot denser than we think. I don't know.

DR. BUDHU (*State University of New York, Buffalo, N.Y.*): The presentations at this session were concerned with the phenomenon of liquefaction and evidence of prehistoric liquefaction. The question that arises is: Is liquefaction likely to occur in a major city in the eastern United States due to a moderate earthquake?

At SUNY Buffalo we attempted to address this question. One of the areas we have selected is a portion of Manhattan in New York City. This area is approximately 7 square miles and is bounded by the Hudson River, 140th Street, the East River, and 90th Street. The 125th Street fault is being mapped by personnel from Lamont-Doherty.

In this study area we collected some three thousand bore hole logs. We placed these bore hole log data in a data base, and selected some results for our analysis. The analysis made use of the standard penetration test number and a probabilistic approach that came out of MIT.

We produced a map of the liquefaction potential of the soils within our study area. The region was broken up into three zones: one where the probability of liquefaction was greater than 50 percent, a second with a liquefaction potential between 10 and 50 percent, and a third area with no possibility of liquefaction.

The greater-than-50-percent area is approximately one square mile and the 10–50 percent areas are approximately three square miles. The areas that we have selected contain a significant amount of fill, and we have not considered this in our analysis. We have only studied those sandy layers that appear to have potential to liquefy, and these layers were within 50 feet from the surface. Consequently, if liquefaction does occur in the deep sand layers, then the fills may spread laterally and may cause a tremendous amount of damage.

UNIDENTIFIED SPEAKER: The example of Mexico City leads me to this question: Is there any evidence in eastern North America that would raise the possibility of soil amplification? Can that type of problem occur here?

DOBRY: I'm glad you asked this question, which is really a question of ground motion rather than liquefaction. Yes, certainly there is evidence of soft clay amplification all over the world—in the western United States as well as in other countries. Usually there is not as much amplification as in Mexico City, but it is still quite significant. And there is some evidence in the eastern United States that this has happened in historic earthquakes. In the Charlevoix 1925 Canadian earthquake there was a small town called Shawinigan, 250 kilometers to the southwest of the epicenter in an area of intensity 5, which suffered damage equivalent to intensity 7. And that town is on about 50 meters of leda clay, a very sensitive clay which is, in some respects, similar to the Mexico City clay. And the same earthquake caused a significant ground motion at an old theater on Boston blue clay, more than 500 kilometers from the epicenter. This theater was a long-period structure, thus showing a similar situation of long-period amplification of a very distant, large earthquake. Therefore, there are some indications that this may happen in the eastern United States, with the situation being worsened by the slow attenuation of the ground shaking characteristic of the eastern and central United States.

IDRISS: If you examine the relationship between peak accelerations for different site condition and peak acceleration on rock sites, you will see that the evidence for Mexico City in 1962 and 1985 is significantly different evidence for soft soil conditions collected from other parts of the world.

Mexico City soils are quite different from many other soft-site conditions. They behave a lot more elastically than we have ever seen any other soil behave that is as soft as the Mexico City soil.

The indications for East versus West is that there should be no particular difference as far as the soils themselves are concerned. Soils are soils. It's just the fact that the frequency characteristics of the ground motions in the East are quite different from those in the West, at least based on the evidence available to date, that the response, especially at the lower end of things, could be quite a bit different. My guess is that it would be higher than what we would expect from the West Coast.

I don't think we should expect things to be behaving as they did in Mexico City just because it's a soft-site condition. We should investigate what the types of soils are, because Mexico City does constitute a very special soil condition, as far as we can tell.

Current Design Spectra:
Background and Limitations[a]

W. J. HALL

Department of Civil Engineering
University of Illinois
Urbana-Champaign
Urbana, Illinois 61801

S. L. McCABE

Department of Civil Engineering
University of Kansas
Lawrence, Kansas 66044

INTRODUCTION

The purpose of this paper is to present some observations about response and design spectra, and about certain selected topics pertaining to seismic design that will need to be developed in further detail if improved design procedures are to occur. The earthquake engineering literature is replete with articles on spectra and their applications in analysis, in part because of the simplicity of the approach offered by this design tool; nonetheless, as pointed out later in this paper, spectra, as currently employed, provide quite limited information. It is our belief that future advances in seismic design procedures will require a somewhat different assemblage and evaluation of input and response information than is currently the case. This paper will offer some observations on this subject on the basis of current research and practice.

RESPONSE AND DESIGN SPECTRA AND GROUND MOTIONS

The response spectrum is defined as a graphic relationship of the maximum response of a damped single-degree-of-freedom (SDOF) elastic system to dynamic motion or forces. In the case of earthquake engineering the dynamic motion is applied at the base of the simple oscillator, as would be the case for a structure supported on the ground. The simple damped mass-spring system normally depicted is shown in FIGURE 1. When excitation is applied to the base,

[a] This paper is based in part on research supported by the National Science Foundation under Grants PFR 80-02582, CEE-8203973 and DFR-8419191 at the University of Illinois at Urbana-Champaign. Any opinions, findings, conclusions or recommendations expressed in this paper are those of the authors and do not necessarily reflect the views of the National Science Foundation. Some of the material presented herein was developed and reported in an earlier paper[13] by the authors.

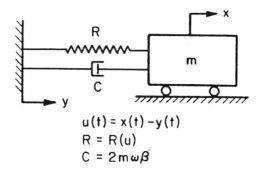

$$u(t) = x(t) - y(t)$$
$$R = R(u)$$
$$C = 2m\omega\beta$$

FIGURE 1. Simple damped mass-spring system.

usually as acceleration versus time in the case of an earthquake, one can calculate any number of quantities that are descriptive of the response of the simple system. Normally the peak value of a response quantity, such as relative displacement between the mass and base, u, pseudo relative velocity between the mass and base, ωu, or pseudo acceleration of the mass, $\omega^2 u$, are plotted as a function of frequency. The plots may be made in any number of ways, but one useful way is that of the tripartite plot that permits presentation of all response parameters together. Examples of two response spectra computed and plotted in this manner are shown in FIGURE 2, one for a long-duration earthquake of strong shaking

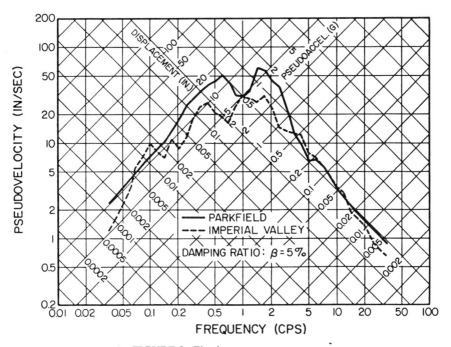

FIGURE 2. Elastic response spectra.

(namely, the Imperial Valley El Centro earthquake of May 18, 1940 [SOOE]), and another for an earthquake with a short burst of energy (the Parkfield earthquake of 27 June 1966, Cholame-Shandom No. 2 Station [N65E]), both for 5 percent damping.

It is not the purpose of this paper to recast the theory of computing and plotting response spectra; the uninitiated reader is referred to the work cited in Refs. 1–5, as well as to other texts, papers and reports, for a description of the details involved. However, one must read and study the literature with care because of the great variation in nomenclature employed, and the subtle differences in intended applications that are sometimes discussed. It is customary to characterize the effects of an earthquake by the peak ground motions, as, for example, peak ground acceleration, velocity and displacement; another means of characterization of the effects of an earthquake is provided by the spectrum. It

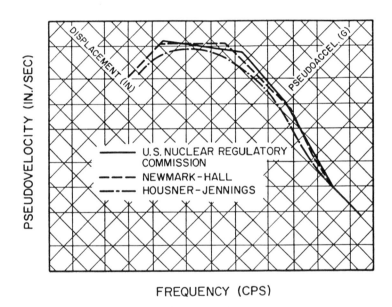

FIGURE 3. Typical shapes of smoothed design spectra.

should be obvious that the spectrum is in one sense a form of transfer function since it provides a relationship between excitation (loading) and simple structural response (resistance).

On the basis of studying response spectra associated with many earthquakes, and of calculations of response from recorded strong ground motion acceleration-time histories, it has been possible to estimate the shape of typical spectra or to arrive at general rules for constructing approximate representative spectra for different values of damping.[1-5] This major development, which seems quite simple now, was of great importance since it permitted a spectrum to be sketched for the purpose of general analysis even though only a few key parameters were known. Typical examples of such "smoothed" response spectrum shapes are shown in FIGURE 3. If these spectra are used for analytic purposes in connection with a seismic design project they would commonly be called "design spectra." Thus a very definite difference exists between response spectra and design spectra; response spectra are associated with computed response for a particular ground motion-time history representing a particular earthquake, whereas a design spectrum is a smoothed characterization of the same general shape intended to capture the essential features of the response spectrum, often for a geological region and class of earthquakes, but scaled to a level believed, in the judgment of the engineer, to be applicable for design purposes.

In some articles and texts the term "effective acceleration" or "effective motion" has been employed to denote the level of acceleration to be used as the high-frequency anchor point (or "zero period" acceleration [ZPA]) for the design spectrum, the shape of the spectrum being determined generally as just described. The effective motion values are chosen by considering many factors, as, for example, the effect on response of sustained long-duration shaking as contrasted to a short burst of energy with a high spike of high-frequency acceleration; the latter normally is not of great consequence in affecting response. Recent research[6] has shown that ground motion occurring at more than twice the structural frequency will have limited effect on the response of simple structures. Distant source motion generally has been so filtered by the ground medium that it would be considered the effective motion. Also, in some cases, the term "effective motion" may signify specified control values of velocity and displacement if these are believed applicable because of site conditions or for other reasons. In any event, the design spectrum (or spectra) must be based upon some specified level of excitation; accordingly, it should be obvious that the reasoning behind "design spectra" and "effective motions" are intertwined and the differences in arriving at the final result are subtle at best, and essentially indistinguishable. For further discussion of these topics the reader is referred to Refs. 1 and 2.

For what purposes are design spectra employed? For a simple SDOF system, or a normal or principal mode of a multiple-degree-of-freedom model of a building, for example, one can enter the spectrum with the appropriate frequency or period and read off a maximum response value (acceleration, velocity, or displacement) and in turn employ that value in the analytic computation in an appropriate fashion.[7-10] Similar computations may be carried out for two or three orthogonal directions of motion and the responses combined appropriately. Or, for two adjacent simple systems, entering with the appropriate frequencies one can estimate displacements of each mass and draw conclusions about relative motions and required clearances. These are but a few examples of many applications that could be cited. Clearly the design spectrum is an approximation to a set of scaled response spectra or equivalent, where the actual peak responses may fall above or below the design spectrum; in light of such realizations one wonders at times about the extreme accuracy that is attributed to such plots by some researchers and designers!

The foregoing discussion sets the stage for the primary point to be made in this section of the paper. Up to this time, in codes and design documents it has been assumed that one design spectrum would suffice for use in design. As more earthquakes are studied in detail around the world, and as more records of ground motions are obtained, it appears that the practice of employing only one design response characterization (one spectrum), even for elastic computations without any concern about nonlinear effects, may have some obvious shortcomings; such observations have not received much attention in the literature in any organized fashion. Several examples of situations where multiple design conditions are needed follow, along with some observations on associated decisions that must be made as well.

Some time ago Hall[8] noted the need in some cases for another spectrum (or alternative motion definition) for a distant earthquake source. In the case cited, the normal design spectrum (broad-banded) was intended to handle near-field to moderate-distance earthquakes from sources with relatively long duration and strong shaking, but that spectrum clearly failed to fulfill the requirement for a distant-source long-period sustained-type motion that could possibly affect high-rise buildings and fluid sloshing in large-diameter tanks, or other long-period items. Such a spectrum is shown in FIGURE 4 as #2 with varying levels of velocity; the normal design spectrum is denoted as #1 there. The shift in spectral amplitude and effects of excitation frequency are apparent. However, even the use of multiple spectra can lead to deficiencies insofar as design parameter definition is concerned, as will be noted a little later herein.

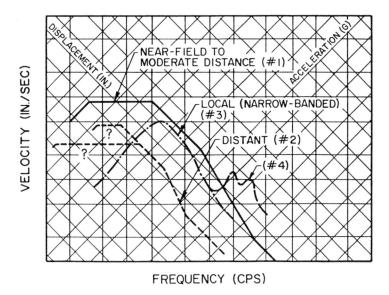

FIGURE 4. Design spectra for different earthquake sources.

It has been observed over the years, especially in the eastern United States, that local earthquakes (for example those of magnitude 4 and 5, which are relatively common) are narrow-banded, as illustrated by spectrum #3 in FIGURE 4. A highly debated question is whether or not such spectra should be used for design solely in lieu of the broad-banded spectrum (#1 in FIGURE 4) that reflects sustained shaking from strong ground motion. How can one be sure that the next earthquake will be so limited in source characteristics as to insure that such a narrow-banded spectrum is appropriate for use? Perhaps the use of several design spectra, coupled with other appropriate criteria, will resolve this problem. We do know that these small earthquakes can lead to sustained, nearly harmonic motion (surface waves) at great distances from the source; even though the amplitudes are measured in tens of microns or less with very long periods, the fact that the motions are repeated for ten or twenty cycles, for example, can lead to significant excitation for long-period items, such as high-rise buildings.

Another interesting item has been detected in connection with some of the local earthquakes, namely, that they may be accompanied by some significant excitation at selected frequencies, and in some cases these frequencies are high. In such cases one often can find another peak at a higher frequency, as noted by #4 in FIGURE 4. Usually such a peak will be associated with a series of moderate-amplitude high-frequency acceleration-type ground motions. At this high frequency the acceleration (force) values are of little structural significance, but they may have an effect on the building contents or equipment. Generally most seismic-qualification shake-table-testing of equipment, by virtue of the properties of the table, leads to testing at modest to reasonably high values of acceleration in the high-frequency range of excitation and thereby provides a measure of resistance in the high-frequency domain; however, this situation is not always the case and this potential for exceedance needs to be addressed in the basic design criteria.

In the last few years there has been increasing interest in utilizing Power Spectral Density (PSD) as a "monitor" of the energy at the frequencies of interest in the ground-motion time-history and thereby the spectrum as well. One suspects strongly that a specified PSD value over a given frequency range may aid in maintaining adequate frequency and amplitude control, but may not be totally sufficient. The variation in input motions and behavior of the item (such as frequency variation, duration, nonlinear behavior, and effects at distance) may serve to make the PSD an incomplete additional criterion. This topic needs further study by researchers and practitioners.

It should be clear by this time that, depending upon the circumstances, one can have a range of conditions that may have to be considered, and it is clear that one response spectrum will not serve to characterize all situations. Even more important is the fact that the spectrum by itself may not represent many of the desired criteria parameters. For example, in the last example given above, for complex equipment incorporating many interconnected elements one may well have to specify the ground-motion time-history (or more likely a set of such values) that is to be used for testing or analyzing the equipment item. Also, because the spectra provide only the maximum response, ignoring the other aspects of the response, such as phasing and amplitude of the non-peak response cycles, the spectrum by itself may not be a sufficient criterion for all applications; this situation is particularly true in the case of nonlinear response.

NONLINEAR RESPONSE, MODIFIED SPECTRA, AND DAMAGE

In those cases where limited nonlinear behavior is to be incorporated in the response through analysis it is possible to rigorously compute the response of a simple oscillator as a function of frequency and thereby arrive at modified response spectra (i.e., spectra that correspond to specific nonlinear resistance-force and deformation relationships). Studies conducted thus far suggest that the use of such spectra may lead to reasonable results, albeit approximate, when the ductility (defined as the ratio of the maximum deformation to the yield deformation) is limited to values of no more than 5 or 6, and more rationally one-half that value or less. Details for constructing such spectra are presented in Refs. 1 and 4, and are not repeated here; a typical modified spectrum is shown in FIGURE 5. This con-

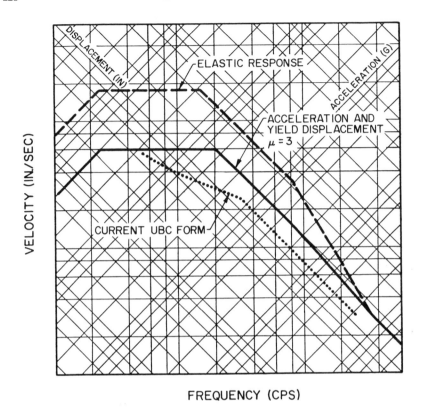

FIGURE 5. Elastic and modified design spectra.

cept was developed initially for monotonic elasto-plastic resistance-type functions subjected to blast-type loadings and was later extended to the earthquake engineering field. A recent definitive study in this area[11] offered suggestions for refining the bounds, but noted that the overall ductility was still bounded fairly well by the maximum deformation achieved, even in the case of cyclic motion. Thus, for an elasto-plastic resistance function, the bounds were essentially the same as developed earlier[1,4,5,8] in terms of reductions, as a function of the ductility factor, reflecting the limit acceleration and yield displacement, as noted in FIGURE 5.

It is customary at present to employ a design spectrum of the general form shown in FIGURE 5 in the design of special structures and facilities. Also shown in that figure, in a relative sense, is the form of the current Uniform Building Code (UBC) seismic coefficient; the reduction in the seismic coefficient is based in part on the same principles as the modified spectra illustrated in the figure, as well as on consideration of risk and economics. Thus, for intelligent application of the UBC procedures, the designer must be fully cognizant of the implicit factors that assume that the designed structure, including the members and especially the connections, can accommodate nonlinear behavior without major distress (loss of strength) to ductility levels of perhaps 4 to 6; this fact is not appreciated by many of those employing the procedures.

Another form of the design spectrum that probably will be employed in new building code provisions in the near future is that shown in FIGURE 6. This form of the spectrum is equally easy to use and is the form adopted in the two recent related national studies.[9,10] These studies had as their goal the development of design procedures wherein modern structural dynamics concepts could be adapted to building code practice in a manner such that the parameters in the analysis were explicitly expressed, as contrasted to the present procedures, wherein many of the principles are implicit.

The latest research that may well form the basis of the next generation of building code provisions deserves some discussion; it will further illustrate the advances that are in progress, and will serve to illustrate some of the limitations on our current procedures. For some time it has been realized that structural behavior was not well represented by the approaches just noted, especially where nonlinear behavior occurred, and a subsequent study[12] undertook to examine the energy considerations and the hysteretic behavior of a SDOF system. In brief: the energy input to a structure subjected to ground motion is dissipated in part by damping and in part by yielding or inelastic deformation in the structural components. Well-designed and well-constructed buildings should be able to absorb and dissipate the imparted energy with minimal damage for moderate shaking, and controlled amounts of damage for more severe shaking; collapse is unacceptable. The energy imparted to a simple structure for a unit mass is given by Equation **1** and the energy absorbed in a simple structure is given by Equation **2**.

$$E_I = \int_0^t \ddot{y}(t)\dot{u}(t)\ dt \tag{1}$$

$$E_I = \int_0^t \ddot{u}(t)\dot{u}(t)\ dt + 2\beta\omega \int_0^t \dot{u}(t)^2\ dt + \int_0^t R[u(t)]\dot{u}(t)\ dt \tag{2}$$

where E is the energy, $\ddot{u}$, $\dot{u}$, and u are the relative acceleration, velocity, and displacement, respectively, $\ddot{y}$ is the base acceleration time-history, β is the critical damping ratio, ω is the natural circular frequency, and $R[u(t)]$ is the structural resistance. In Eq. **2** the first term represents the kinetic energy, the second term

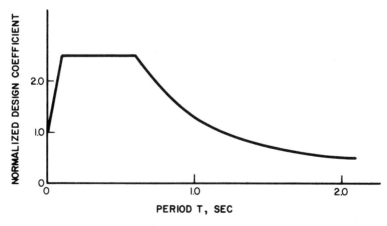

FIGURE 6. Proposed normalized response spectrum for use in building codes.

represents the energy dissipated by viscous damping, and the third term represents the sum of the hysteretic energy plus the strain energy.

Two examples of the energy, time, and yielding sequence are shown in FIGURES 7 and 8 for a simple structure subjected to the 1940 El Centro earthquake and the 1966 Parkfield earthquake identified earlier; in the former case there was sustained long-duration shaking and energy input, and in the latter case a short burst of energy was input into the structure. It will be observed that the energy input curve for a structure subjected to long-duration motion has a large number of peaks and troughs as compared to that for short-duration motion. Also, one should note the number of yield excursions: 15 in the case of the long-duration El Centro earthquake and 4 in the case of the shorter Parkfield earthquake. Many other observations arose out of this study pertaining to energy considerations as a function of structural frequency, including such items as effective times for energy absorption, effective motions, the observation that acceleration by itself is often a poor indicator of damage potential, and energy spectra; the study offered some limited insight into steps that might be taken to evaluate damage in manners not previously addressed by researchers.

This latter study led to a second study just completed[6] dealing with development of structural damage evaluation criteria for seismic loading situations

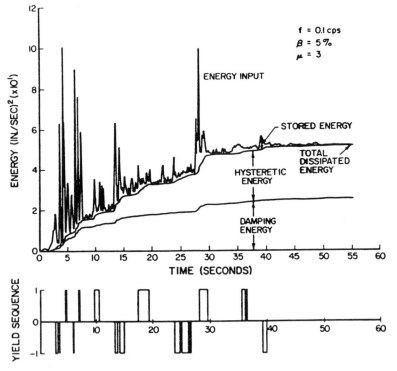

FIGURE 7. Energy versus time and yield sequence for a simple structure subjected to El Centro ground motion.

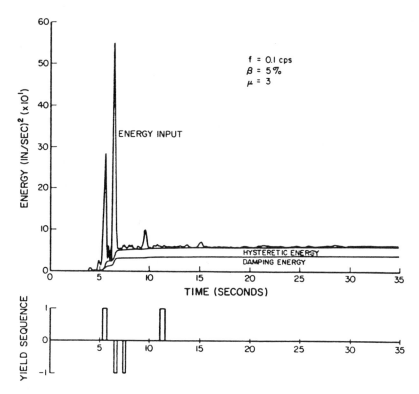

FIGURE 8. Energy versus time and yield sequence for a simple structure subjected to Parkfield ground motion.

wherein low-cycle fatigue concepts are adapted to the cyclic deformation process that occurs in structural members during an earthquake. An intensive examination of the low-cycle fatigue damage theories suggested that several of the plastic strain rules have potential application in the seismic response field if modified to represent energy absorption instead of cycles to fracture; it was apparent that the mechanisms involved in fatigue fracture and cyclic earthquake motion were quite similar in concept. One such expression, developed by Prof. J. D. Morrow of the University of Illinois Theoretical and Applied Mechanics Department, has the following form:

$$\frac{\Delta\varepsilon_p}{2} = \varepsilon'_f(2N)^{-0.6} \tag{3}$$

where $\Delta\varepsilon_p/2$ is the plastic strain amplitude ε'_f is the fatigue ductility coefficient, and $2N_f$ is the number of reversals to failure.

In turn, this general form of representing cyclic fatigue behavior can be re-structured into a form to reflect the structural behavior associated with monotonic plastic ductility and effective hysteretic plastic ductility. Also, the theory reflects the fact that the straining cycles in the member can be tensile and compressive, not necessarily equal cycles in each domain, and accounts for partial cycles. And finally it has been possible to demonstrate the applicability in simple structures for different types of seismic excitation and to check the developed relationships against published experimental evidence. Other potential applications of the theory demonstrated in the study[6] pertain to construction of damage-consistent inelastic spectra, and to a new tentative approach to drift limits for buildings.

This important development should point the way to permit us to gain some insight into how one might incorporate realistic damage concepts into design spectra or alternatively other design criteria reflecting the ability of the structure to withstand cyclic loading. Also, it may provide a basis in some part for helping to evaluate the damage sustained by a structure in an earthquake, the remaining margin of safety, and in turn to reduce the large number of structures that are deemed unusable after an earthquake and thereby torn down.

In conclusion, one final observation needs to be made. Rational modeling of structures for purposes of analysis, no matter how approximate, requires one to be able to predict the behavior of the elements well into the nonlinear range (including also connections, components, and cladding, for example) making up the structure, as well as the global behavior of the structure. Our reservoir of experimental information in this area is indeed small and more studies are needed to develop the entire strength relationship and to arrive at better design in the sense of use of materials and detailing. Such studies also are needed to provide a better basis for estimating serviceability limits and remaining margins of strength. Experimental laboratory studies, and studies of undamaged and lightly damaged structures in earthquakes, the ultimate laboratory in this case, should do much to increase our knowledge in this area. When this information is coupled with studies of the functional relationships (of which spectra are one example) for combining the effects of input motion and response (or behavior) in even a more meaningful manner, we should be able to provide a basis for design analysis that is much improved over that used at present. In short, much work remains to be done.

REFERENCES

1. NEWMARK, N. M. & W. J. HALL. 1982. Earthquake Spectra and Design. Earthquake Engineering Research Institute. Berkeley, CA.
2. HOUSNER, G. W. & P. C. JENNINGS. 1982. Earthquake Design Criteria. Earthquake Engineering Research Institute. Berkeley, CA.
3. CHOPRA, A. K. 1981. Dynamics of Structures—A Primer. Earthquake Engineering Research Institute. Berkeley, CA.
4. HALL, W. J. 1988. Vibrations of structures induced by ground motion. In Shock and Vibration Handbook, 3rd ed. C. M. Harris & C. E. Crede, Eds. 24-1–24-19. McGraw-Hill. New York.
5. NAU, J. M. & W. J. HALL. 1984. Scaling methods for earthquake response spectra. J. Struct. Engr. ASCE. 110: 1533–1548.
6. McCABE, S. L. & W. J. HALL. 1987. Evaluation of Structural Response and Damage Resulting from Earthquake Ground Motion. Civil Engineering Studies Structural Research Series No. 538. Department of Civil Engineering. University of Illinois. Urbana-Champaign, IL.
7. CRUZ, E. F. & A. K. CHOPRA. 1986. Simplified procedures for earthquake analysis of buildings. J. Struct. Engr. ASCE. 112: 461–480.
8. HALL, W. J. 1982. Observations on some current issues pertaining to nuclear power plant seismic design. Nucl. Engr. Des. 69: 365–378.
9. APPLIED TECHNOLOGY COUNCIL. 1978. Tentative Provisions for the Development of Seismic Regulations for Buildings. Report ATC 3-06. Palo Alto, CA.

10. FEMA. 1985. NEHRP-Recommended Provisions for the Development of Seismic Regulations for New Buildings. Earthquake Hazards Reduction Series Report No. 17. Federal Emergency Management Agency.
11. RIDDELL, R. & N. M. NEWMARK. 1979. Statistical analysis of the response of nonlinear systems subjected to earthquakes. Civil Engineering Studies Structural Research Series Report No. 468. University of Illinois. Urbana-Champaign, IL.
12. ZAHRAH, T. F. & W. J. HALL. 1984. Earthquake energy absorption in SDOF structures. J. Struct. Engr. ASCE. **110:** 1757–1772.
13. HALL, W. J. & S. L. McCABE. 1986. Observation on spectra and design. Proceedings of the Third U.S. National Conference on Earthquake Engineering, Charleston, SC. Earthquake Engineering Research Institute. El Cerrito, CA. Vol. II: 1117–1127.

An Assessment of Uncertainties in Seismic Response[a]

M. SHINOZUKA

Department of Civil Engineering and Operations Research
Princeton University
Princeton, New Jersey 08544

K. MORIYAMA

Taisei Corporation
Tokyo, Japan

INTRODUCTION

Loads, man-made or natural, acting on structures are usually difficult to predict in terms of their time of occurrence, duration, and intensity. The temporal and spatial characteristics needed for detailed analysis are also subject to considerable uncertainty.

With respect to earthquake engineering, there are many ways in which strong-motion earthquake phenomena can be modeled from the engineering point of view. Each model consists of a number of component models which address themselves to particular phenomena of seismic events. For example, a succession of arrival times of earthquakes at a site may be modeled as a stationary or nonstationary Poisson process, and the duration of significant ground motion in each earthquake may be modeled as a random variable with its distribution function specified. Also, temporal and spatial ground-motion characteristics may be idealized as a trivariate and two-dimensional nonstationary and nonhomogeneous stochastic process with appropriate definitions of intensity. An alternative approach would consider, for example, at differing levels of sophistication, models for locations of seismic source, source mechanisms with appropriate definitions of magnitude, seismic wave propagation paths and attenuation, amplification, and soil-structure and stucture-to-structure interaction.

The difference in ground motion and resulting structural response estimates arising from the use of various models represents modeling as well as parametric uncertainties, since each component model contains a certain number of parameters for which appropriate values must be assigned for numerical analysis. Hence, the total uncertainty consists of modeling uncertainty and parametric uncertainty.

A number of methods are available and in fact have been used for identifying the extent of uncertainty of parametric origin. As for modeling uncertainty, what can be done appears to be limited to the extent that we are able to construct as many plausible models as practical and to examine the variability of the results

[a] This research was supported by the National Center for Earthquake Engineering Research (NCEER) under Contract No. NCEER-87-1006A under National Science Foundation Contract No. ECE-86-07591 as well as by the Taisei Corporation.

from these different models. The degree of variability expressed in terms of range or any other meaningful quantity may be seen as representative of modeling error when, for example, "best estimates" are used for parameters within each model.

Similarly, with respect to structures, irregularity in structural shape and dimensions and variability in the spatial distribution of mass, rigidity, damping and strength parameters give rise to uncertainties in dynamic as well as static structural behavior, particularly in the nonlinear range. The situation is much worse when we deal with soil and soil-structure interaction problems.

Partly because of the limited space allotted for the treatment of uncertainty, an extremely complex subject, and partly in view of other existing reviews (e.g., Ref. 1) that deal more generally with uncertainty issues, specific cases of seismic response evaluation and the uncertainties associated therewith are considered below.

RESPONSE MODIFICATION FACTOR (RMF)

One of the more design-oriented and risk-related quantities in the seismic response issue is the response modification factor (RMF). In the eastern United States, where seismic hazards are perceived to be of a lesser degree than those, for example, of California or Alaska, structures are designed and constructed accordingly. Therefore, the effect of even a moderate seismic event, say in the range of magnitude 6, may be significant from the points of view of damagability and seismic risk,[2] particularly for old structures which might be subject to considerable strength degradation due to material deterioration and lack of proper maintenance. In this respect, the issues of an RMF and uncertainty associated therewith are also important in the design and construction of structures in the eastern United States. The RMF factor is often estimated as a function of the ductility factor μ so that, given a ductility or maximum nonlinear deformation capability, the corresponding maximum elastic response is determined. This maximum elastic response is then used as the failure-related limit state in safety evaluation as well as in structural design in which only linear response analysis is required, thus expediting the process of analysis and design.

The concept of an RMF was originally developed for single-degree-of-freedom (SDOF) systems.[3,4] Application of the same concept to multiple-degree-of-freedom (MDOF) systems is not straightforward. It can, however, be dealt with by introducing the notion of a system response modification factor.[5,6] To accomplish this, we utilize Monte Carlo simulation techniques by idealizing the ground acceleration $\ddot{z}(t)$ as a product of a Gaussian stationary process $g(t)$ and a deterministic nondimensional envelope function $f(t)$.

$$\ddot{z}(t) = g(t)f(t) \tag{1}$$

The function $f(t)$ has a trapezoidal shape with rise time of 2.5 sec, unit height of 10-second duration and decay time of 2.5 sec. The process $g(t)$ is assumed to have the Kanai-Tajimi spectrum[7]

$$S(\omega) = S_0 \frac{1 + 4\zeta_g^2 \left(\frac{\omega}{\omega_g}\right)^2}{\left(1 - \left(\frac{\omega}{\omega_g}\right)^2\right)^2 + 4\zeta_g^2 \left(\frac{\omega}{\omega_g}\right)^2} \tag{2}$$

in which ω_g and ζ_g are the characteristic ground frequency and damping ratio of the surface soil layer, respectively, and S_0 is the intensity of the white noise acceleration at bedrock and is related to the peak ground acceleration. Sample functions of $g(t)$ can be generated with the aid of the following form[8]

$$g(t) = \sqrt{2} \sum_{k=1}^{N_f} \sqrt{G(\omega_k)\, \Delta\omega}\, \cos(\omega_k t - \phi_k) \tag{3}$$

where $\omega_k = k\Delta\omega$, $G(\omega_k) = 2S(\omega_k)$, $N_f\Delta\omega$ = upper cut-off frequency, and ϕ_k = uniformly distributed random phase angle. Different sets of ϕ_k yield different sample functions of $g(t)$. A sample function of $\ddot{z}(t)$ with $\omega_g = 8\pi$ rad/sec, $\zeta_g = 0.6$, and $S_0 = 1$ (with proper dimensions) is shown in FIGURE 1.

The MDOF systems considered here are stick models of the shear beam type with n lumped masses and bilinear elasto-plastic shear springs.

$$M\ddot{y} + C\dot{y} + Q = -MI\ddot{z} \tag{4}$$

in which M = diagonal mass matrix, C = Rayleigh damping matrix based on the corresponding linear system, Q = restoring force vector and I = influence vector.

The story ductility factor μ_i at the ith story is defined as $\mu_i = U_{ni}/U_{yi}$ in which U_{ni} indicates the absolute maximum interstory displacement at the ith story obtained from inelastic response analysis and U_{yi} the yield displacement of the same story. The RMF R_i of the ith story is now defined as

$$R_i = Q_{1i}/Q_{ni} \tag{5}$$

where Q_{ni} is the maximum interstory shear force at the ith story obtained by numerically solving Eq. 4 under $\ddot{z}(t)$ generated by Eq. 1 and Q_{1i} is the maximum interstory shear force at the same story of the corresponding linear elastic system.

For an SDOF system, some approximate relationships were proposed. Among them, the following form is well known[3]:

$$R = \sqrt{2\mu - 1} \tag{6}$$

where R and μ are the RMF and ductility factor for SDOF systems. When an MDOF system is considered, inelastic deformations tend to concentrate at certain stories, depending on the way in which mass, stiffness, and damping are distributed in the system and, as a result, each story tends to have its own relationship between R_i and μ_i.

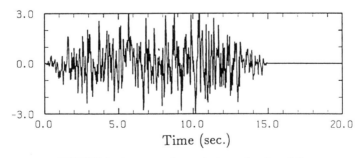

FIGURE 1. A generated sample ($\omega_g = 8\pi$, $\zeta_g = 0.6$).

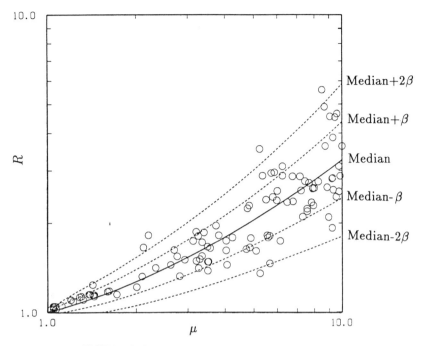

FIGURE 2. System $R - \mu$ relationship (5% damping).

From the system reliability point of view then, it is convenient to consider, as a measure of the system's deformational capacity, the ductility factor μ_j of the jth story, which happens to be the maximum value among μ_i ($i = 1, 2, \ldots, n$) under a particular ground acceleration history $\ddot{z}(t)$. The corresponding RMF R_j can also be computed under the same ground acceleration. These μ_j and R_j are defined as the system ductility factor and system RMF, respectively, and, in this sense, the subscript j will be dropped hereafter.

Since the system ductility factor and RMF depend on the ground acceleration history, a Monte Carlo simulation study has been performed by generating a large number of ground acceleration histories in the form of Eq. **1** and evaluating μ and R corresponding to each history. A Monte Carlo simulation result for $\omega_g = 8\pi$ rad/ sec, $\zeta_g = 0.6$ and damping ratio $h = 0.05$ (for the first two linear modes) is shown in FIGURE 2. This result is based on a total sample size of 200 (simulated acceleration histories) with a sample size of 40 each for each value of $S_0 = 0.16, 0.4, 1.0,$ 2.5 and 6.25 ft^2/sec^3. These values of S_0 are used to cover the range of acceleration intensity necessary to produce μ in the range up to 10. The stick model considered here idealizes an auxiliary shear wall building in the Zion nuclear power plant as shown in FIGURE 3.[9] Note that in this exercise of Monte Carlo simulation, the structural models are considered deterministic.

The result in FIGURE 2 is analyzed to obtain the least-square fit between R and μ in the following form (similar to that proposed in Ref. 4):

$$\check{R} = (p\mu - p + 1)^r \qquad (7)$$

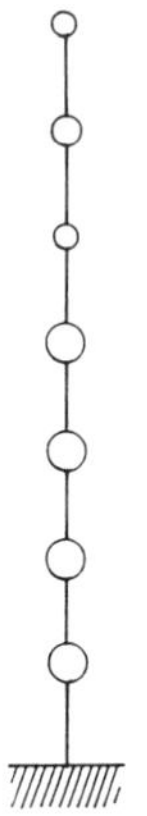

	Mass (ks^2/ft)	Elastic Stiffness (k/ft)	Yield Force (k)	Yield Story Displacement (ft)
	381.0			
		7.16×10^6	76,350	.01003
	465.2			
		23.63×10^6	126,100	.00534
	368.9			
		28.68×10^6	158,800	.00554
	753.9			
		12.90×10^6	137,800	.01068
	917.3			
		30.23×10^6	245,000	.00810
	870.2			
		23.10×10^6	258,500	.01119
	914.2			
		31.28×10^6	304,800	.00974

FIGURE 3. Mechanical properties (Zion auxilliary shear building).

where $\check{R}$ is interpreted as the median system RMF and p and r are parameters whose values are to be determined by means of the least-square method. Note that $p = 2$ and $r = 0.5$ reduce Eq. 7 to Eq. 6. The scatter of simulated $R - \mu$ points observed in FIGURE 2 is then interpreted as indicating the extent of uncertainty of the $R - \mu$ relationship. Under the assumption that R is log-normally distributed at each value of μ, a quantity β is evaluated in approximation as

$$\beta^2 = (\ln R - \ln \check{R})^2 = s(\mu - 1)^t \qquad (8)$$

and is interpreted as the log-standard deviation of R given μ. The parameters s and t in Eq. 8 are again evaluated by means of the least-square method. In FIGURE 2, the median relationship given by Eq. 7 is shown by a solid curve, while the relationships for the median $\pm\beta$ and $\pm2\beta$ are shown by dashed curves. In FIGURE 4, the median curves under different values of h are plotted. The relationship $R = \sqrt{2\mu - 1}$ is also plotted to suggest that except for the case $h = 0.5\%$, the relationship $R = \sqrt{2\mu - 1}$ tends to underestimate the maximum nonlinear shear force. The results shown in FIGURES 2 and 4 also suggest that if the nonlinear effect is expressed in terms of the RMF in the form $\varepsilon \sqrt{2\mu - 1}$, as is often done in PRA studies, ε must be a function of μ and h.

While these results are useful, they are in principle unique to a particular building design and specific form of ground acceleration idealization. Hence, a Monte Carlo simulation study must be performed on each set of design and ground acceleration idealizations to estimate the uncertainty in the RMF as well as the median $R - \mu$ relationship. In this context, Monte Carlo simulation studies are not at all as expensive as one might have been led to believe, particularly when structural systems are as simple as the ones considered here. In fact, since computer codes are almost invariably available for deterministic response analysis, for this type of structure, the cost of performing Monte Carlo simulation studies is trivially low, and yet such studies provide a great deal of uncertainty-related information.

Nevertheless, it is intriguing and interesting to know under what conditions a design (determination of mass and stiffness distribution) for shear buildings may be considered optimal in the sense that the median $R - \mu$ relationship and nature of the uncertainty are the same for all stories. In this respect, consider the optimal design of a shear building in which each story will be subjected to an identical linear interstory displacement under specified deterministic design seismic conditions. Equivalent to such a design, a shear building can be optimally designed[10] in such a way that each story will have identical maximum values for the standard deviation of interstory displacement, if the ground acceleration is idealized as a nonstationary Gaussian process as introduced in Eq. 1. In fact, three-, five- and ten-story shear buildings are designed to meet this requirement in Ref. 10 by determining the stiffness distribution by a trial and error method. Each shear building is assumed to have an identical mass at each floor and also identical story height. The ground acceleration is modeled in the same way as before, except for the envelope function $f(t)$; in this case, $f(t) = (e^{-\alpha t} - e^{-\beta t})H(t)$, where $\alpha = 0.25/$ sec and $\beta = 0.5/$sec and $H(t)$ = the Heaviside unit step function. FIGURE 5 shows the optimal stiffness distribution for three buildings (see TABLE 1 for details) and FIGURES 6–8 plot simulated $R - \mu$ points for the first, second, and third story, respectively, of a three-story building. The median $R - \mu$ relationship and the nature of uncertainty at each story in these figures show considerable similarities usually not observable unless the type of optimization performed here is imple-

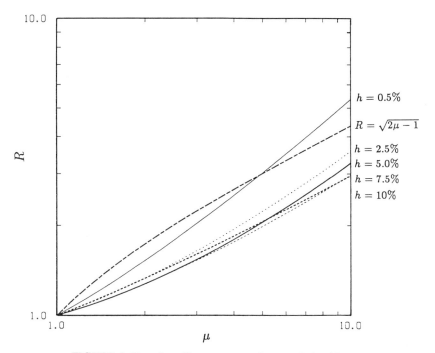

FIGURE 4. Damping effect on system $R - \mu$ relationships.

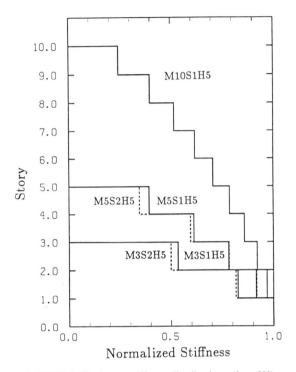

FIGURE 5. Optimum stiffness distributions ($h = 5\%$).

TABLE 1. Building Properties[10]

Case Name	Number of Mass	Earthquake Spectrum	Damping Ratio	Strain Hardening	Stiffness[a] of First Story	Yielding[b] Displacement
M3S1H5	3	S_1	0.05	0.0	999.01	0.060, 0.040
M3S2H5	3	S_2	0.05	0.0	1018.3	0.050, 0.035
M3S1H2	3	S_1	0.02	0.0	998.53	0.090, 0.060
M3S2H2	3	S_2	0.02	0.0	1009.0	0.060, 0.040
M3S1A1	3	S_1	0.05	0.1	999.01	0.060, 0.040
M3S2A1	3	S_2	0.05	0.1	1018.3	0.050, 0.035
M5S1H5	5	S_1	0.05	0.0	1319.2	0.060, 0.040
M5S2H5	5	S_2	0.05	0.0	1336.5	0.060, 0.040
M5S1H2	5	S_1	0.02	0.0	1320.9	0.080, 0.050
M5S2H2	5	S_2	0.02	0.0	1324.0	0.080, 0.050
M5S1A1	5	S_1	0.05	0.1	1319.2	0.060, 0.040
M5S2A1	5	S_2	0.05	0.1	1336.5	0.060, 0.040
M10S1H5	10	S_1	0.05	0.0	2324.8	0.050, 0.035
M10S1H2	10	S_1	0.02	0.0	2318.2	0.060, 0.040
M10S1A1	10	S_1	0.05	0.1	2324.8	0.050, 0.035

NOTE: S_1: rock or stiff soil ($S_0 = 1.0 ft^2/sec^3$, $\omega_g = 8\pi rad/sec$, $\zeta_g = 0.6$).
    S_2: soft to medium clay and sand ($S_0 = 1.0 ft^2/sec^3$, $\omega_g = 2.4\pi rad/sec$, $\zeta_g = 0.85$)
[a] (k/ft)
[b] (ft)

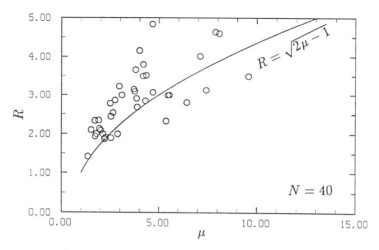

FIGURE 6. $R - \mu$ relationship (first story, case M3S1H5).

mented. FIGURE 9 shows the median $R - \mu$ relationship and uncertainty of the combined data for the three stories. It appears that, in this case, the median $R - \mu$ relationship and uncertainty of the $R - \mu$ relationship may be established on the basis of combined data. Not only the same trend can be observed for five- and ten-story buildings, but also the median $R - \mu$ relationship and nature of the uncertainty are practically identical for all these optimized buildings regardless of the number of stories involved (see FIGURES 10 and 11).

The results shown above involve uncertainties arising from a number of sources. With respect to ground motion, the shape of the spectral density function of the ground acceleration in Kanai-Tajimi form may not always be representative

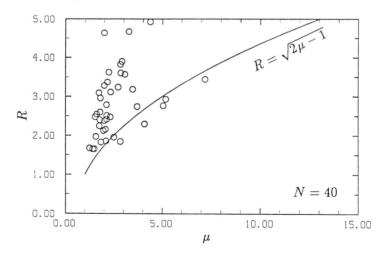

FIGURE 7. $R - \mu$ relationship (second story, case M3S1H5).

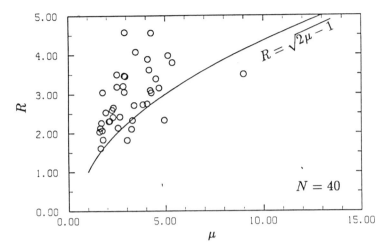

FIGURE 8. $R - \mu$ relationship (third story, case M3S1H5).

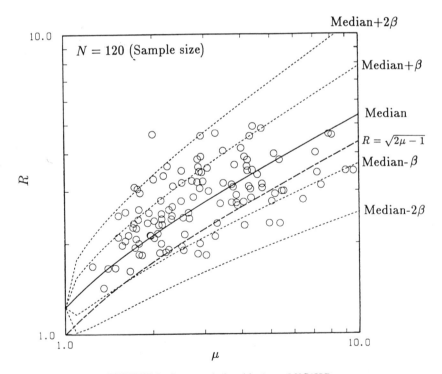

FIGURE 9. $R - \mu$ relationship (case M3S1H5).

of the actual acceleration of all future earthquakes to which the structure will be subjected. This is in spite of the fact that considerable latitude in the spectral model exists to alter its shape by assigning different values to ω_g and ζ_g. At the same time, the envelope function may not be adequate enough. As a matter of fact, this simple form of a nonstationary stochastic process model for ground acceleration, constructed as a product of a stationary process and an envelope time function, disregards the potentially important influence of the time-varying spectral contents. A nonstationary process representation of the ground acceleration by means of evolutionary power may be more appropriate. Furthermore, seismic intensity is represented by the parameter S_0 in this model. In a more

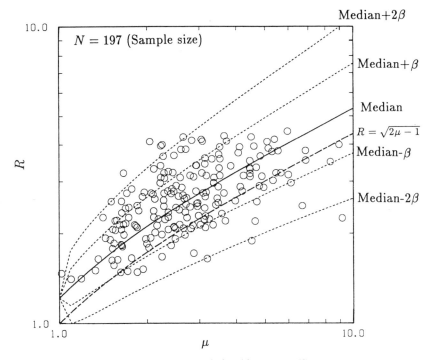

FIGURE 10. $R - \mu$ relationship (case M5S1H5).

global risk assessment of a building structure, the seismic hazard curve is brought into the analysis by interpreting the peak ground acceleration (PGA) considered in the construction of the hazard curve as equal to the expected maximum ground acceleration obtained from the assumed stochastic process model, or, PGA = $p\sigma_{\ddot{z}}$, where p is a peak factor depending on the duration, spectral contents, and nonstationary characteristics of the process and is usually somewhere around 3.0. It is well known that a seismic hazard curve is full of uncertainty, not to mention the fact that characterization of seismic hazard only in terms of PGA leaves a lot to be desired. While it is a straightforward detail, the value to be used for the peak factor is not necessarily easy to determine; unfortunately, the peak factor has a direct and probably strong impact on the limit state probability to be estimated for the ensuing reliability analysis.

As opposed to a method using stochastic process models, observed records or time histories generated to satisfy specified response spectra can be directly used in analysis and design with a scaling factor applied as appropriate. This approach may appeal more to the practicing segment of the profession and is preferred by some researchers over the stochastic process approach. A combination of these two methods may provide conclusions with more engineering confidence, however.

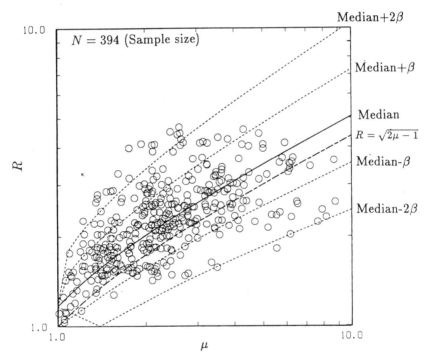

FIGURE 11. $R - \mu$ relationship (case M10S1H5).

With respect to nonlinear structural characteristics, the elasto-plastic models used above for reinforced concrete column and beam members are obviously idealizations of rather complex reinforced concrete behavior. More sophisticated models are available to account for the more detailed and often quite important dynamic behavior. Note that nonlinear structural behavior produces a considerable degree of response uncertainty, particularly in comparison with that of a linear structure; the variability in the peak response value is not so significant for a linear structure subjected to time histories derived from an identical power spectral density function. Finally, it is noted that when an RMF is used in a reliability analysis, the effect of energy absorption is considered only indirectly. Indeed, a number of models have been proposed to incorporate more explicitly such an effect into damage estimation, which forms the basis for the ensuing reliability analysis. A comprehensive review of these models is made in Ref. 11.

EQUIPMENT FRAGILITY

Equipment often represent functionally critical components of large-scale systems, particularly under seismic ground motion. Also, in general and particularly in the eastern United States, potential seismic problems associated with equipment are less seriously dealt with than those of ordinary buildings and structures, if indeed these problems are perceived at all.

It is well known, for example, that an emergency power supply device is one of the most critical components for the integrity of nuclear power plant operation under seismic conditions, which might induce an off-site electric power outage. Other examples include: a safe shut-down device for a water transmission system which must be operational at the time of seismically induced pipe breaks, that is, when it is most needed; communication and utility equipment housed in a telephone exchange building whose seismic integrity is a necessary condition for the serviceability of the telecommunications system of which it is an integral part; and a computer system whose malfunction due to a seismic event might disrupt regional, national, or even international business transactions, resulting in substantial financial loss to related industries and ultimately to the general public.

The content of this section represents a brief outline of the work performed by Bandyopadhyay et al.[12] Although this work considers the seismic fragility of nuclear power plant equipment such as a motor control center, switchboard, panel board and power supply, on the basis of fragility as well as qualification test data, the method for estimating the seismic fragility developed therein is in principle applicable to equipment housed in or attached to other types of primary structures.[12]

In this work of Bandyopadhyay et al. most of the test data represent the results of random multifrequency biaxial (vertical and horizontal) inputs, and hence these inputs and corresponding TRS (test response spectra) at 2% damping are considered standard. Test data based on other than biaxial inputs and TRS at different damping ratios are converted into standard form in the proper fashion. For the purpose of developing fragility data that can be used in seismic margins studies and probabilistic risk assessment, a median g-value and associated log-standard deviation are derived from these test data under the assumption that the fragility is log-normally distributed. To achieve this, however, it is necessary that a g-value is selected to represent the TRS associated with each qualification and fragility test. A sample of these g-values is then treated statistically for estimation of the median and log-standard deviation.

As a representative g-value for each TRS, the zero-period acceleration (ZPA) and average spectral acceleration (ASA) are used, where the ZPA is equal to the peak input acceleration value and the average spectral acceleration is obtained by dividing the area under a portion of each TRS curve (plotted in terms of g-value versus frequency both in arithmetic scale) by a frequency band representing the frequency range of significance for the equipment under consideration (the 4 ~ 16 Hz range in Ref. 12).

Listed in TABLE 2 are those ZPA and ASA values associated with qualification and fragility tests performed on motor control centers for the functional failure mode arising from "contact chatter." Although TABLE 2 does not identify which

TABLE 2. Input Data for Statistical Analysis–MCC[12]

Failure Mode	ZPA (g)		ASA (g) at 2% Damping	
	Qualification	Fragility	Qualification	Fragility
Contact chatter	0.9, 0.9, 1.1	0.9, 1.0, 1.0	2.0, 2.0, 2.6	2.1, 2.1, 2.1
	1.2, 1.3, 1.3	1.0, 1.0, 1.0	2.7, 2.7, 3.2	2.1, 2.1, 2.2
	2.1, 2.1, 2.1	1.0, 1.1, 1.1	5.6, 5.6, 5.6	2.2, 2.2, 2.2
	2.4	1.1, 1.1, 1.1	5.6,	2.2, 2.2, 2.2
		1.1, 1.1, 1.1		2.3, 2.3, 2.3
		1.1, 1.1, 1.1		2.3, 2.6, 2.7
		1.1, 1.1, 1.1		2.8, 2.8, 2.8
		1.1, 1.1, 1.1		2.8, 3.0, 3.0
		1.1, 1.1, 1.1		3.0, 3.0, 3.0
		1.2, 1.2, 1.2		3.1, 3.2, 3.2
		1.2, 1.3, 1.4		3.2, 3.2, 3.2
		1.5, 1.5, 1.5		3.2, 3.3, 3.3
		1.5, 1.5, 1.5		3.3, 3.3, 3.4
		1.5, 1.5, 1.5		3.5, 3.5, 3.6
		1.5, 1.6, 1.6		3.7, 3.7, 3.7
		1.6, 1.7, 1.7		3.7, 3.7, 3.7
		1.8, 1.8, 2.4		4.1, 4.9, 7.4

data belong to which specimen, the database therein involves a number of specimens produced by different manufacturers. Reference 12 interprets the statistical scatter in the g-values associated with a particular specimen as primarily representing randomness in the input motion.

The median fragility level and total log-standard deviation (β_c) are separately estimated for ZPA and ASA data involving all the specimens. However, the log-standard deviation of the randomness within a specimen is estimated from the ZPA and ASA values associated with each specimen. A weighted average (weight-proportional to number of tests involving each specimen) of these log-standard deviations with respect to all the specimens in the database then provides the log-standard deviation (β_r) representing the randomness within an individual specimen in accordance with the interpretation given above. The statistical analysis here is quite analogous to the analysis of variance: β_c represent the total variation and β_r the intra-specimen variation. Hence, if β_u is computed as

$$\beta_u^2 = \beta_c^2 - \beta_r^2 \tag{9}$$

it represents that interspecimen variation interpreted in this case as the log-standard deviation of the uncertainty.

The median fragility values, β_r and β_u, are computed by the moment and maximum likelihood methods (Appendix A of Ref. 12). When the former method is employed, only fragility data are used for computation, while both fragility and qualification data are utilized in the maximum likelihood method.

With the aid of these fragility parameters, we can estimate the probability of failure of equipment with a confidence statement. Following Budnitz et al.,[13]

$$\text{HCLPF} = \text{Median} \times \exp[-1.645(\beta_r + \beta_u)] \tag{10}$$

where, in this example, HCLPF (High Confidence Low Probability of Failure) indicates a fragility value corresponding to a (low) probability of failure of 5% at a (high) confidence level of 95%. HCLPF values based on the data are shown in TABLE 3.

The expression for the HCLPF value in Eq. **10** shows how the median fragility value must be reduced by a factor $\exp(-1.645\beta_r)$ to account for the effect of response randomness, and once more by a factor $\exp(-1.645\beta_u)$ for the effect of uncertainty.

The probabilistic concept from which Eq. **10** emerged was developed over the years and formed the basis for seismic probabilistic risk assessment (PRA) procedures for nuclear power plant structures. Equation **10** is used to determine the degree of the safety margin with respect to the response, computed on the basis of an input motion consistent with the design seismic conditions stipulated by design codes.

BOUNDING TECHNIQUES

When uncertainty problems cloud the process of estimating the structural response, the use of bounding techniques permits estimation of the maximum response, which only depends on one or two key parameters of design and analysis. The maximum response thus estimated provides some idea as to the range of the structural response variability, although the applicability of bounding techniques is at this time limited to relatively simple load and structural models.

Maximum Response under Excitations with Identical Total Energy or Fourier Amplitude

Consider a linear structure subjected to a class of excitations $x(t)$ with total energy less than or equal to M^2;

$$\int_{-\infty}^{\infty} x^2(t)\, dt \leqq M^2 \tag{11}$$

If the impulse response function of this structure is $h(t)$, we can show[14,15] that the structural response $y(t)$ is bounded from above as shown in the following equation:

$$\max_t |y(t)| \leqq I \leqq MN \tag{12}$$

TABLE 3. MCC Fragility Analysis Results[12]

Failure Mode	Indicator	Method[a]	Median (g)	β_u	β_r	HCLPF (g)
Contact chatter	ZPA	1	1.3	0.18	0.10	0.8
		2	1.3	0.24	0.09	0.8
		Recommended	1.3	0.20	0.10	0.8
	ASA at 2%	1	2.9	0.25	0.06	1.8
		2	3.1	0.31	0.06	1.7
		Recommended	3.0	0.27	0.06	1.7

[a] 1 = method of moments; 2 = method of maximum likelihood.

where

$$I = \frac{1}{2\pi} \int_{-\infty}^{\infty} |X(\omega)|\, |H(\omega)|\, d\omega \tag{13}$$

and

$$N^2 = \int_{-\infty}^{\infty} h^2(t)\, dt \tag{14}$$

where $H(\omega)$ is the Fourier transform of $h(t)$ or the frequency response function. The excitation that will produce response MN can be shown to be[14]

$$x(t) = (M/N)h(-t) \tag{15}$$

Therefore, if the total energy as defined in Eq. 11 is the only input information available, then MN is the worst response we can expect subject only to the uncertainties existing in the estimation of M and N. On the other hand, if we have some knowledge of the Fourier amplitude $|X(\omega)|$, or at least of its envelope, our upper bound for the structural response is I, which is usually smaller than MN. In fact, it is known[14] that MN is usually too large to be useful. However, use of an appropriate Fourier amplitude envelope[15] will provide an upper bound I that is much smaller than MN (sometimes less than one-tenth of MN). We observe that information on $X(\omega)$ is readily available, particularly for ground acceleration records, and hence estimating I does not require anything extraordinary. This bounding technique, therefore, involves uncertainties only in the estimation of the envelope of the Fourier amplitude and N.

Maximum Response under Uncertain Structural Properties

While it does not deal with dynamic problems, this subsection deals with the maximum structural response under uncertainty of spatial distribution of the material and geometric properties of structures.

For the purpose of illustration, we consider the simplest example possible in which we deal with a straight bar with length L and extensional rigidity of EA. The bar is fixed at $x = 0$ and subjected to a force F at $x = L$ as well as to distributed force $p(x)$ per unit length. Furthermore, we consider that the flexibility $1/(EA)$ is not deterministic and is expressible as

$$\frac{1}{EA} = \frac{1}{(EA)_0} [1 + f(x)] \tag{16}$$

where $1/(EA)_0$ is the expected value of $1/(EA)$ and $f(x)$ is a zero-mean homogeneous stochastic field with spatial correlation function $R_{ff}(\zeta) = E[f(x + \zeta)f(x)]$, where $E[\cdot]$ represents the expected value.

Under these conditions, we can show that, for example, the displacement $u(x)$ of the bar has the expected value and variance

$$E[u(x)] = \frac{1}{(EA)_0} \{Q(x) + Fx\} \tag{17}$$

$$\text{Var}[u(x)] = \frac{1}{(EA)_0^2} \int_0^x \int_0^x R_{ff}(\zeta_1 - \zeta_2) V(\zeta_1) V(\zeta_2) \, d\zeta_1 \, d\zeta_2 \tag{18}$$

where

$$Q(x) = \int_0^x P(\zeta) \, d\zeta_1, \quad P(x) = \int_x^L p(\zeta) \, d\zeta, \quad V(x) = P(x) + F \tag{19}$$

We can show that the following upper bounds for the variance can be derived[16]:

$$\text{Var}[u(x)] \leq \frac{\sigma_{ff}^2}{(EA)_0^2} \left[\int_0^x |V(\zeta)| \, d\zeta \right]^2 \tag{20}$$

and

$$\text{Var}[u(x)] \leq \frac{\sigma_{ff}^2}{(EA)_0^2} |V(\kappa^*)|^2 \tag{21}$$

where $\overline{V}(\kappa) = \int_0^x V(\zeta) e^{i\kappa\zeta} \, d\zeta$ and κ^* is the wave number at which $|\overline{V}(\kappa^*)|^2 \geq |\overline{V}(\kappa)|^2$. Equations **20** and **21** respectively indicate the upper bounds based on space and wave number domain consideration.

If the autocorrelation function $R_{ff}(\zeta)$ is given by

$$R_{ff}(\zeta) = \sigma_{ff}^2 \frac{b^4(b^2 - 3\zeta^2)}{(\zeta^2 + b^2)^3} \tag{22}$$

and $p = 2F/L$, $(EA)_0 = 1.57 \times 10^9 N$, $L = 1$ m, $F = 10^7$ N and $\sigma_{ff} = 0.10$, the maximum value of $\text{Var}[u(x)]$, say, at $x = 0.75L$ turns out to be 2.2×10^{-8} m^2 when $b \doteq 0.8$ m. The upper bounds obtained from Eqs. **20** and **21** are respectively equal to 3.9×10^{-8} m^2 and 2.8×10^{-8} m^2, which are very close to the actual maximum, particularly in view of the fact that they are obtained with knowledge of σ_{ff} but without knowledge of the autocorrelation function. Hence, the uncertainty issue associated with material and geometric properties may be reduced to that associated with σ_{ff} only. Since it is rather difficult to experimentally estimate the autocorrelation function of spatially varying flexibility, the upper bounds thus established are of particular importance. Extension of these bounding techniques to more complex structures appears possible.

CONCLUSION

The uncertainty associated with seismic response estimation is difficult to handle, primarily because the degree of such uncertainty can only be established by a balanced exercise of objective analysis of the available facts and subjective judgment based on seasoned professional experience. This paper attempts (1) to illustrate how complex the analytical process becomes to establish the degree of such uncertainty even when structural uncertainty is disregarded, by considering, as an example, the response modification factor; (2) to outline briefly the HCLPF approach for estimating equipment fragility in which the degrees of uncertainty and randomness are rather clearly identified; and (3) to demonstrate some bounding techniques in which the source of the uncertainty can be reduced to only a very small number of parameters, facilitating the uncertainty analysis that follows.

Modeling uncertainties may be handled by developing alternative models and comparing results, although many analytical and numerical methods are available (e.g., Ref. 17) for estimating the response uncertainty arising from parametric uncertainties. Human error is extremely difficult to quantify and this issue has just begun to be addressed by the profession (e.g., Ref. 18). In this respect, further study is encouraged.

REFERENCES

1. WORKING GROUP ON QUANTIFICATION OF UNCERTAINTIES, COMMITTEE ON DYNAMIC ANALYSIS OF THE COMMITTEE ON NUCLEAR STRUCTURES AND MATERIALS. 1986. Uncertainty and Conservatism in the Seismic Analysis and Design of Nuclear Facilities. STD, ASCE.
2. SHINOZUKA, M., H. HWANG & H. USHIBA. 1987. Proceedings of US-Korea Joint Seminar/Workshop on Critical Engineering System. C-K. Choi & A. H-S. Ang, Eds. Vol. 1: 331–348.
3. VELETSOS, A. S. & N. M. NEWMARK. 1960. Proceedings of the 2nd World Conference on Earthquake Engineering, Tokyo, Japan: 895–912.
4. RIDDLE, K. & N. M. NEWMARK. 1979. Statistical Analysis of the Response of Nonlinear Systems Subjected to Earthquakes. Department of Civil Engineering Report, UILU 79-2016. University of Illinois. Urbana-Champaign, IL.
5. SHINOZUKA, M. et al. 1986. Paper presented at the ASCE National Convention, Seattle, Washington.
6. MIZUTANI, M. et al. 1987. Elasto-plastic response of multi-story shear wall structures. Proceedings of Structural Mechanics in Reactor Technology-9, M: 441–446.
7. TAJIMI, H. 1960. Proceedings of the 2nd WCEE, Vol. II: 781–798.
8. SHINOZUKA, M. 1987. Stochastic fields and their digital simulation. In Stochastic Mechanics. Vol. I: 1–43, Columbia University. New York.
9. COMMONWEALTH EDISON CO. 1981. Zion Probabilistic Safety Study. Docket 50295. U.S. Nuclear Regulatory Commission. Washington, D.C.
10. TAKADA, T. & M. SHINOZUKA. 1988. Response Modification Factors for MDOF Systems. NCEER Technical Report (under NCEER Contract Nos. SUNYRF-NCEER-86-3033 and SUNYRF-NCEER-87-1006A). In preparation.
11. CHUNG, Y-S., C. MEYER & M. SHINOZUKA. 1987. Seismic Damage Assessment of Reinforced Concrete Members. Technical Report No. NCEER-87-0022. National Center for Earthquake Engineering Research. Buffalo, NY.
12. BANDYOPADHYAY, K. K., C. H. HOFMAYER, M. K. KASSIR & S. E. PEPPER. 1987. Seismic fragility of nuclear power plant components (phase II). NUREG/CR-4659, U.S. Nuclear Regulatory Commission. Washington, D.C.
13. BUDNITZ, R. J. et al. 1985. An approach to the quantification of seismic margins in nuclear power plants. NUREG/CR-4334. U.S. Nuclear Regulatory Commission. Washington, D.C.
14. DRENICK, R. R. 1968. Functional analysis of effects of earthquakes. Collection of Pre-Printed Papers for the 2nd Joint US-Japan Seminars in Applied Stochastics. Washington, D.C.
15. SHINOZUKA, M. 1970. J. Eng. Mech. ASCE 96: 729–738.
16. SHINOZUKA, M. 1987. J. Eng. Mech. ASCE 113: 825–842.
17. WONG, F. S. 1985. J. Comp. Struct. 20: 779–791.
18. NOWAK, A. S. 1986. Modelling human error in structural design and construction. NSF Workshop Proceedings, ASCE. Ann Arbor, MI.

Lessons Learned from Recent Moderate Earthquakes

FRANK E. McCLURE

Lawrence Berkeley Laboratory
Berkeley, California 94720

INTRODUCTION

The Earthquake Engineering Research Institute (EERI) has, since its inception in 1949, conducted postearthquake investigations to improve the science and practice of earthquake engineering. The EERI Learning from Earthquakes (LFE) Program activities in the past 11 years have resulted in the development of extensive procedures for mobilizing investigating teams, for conducting postearthquake disaster investigations, and disseminating these lessons to researchers, design practitioners, educators, government officials, and the public. The EERI LFE Program has produced significant input into programs for mitigating earthquake hazards.

These multidisciplinary studies have been conducted by geoscientists, engineers, architects, planners and social scientists. These studies have added to the cumulative knowledge base on geology, seismology, engineering seismology, geotechnical engineering, structural engineering, architecture, urban planning, social sciences, and emergency response. Many important lessons have been extracted from this continuously expanding body of knowledge. Also, significant gaps in knowledge have been identified and used to guide planning for future earthquakes and research needs.

These lessons are presented in an EERI Report, *Reducing Earthquake Hazards: Lessons Learned from Earthquakes.*[1] Extensive use of the material in this EERI Report and the EERI Continuing Education Committee's "Slides on Learning from Earthquakes" text were used in the preparation of this paper.

PHILOSOPHY OF POSTEARTHQUAKE INVESTIGATIONS

While a great deal can be learned about earthquake hazard mitigation through laboratory and analytical studies, the most effective "teacher" is to witness the impact of an earthquake on a full-scale city. No structure or design method can be proved fully adequate except by such full-scale field tests. No theory of the cause of earthquakes or mathematical model can be accepted unless it correctly explains what happens in nature. No seismic disaster preparedness plan can be confidently implemented unless its principles have been tested through actual use.

The process of learning from earthquakes typically involves three major steps: (1) discovery, (2) theoretical and experimental evaluation and development of the discovery, and (3) verification of theoretical developments made. Thus, a mere anecdotal observation (discovery) from a single earthquake generally does not yield a lesson learned. Verification is usually required to build consensus among the earthquake professionals involved. Once the three-step process is completed, and there is a consensus that a lesson has been learned, implementation in the form of a code revision, policy development, or acceptance of a new technology tool is usually fairly rapid.

PURPOSE AND SCOPE OF POSTEARTHQUAKE INVESTIGATIONS

The purpose of postearthquake investigations is to observe and document earthquake processes and the effects of earthquakes on man and man-made works. Progress in earthquake engineering has been influenced significantly through experiences in destructive earthquakes. Such earthquakes provide the only true tests of the abilities of engineered works to withstand seismic forces and of people to react and survive.

The scope of EERI postearthquake investigations is best presented in terms of the breadth and depth of coverage. The breadth of investigative activities includes all disciplines involved in the earthquake problems.

A representative listing of disciplines and topics included in postearthquake investigations might include the following: (1) geosciences (geology, seismology, strong ground motion); (2) geotechnical engineering; (3) structural engineering (engineered buildings, nonengineered buildings, special engineered structures); (4) architecture and urban planning; (5) lifeline engineering (power and communications, gas and liquid fuels, transportation, water and sewage); (5) social sciences and public policy; and (6) emergency response.

The following sections will discuss some of the things we have learned from earthquakes in some of the above disciplines as interpreted and considered to be of importance to a practicing structural engineer.

LESSONS FROM GEOLOGIC STUDIES

☐ Studies of active fault zones have shown that some faults tend to rupture with characteristic earthquakes, whereas some very long fault zones appear to be segmented. The length of the fault segment is a better guide to identifying the maximum earthquake than the total length of the fault zone.

☐ The tectonic environment—a combination of the earthquake source mechanism and the regional attenuation of seismic energy—controls the primary characteristics of the seismic waves and causes regional differences in ground motion.

LESSONS FROM ENGINEERING SEISMOLOGY

☐ Free-field strong motion data acquired in and since the 1971 San Fernando earthquake has increased understanding of (1) the variability in seismic waves resulting from local soil and building foundation conditions, (2) the focusing of ground motion in preferred directions, and (3) the importance of frequency content and duration along with amplitude in characterizing damage capability of ground motion.

☐ Analyses of strong motion data have shown that even small earthquakes can produce high levels of peak ground accelerations. However, these accelerations are not necessarily damaging, especially if they have single high-frequency pulses.

□ Peak ground acceleration seems to be a poor indicator of damage close to the earthquake source. This parameter is relatively independent of earthquake magnitude and Modified Mercalli intensity within 10–20 km of the fault.

□ Close to the earthquake source, amplitudes of high-frequency ground motion (causing the peak acceleration) are relatively independent of local geologic conditions. Notable exceptions have occurred at sites located on shallow soil deposits or in areas of steep topography or on crests of ridges where significant amplification has occurred.

□ Low-frequency amplitudes (causing the peak velocity and displacement) are generally lower on rock than on soil.

□ Earthquakes whose rupture front propagates in a relatively unilateral coherent manner can substantially enhance strong ground motion in the direction of propagation.

LESSONS FROM GEOTECHNICAL ENGINEERING STUDIES

□ The characteristics of free-field ground motions are influenced by three major factors: source, travel path and local soil conditions.

□ The travel path and the local soil conditions modify the basic characteristic of strong ground motion by acting, respectively, as low-pass and band-pass filters.

□ Depending on the geometry of the source-path-local site and the size of the earthquake, variations in ground motion due to source effects can overshadow the effects of local soil conditions or the effects of the local soil conditions can overshadow the source effects.

□ Near-field records can contain a low-frequency pulse corresponding to the "fling" of the fault, a unidirectional ground displacement near the rupturing fault.

□ Cyclic loads, including uplift forces, weakened the soft clay supporting the friction piles during the 1985 Mexico earthquakes, allowing the loaded piles to penetrate and in some cases, pull out of the ground. These effects led to settlement, tilting, and failure of the superstructure.

□ Most retaining structures that have collapsed or experienced damage during past earthquakes are waterfront structures extending below the water surface. Only a few cases of damage have been reported for retaining structures constructed above the ground-water table. Few reports have been made of damage to retaining walls caused directly by ground-shaking.

LESSONS FROM BUILDING BEHAVIOR

□ Correlation of analytical and experimental predictions of damage with observed patterns of damage has led to a capability for explaining almost all building damage and modes of failures. This success has lead to many changes in earthquake-resistant design provisions of building codes.

□ Designing to code does not always safeguard against excessive damage in a severe earthquake. Enforcement of code compliance and quality of construction are the two factors that most affect the degree to which code design meets the intent of the code, which is to prevent collapse.

□ The difference between good and poor building performance in an earthquake can be attributed to a small number of simple design measures. They include: (1) incorporation of ductility, strength, and redundancy to provide a margin of safety against collapse; (2) avoidance of stiff elements not considered in the design, (e.g., frames with stiff masonry infill walls) to keep the dynamic properties of the structural system from changing during the earthquake; (3) incorporation of well-connected horizontal diaphragms (roof and floor systems) to transfer the seismic forces to the vertical lateral force resisting building elements (columns or walls); (4) incorporation of stiffness in the lateral force-resisting system to enhance performance as a general rule; (5) avoidance of irregularities in plan and elevation to minimize sudden changes (discontinuities) in stiffness between the adjacent stories; and (6) avoidance of soft stories to prevent forces from being directed to places where strength is minimal.

□ Well-designed, well-detailed, and well-constructed buildings will resist large earthquake-induced lateral forces without collapse or excessive damage.

□ Unreinforced masonry buildings usually perform very poorly, whereas reinforced masonry buildings usually perform well.

□ Precast and prestressed concrete elements must be tied together well to ensure good performance.

□ Performance of cast-in-place reinforced concrete buildings depends on the type of structural system and the quality of the detailing.

□ Steel buildings generally perform well, but the 1985 Mexico earthquake showed that they will collapse when the forces, displacements, and duration of shaking are large.

□ Wooden buildings are intrinsically tough when irregular configurations and undesirable combinations with other materials are avoided.

□ The weakest links in building systems are typically the connections between structural elements.

□ Developing adequate connections between structural elements is frequently more difficult than providing strength within the members.

□ Collapse may occur if the strength of nonductile elements is insufficient.

□ Corner columns in moment-resisting frame structures are more vulnerable to severe seismic damage than the other columns.

□ Exterior panels and parapets need strong anchorage to protect life safety.

□ Code design does not, and is not intended to, safeguard against inelastic deformations in a severe earthquake. Potential failure modes should be identified in the design process, and ductility should be provided at all locations where inelastic deformations are anticipated.

□ The uncertainties in seismic input and structural response require an emphasis on detailing and the conceptual aspects of design rather than the analytical aspects of design.

□ Discontinuities in the path over which inertia forces are transferred through the structure to the foundation are a common cause of collapse.

□ Irregularities of mass, stiffness, and strength in plan or elevation give rise to demands in force and deformation that are difficult to evaluate and accommodate in the design process.

□ Redundancy provides for redistribution of forces and can prevent collapse when individual structural members deteriorate in an earthquake.

□ The stiffness and strength of structural elements that are not considered part of the lateral force-resisting system may strongly affect the response of the building.

□ Inadequate building separation may result in severe pounding damage and failure.

□ Many failures have been caused by inadequate detailing of connections. Structural elements must be well tied together. Brittle connections should be designed for the maximum forces that can be generated in a severe earthquake.

☐ The damage to nonstructural elements and building contents is often more severe than the damage to the structural elements. Inadequate anchoring of the nonstructural elements and building contents poses a safety hazard.

☐ Poor quality control of construction has caused severe damage and failure.

☐ Some observations from past earthquakes still cannot be put in a lesson format. These observations include the following: (1) Damage correlates poorly with magnitude or other global measures of earthquake size. (2) Local ground-shaking often varies greatly from location to location, even within a small geographic region. Many examples of severe damage in individual buildings have been reported when adjacent buildings of equal or poorer quality withstood the ground-shaking without distress.

LEARNING FROM ARCHITECTURAL STUDIES

☐ The influence of building configuration (size and shape) on seismic performance has been recognized for many years. Examples of such influences include the following: (1) Buildings having irregular shapes (e.g., L- or T-shaped), triangular plan layouts, and set-back elevations generally always perform poorly. (2) Buildings having soft first stories consistently perform poorly and nearly collapse.

☐ Buildings having a triangular plan are prone to torsion effects that can lead to severe structural damage.

☐ Elevations where the resisting elements are distributed nonuniformly around the building perimeter have experienced severe torsion and performed badly.

☐ Severe pounding damage can result when adjacent buildings do not have adequate separations and space to move without touching.

☐ Structures with adequate shear walls have consistently performed well except where planning or image requirements have created discontinuities in the shear walls.

☐ The weak column/strong beam configuration has resulted in severe shear stresses in short columns, leading to failure.

☐ Well-coordinated architectural and engineering design requires that the engineer is brought into the conceptual design process early.

☐ Configuration is always important in determining building seismic performance, but there is still a need for a systematic analysis of a large number of modern buildings in order to fully understand its relative importance.

☐ Recent moderate earthquakes suggest a potential problem when property losses in modern buildings resulting from damage to architectural systems and components may greatly exceed losses due to structural damage.

☐ Modern buildings are relatively flexible and must be designed to limit component interactions.

☐ Unreinforced unit masonry partitions are very prone to damage.

☐ Plaster-sheathed partitions are very prone to cracking and spalling, especially in tall flexible buildings.

☐ Dry wall partitions on steel or wood studs with proper details perform well.

☐ Partitions having controlled movement capabilities offer promise for reducing damage; however, they have not been tested by major earthquakes.

☐ The primary cause of partition damage is cracking from distortion or drift in the primary structural framing system. The factors controlling damage are directly related to the adequacy of the connections and the brittleness and stiffness incompatibility of the partitions.

□ Supported ceiling systems are particularly vulnerable to horizontal and vertical ground shaking.

□ Exposed "T-bar" suspended ceilings with lay-in tiles have performed poorly, especially where heavy light fixtures are supported by the "T-bar" ceiling members.

□ Concealed-spline suspended ceilings perform better than lay-in systems.

□ Proper detailing and construction of ceiling systems and components to control horizontal and vertical shaking offers promise for reducing damage.

□ Heavy precast concrete components must have careful design of their connections in order to perform well.

□ Stiff unit masonry filler walls without adequate joint detailing consistently suffer damage.

□ Brick or stone veneers often crack and sometimes collapse unless adequate detailing is performed.

□ Exterior cladding systems and components adequately detailed and installed to permit realistic interface movement generally perform well.

□ Appendages and parapets often perform poorly because of inadequate or deteriorated anchorage.

□ When appendages are located high on a building, they are subjected to large accelerations. Their failure may cause casualties and additional damage to their own as well as to adjacent buildings.

□ Stairs are vulnerable to damage and they may act as diagonal braces or struts introducing additional stiffness that may not have been considered in the overall design.

□ Stair exits and egress points in old buildings are potentially vulnerable to blockage when they are enclosed with unreinforced unit masonry walls.

□ Building contents may be severely damaged in earthquakes having magnitudes of 5 and greater if equipment, fixtures, and shelving are not anchored to prevent falling, sliding, and overturning.

□ Fires may be caused by ruptured gas lines, disturbed chemicals, and the restoration of electrical power.

□ Recent studies show that motion of some modern building types create great occupant concern for their safety and ability to exit the building. Building exit design must reflect the actual behavior of people during evacuation following earthquakes.

LESSONS FROM SOCIAL SCIENCE STUDIES

□ Social science investigations of an earthquake normally focus first on actual behavior and then seek to understand the causes of that behavior.

□ People living in seismically active regions having relatively frequent damaging earthquakes (e.g., California) generally believe that earthquakes can occur in their lifetime and have an impact on their community.

□ People living in areas of low seismic activity but high chance of loss (risk) (e.g., the Central United States) are surprisingly aware that earthquakes could disrupt their community sometime in the near future.

□ More research is needed on the awareness and attitudes of people living in Eastern United States to the hazards from earthquakes.

□ However, citizens typically are reluctant to do anything about the risk to which they are exposed regardless of where they live.

□ Community decision-makers, although aware of the threat to their communities, may not be willing to take steps to reduce the risk.

□ Administrative or appointed key actors (as opposed to elected officials) are much more aware of earthquake hazards, concerned about the consequences of a large earthquake, and supportive of taking preparatory action toward mitigation of the effects of an earthquake. However, these perceptions and sentiments do not result, in most cases, in actions to reduce their communities' exposure to seismic risk.

□ Technical knowledge, although important, is never sufficient in itself to drive the process of earthquake preparedness in a community.

□ The knowledge utilization pyramid consists of (1) the base of technical knowledge, (2) trained, concerned, and committed people (i.e., advocates, champions), (3) coordinated, integrated, and focused programs, (4) earthquake experience (sometimes another natural disaster can act as a "surrogate" earthquake) to open the "policy window of opportunity," and (5) implementation of loss-reduction measures.

□ The basic question is whether loss-reduction measures can be implemented without an earthquake or its surrogate.

□ Experience suggests that effective implementation leading to changes in design, construction, and land-use practices occurs mainly after an earthquake.

□ Precursory phenomena to damaging earthquakes, such as foreshocks or predictions of the size, place, and time, increase public awareness and can lead to public actions to reduce potential losses and enhance preparedness.

□ Public awareness of earthquake hazards and risk may increase because of personal experience, public education, and public information; however, the effect of education and public information on increased public action has not been established for the general public.

□ Following a disaster, a dramatic need for information exists and people will seek information from a myriad of sources.

□ By focusing on the spectacular "newsworthy" aspects, the media can overemphasize the nature and extent of earthquake effects.

□ "Occupant behavior," a new area of study, indicates the ways in which occupants of a building are affected by: (1) whom they are with, (2) prior experience with earthquakes, and (3) previous emergency training.

□ People interact with the structural and nonstructural elements of their building in predictable ways that correlate with prearranged escape routes, habits, and the role being enacted at the time of the earthquake.

□ Most communities do become safer and less vulnerable to earthquake hazards as postdisaster mitigation policies are implemented in reconstruction.

□ The primary factors that limit implementation of mitigation policies in the reconstruction phase are: (1) availability of funds, (2) desire for a "return to normalcy," rebuilding "just as" and "just where" everything was before the disaster, (3) pre-existing community trends (e.g., economic expansion or decline) and competing interest (e.g., code vs. no code), and (4) presence of undamaged structures (impedes rezoning for less hazardous use).

□ Adoption of mitigation policies declines fairly rapidly as time passes after a disaster.

□ Great opportunities for mitigation policies exist after an earthquake disaster. Awareness is high and the earthquake problem has high priority on political agendas. The political climate at local, state, and federal levels may be receptive to change.

□ The knowledge base of enhancing mitigation policies after a disaster is growing, but the knowledge base on developing mechanisms that produce effective long-term implementation of these policies is conspicuously absent.

☐ Mitigation refers to actions taken before and after a damaging earthquake to reduce the effects on the built environment and the human community.

☐ In cases where mitigation policies (e.g., ordinances to control unreinforced masonry buildings, land-use regulations) have been enacted and implemented, anticipated negative consequences have not arisen.

☐ Adoption of mitigation policies does not assure continuous rigorous implementation. As time passes, communities may even rescind mitigation policies.

CONCLUSION

The lessons learned and the results of research on earthquake engineering are of value to a community only when they are used. The experience accumulated since 1964 shows that social, economic, and political factors are critical to determining where and when utilization occurs. Some of the factors include: (1) the relative rarity of damaging earthquakes as compared to floods and hurricanes; (2) the relative power of interest groups; (3) the public's awareness of the earthquake hazard; (4) existence of credible internal advocates and/or external champions; (5) economic incentives; and (6) a recent natural disaster that calls renewed attention to the value of prudent preparedness and mitigation activities and opens the "policy window" or "window of opportunity."

Although technical knowledge of the geosciences, geotechnical and structural engineering, architecture, and urban planning components of the earthquake problem is very important and necessary, implementation of preparedness and mitigation activities may depend as much or more on understanding the social, economic, and political factors as on understanding the technical factors. The required understanding is gained slowly over time.

REFERENCE

1.   1981. Reducing Earthquake Hazards: Lessons Learned from Earthquakes. EERI Publication No. 86-02. Earthquake Engineering Research Institute.

General Discussion: Part V

PETER GERGELY, CHAIRMAN: It was noted yesterday that base isolation of structures is gaining interest worldwide. As a structural engineer, I recognize the following characteristic of such systems. First, the very low-frequency content of the input motion is of particular interest. Second, the characteristic is the elastic response of the overall isolated structure, and, third, the failure mechanism is quite different. Would any of the earlier speakers, or perhaps some seismologists, care to comment on what kind of rethinking might be appropriate for defining motions for such systems?

RESPONSE: You are correct that many of the considerations that go into designing a base isolated structure are somewhat different from those that would do in a conventional structure. However, they are not necessarily much different from those that you would consider in a structure that might be highly nonlinear. In both cases you're trying to isolate the forces that are transmitted into the building; in one case by limiting the capacity of the structure to resist forces and, in the second, by limiting the forces that can go into the structure itself. But in both of those cases what you're worried about is the ground displacements that you have to isolate, not so much the peak ground accelerations. And so, by picking the isolator characteristics or the strength of the structure, you are going to more or less control the forces; nevertheless, you have to be concerned about long-period motions, and, if you're close to a causative fault, the maximum displacements. I have not touched on some of the other issues, but I think these comments are pertinent to the specific question.

PAUL SOMERVILLE (*Woodward-Clyde Consultants, Pasadena, Calif.*): To add a seismological perspective, modern digital strong-motion instruments allow us to integrate reliably to ground-motion displacement. This was done, for example, for the Michoacan, Mexico earthquake of 1985, and these records show very large static ground-motion displacements—in that case on the order of one meter of ground-motion displacement above the fault zone occurring over a 10-second time interval. These very large displacements need to be thought about very carefully. If they occurred in San Bernadino, for example, the building would effectively hit the stops. A lot of thought must be given to how large these static ground motions can be.

CARL TURKSTRA (*Polytechnic University, Brooklyn, N.Y.*): I would like to ask a question of seismologists that relates to risk analysis and engineering.

We have a consensus that the response of structures in the nonlinear range cannot be scaled by the peak ground acceleration of a record. That means that our present methodology, which gives us a map of peak ground accelerations with a certain probability of exceedence in a reference time period, such as fifty years, is no longer a sufficient or perhaps even a very useful basis for zoning or for risk analysis in general.

As a result, we have to go to a more meaningful measure, a more meaningful scaling procedure for damage potential in earthquake records. The seismologists speak of things like seismic moments, and they have other parameters which are of no interest to structural engineers whatsoever. In fact, the structural engineer is concerned with the shape of response spectra for ductility demand or a damage index among other things—and he wants to scale the spectra to an appropriate earthquake magnitude. That means that we have to have a new mapping procedure.

We now use the Cornell-McGuire approach where we establish zones to which we attach seismicity, laws for attenuation and so on, all of which we relate to peak ground acceleration.

I would like to ask the seismologists what they can give us as an alternative basis for mapping that will allow us to do a more realistic hazard analysis for structural engineering. Are there approaches to mapping that will give us the lifetime probability distribution functions of other earthquake parameters that are better damage estimators?

ALLIN CORNELL (*Stanford University, Stanford, Calif.*): I'd like to reinforce Steve Mahin's use of looking at nonlinear behavior of simple systems as a way of characterizing ground motions. This comment fits in, not only with Carl Turkstra's recent comments, but also with those who ask why we are not moving forward further in considering duration of records in our hazard assessment.

UNIDENTIFIED SPEAKER: I would like to present some results that are related to previous presentations and to the comments just made. We have done studies that are very similar to those of Professors Hall, Mahin, and Bertero.

We have selected two earthquakes of interest for different reasons—the El Centro and the Mexico City earthquakes.

If you examine the pseudo-acceleration spectrum for El Centro with 50% damping, you will find that the maximum pseudo-acceleration is of the order of .9 and occurs at about .5 seconds. If we look at the lower end of the curve for zero period, the curve will tend to the maximum acceleration, which, for El Centro, is 0.37.

If we compare that to a similar spectrum for Mexico City, it is interesting to note that the maximum pseudo-acceleration is also of the order of 1. So the spectra have that similarity even though the maximum acceleration in Mexico City for the records that we used was of the order of .17.

Now, if we look at the hysteretic energy demand, which is exactly what Drs. Cornell and Mahin presented earlier, we see an interesting difference. For similar assumptions, the maximum hysteretic energy demand for El Centro is about .05. If we compare this to Mexico City, we see that the peak hysteretic energy demand is about .45. This is an order of magnitude greater than that of El Centro, even though the maximum peak elastic response is the same.

So, if we just look at the elastic spectra and then reduce them for ductility demand, they might have similar peaks, but the structures in Mexico City would have required substantially larger energy dissipation properties to resist the same accelerations as in El Centro.

Of course we're talking about different periods. But nevertheless, when we just look at the peaks there are basic differences.

We have also studied the area's intensity and some of the other parameters that Dr. Mahin presented, and basically we obtained similar results. So, I think these results point to the need to look at different ways of measuring response of structures.

CHRISTIAN MEYER (*Columbia University, New York, N.Y.*): I would like to comment on the basic question that was raised in the first two papers by Professors Turkstra and Mahin. It's really to look at the design problem from a very broad view. Generally, when we design structures, we have to assure that the demand never exceeds the capacity. Now, in earthquake-resistant design, this simple inequality is extremely complicated because the capacity is really interdependent with the demand and the nature of the earthquake. Just now we've heard the advice: make sure your ductility capacity is 4 and doesn't get exceeded. But, unfortunately it's not as simple as that.

In a reinforced concrete frame, for example, with regard to cyclic loading its strength and stiffness deteriorate. In other words, we are talking about a low cycle fatigue phenomenon. The capacity of a member or a structure to dissipate energy, the energy dissipation capacity, is a finite quantity, and to cast this information into a simple design spectrum will be very difficult. I don't see yet how we will be able to resolve this kind of complication. I know that Professor Hall alluded to this, but I don't see how we can come up with simple design spectra that realistically take this interaction between capacity and demand into account so that we can assure our designs to be safe.

ERIC VAN MARCKE (*Princeton University, Princeton, N.J.*): Dr. Hall has mentioned that perhaps the time is right to move away from response spectra as the basic characterization of earthquake ground motion for engineering purposes; he also noted the need for multiple response spectra, reflecting different kinds of earthquakes that have very different frequency content.

In attempting to look at many different kinds of measures of nonlinear response or liquefaction potential such as low cycle fatigue, or in other words, duration-sensitive measures of response that show a great variety of types of sensitivity to duration, it is not really efficient to try to capture the effect of duration in all these different kinds of response. Period-dependence also enters into the picture. So, to have risk defined explicitly in terms of nonlinear response measures is quite inefficient.

Why not explicitly recognize duration in the characterization of earthquake ground motion? And the most logical way to do this is in terms of the spectral density function combined with something like a strong-motion duration. Certainly, methods exist based on either the vibration theory or simulation to relate the combination spectral density function duration not only to linear response spectra, but also to multi-degree response, the response of various kinds of simple nonlinear systems, as well as liquefaction prediction.

In addition, of course, the spectral density function–based approach ties in very nicely with the newly available methodologies coming from geophysics to predict the character of the Fourier amplitude spectra as a function of magnitude and distance. So I very much favor an approach in terms of a design specification that is not expressed in terms of a single spectral content, but a spectral content associated with different characteristic earthquakes that may cause damage to the structure.

Wind versus Seismic Design[a]

GUY J. P. NORDENSON

Ove Arup & Partners
260 Fifth Avenue
New York, New York 10001

INTRODUCTION

While there is a growing recognition of the potential for damaging earthquakes in the eastern United States, the general opinion among building designers and the public is that modern structures designed according to the code-specified wind loads should prove safe in the moderate-level earthquakes that are considered possible. In New York City (NYC) the code requires that buildings over 20 stories be designed for wind pressures up to 30 and even 40 pounds per square foot (psf). A typical 30-story apartment building would then be designed for a total lateral wind load equivalent to between 1.5 and 2 percent of its weight. This is, in fact, comparable to the design loads specified in current seismic design codes[1,2] for such a building in NYC. Given the infrequency of even moderate earthquakes in the NYC area (only one with a Richter magnitude slightly over 5 is known to have occurred to date), there would appear to be little cause for concern.

If one examines more closely the assumptions underlying the loads specified in the building codes for wind- and earthquake-resistant design in relation to the actual response of buildings to earthquakes, the comparison is in fact less simply made. Further, it has become clear from recent seismological research that earthquakes of at least one magnitude greater than the historical record indicates must be considered possible. In this paper an attempt is made to clarify the difference between wind and seismic design assumptions and results through a review of recent research work and several simple parametric studies. Implications regarding the seismic hazards and risk in NYC are discussed in the conclusions.

THE ACTION OF WINDS AND EARTHQUAKES ON BUILDINGS

The overall effects of winds and earthquakes on building are well understood and incorporated in building codes and design practice. The major uncertainties arise when an effort is made to predict the specific character and frequency of extreme and potentially damaging hurricanes or earthquakes. There is, as well, a fundamental difficulty inherent in any structural analysis in precisely characterizing the properties and interaction of the "structural" and "nonstructural" materials and elements that make up a building to arrive at a complete representation of its behavior. How is one to give an exact account of the variation in stiffness with increasing load of an unreinforced brick masonry wall built as infill to a

[a] The work described was done under a research contract from the National Center for Earthquake Engineering Research (SUNY at Buffalo) at Weidlinger Associates, New York.

riveted steel frame, up to and past the "elastic" limit? How can one model the elastoplastic, interactive behavior of concrete-encased steel girders?

The physical effects of strong winds and earthquakes differ in a number of essential ways:

1. The wind velocity at a given location during a storm includes both a mean component that increases gradually over the course of hours and a gust component that fluctuates about that mean in periods of seconds or less. The gust component is usually less than 40 percent of the magnitude of the mean wind pressure value. In contrast, earthquakes last from 10 to 90 seconds with complete reversals of ground motion occurring in periods from a 10th of a second to 2–3 seconds.

2. For all but tall slender structures, wind effects are essentially static loadings, whereas in earthquakes the accelerations reverse rapidly and induce significant inertial effects. Wind loads are proportional to the building shape and exposed area, earthquake loads to the building mass and stiffness.

3. Extreme winds (e.g., hurricanes) generally give some warning, while earthquakes give none that are dependable.

4. The wind loads specified in building codes are tied to the "50-year wind velocities," that is, those estimated from the historical record to have less than a 2 percent annual probability of being exceeded (annual probability of exceedence [APE]). Important structures may be designed for the "100-year wind" or 1 percent APE. The design loads in current seismic design codes are derived from ground motions with a 10 percent probability of exceedence in 50 years (i.e., a 475-year return period of 0.2 percent APE).

5. As tragically demonstrated in the 1985 Mexico City earthquake, the type of soil underlying a building can have a profound and determining effect on the structure's response in both amplifying the signal and shifting the input frequency content to that resonant with the structure's natural frequency. The wind's effect is unaffected by soil conditions.

6. Structures are typically designed to remain elastic in extreme winds, whereas in an earthquake they are expected to locally deform well into the material plastic range in order to dissipate in work the seismic energy input. Wind design is largely concerned with achieving sufficient stiffness and damping to keep horizontal vibrations below perceptible levels.

THE WORST CASE

The probable hurricane wind speeds measured at 30 feet above ground have been estimated for various mean recurrence intervals in Ref. 3. For 50-, 100-, and 2000-year intervals (2.0, 1.0, and 0.05 percent APE) the estimates are 86, 97, and 134 miles per hour (mph) fastest mile wind speed. The wind loads specified for use in the NYC area in current codes[4] are based on the "50-year" wind speed of 90 mph. In a hurricane with wind velocities as high as 134 mph the surface pressures on a building could exceed the code design basis by 122 percent (pressure varies with the square of the velocity). Since structural elements are typically designed to allowable stress levels of 80 percent of the tensile yield stress when considering wind effects, they would at most be overstressed by slightly over 75 percent. Furthermore, since wind bracing is often sized to meet lateral stiffness requirements (and is therefore at lower than allowable stress levels under design loads), it is likely that only a few elements would in fact yield in even the worst hurricane.

In the case of earthquake effects the analogous ratio of design loading to the maximum conceivable in the extreme instance can be far greater. In the Los Angeles area the largest "credible" earthquake (of magnitude 8.0) for the South San Andreas fault might generate peak ground accelerations (PGA) at a site 10 kilometers (km) from the fault of 70 percent of gravity (g).[5] The average return period for such an earthquake is believed to be less than 1000 years. For a 500-year return period the PGA is estimated to be approximately 64 percent g. For a site in Santa Monica, California, further from the San Andreas fault, the corresponding PGA for 1000- and 500-year average return periods are estimated at 52 and 42 percent g, respectively.[5] The lateral loads in the current Los Angeles seismic code[i] are derived from ground motions with an effective peak acceleration of 0.40 g. In the worst conceivable event, at a site close to the San Andreas fault, this design basis PGA may be exceeded by up to a factor of 1.75. The increased demand on the structure's energy-dissipating capacity will undoubtedly cause extensive damage to some structures. For a good number of the structures designed to the current code this increased demand may well be accommodated without collapse.

A similar comparison of the proposed code[1] design basis PGA to the maximum credible values, for NYC, illustrates the essential difference in the nature of seismic hazards in zones of high and moderate seismicity. Our knowledge of the seismic hazard in the East is not as good as for the western United States because of the relative infrequency of damaging events and the comparatively small extent of our history. The historical record for the East dates back to the 1638 St. Lawrence River, Canada earthquake. There have been sizable earthquakes with Richter magnitudes estimated at or in excess of 7.0 in 1755 (Cape Ann near Boston, Massachusetts), 1812 and 1813 (New Madrid, Missouri), and 1886 (Charleston, North Carolina). In 1884 an earthquake with an estimated Richter magnitude of 5.0 to 5.5 occurred off-shore near Coney Island, New York causing a few large cracks in masonry walls in Jamaica, Queens. There have been numerous small earthquakes as well. A recent probabilistic hazards evaluation undertaken for the NYC area[6] based on available historical and geological data projects that for an average return period of 400 to 500 years a PGA value around 7 percent g. This corresponds roughly to an event with a Richter magnitude near 5.2. A clarification issued by the U.S. Geological Survey in response to a query from the Nuclear Regulatory Commission has suggested that the risk may in fact be even greater.

> Because the geologic and tectonic features of the Charleston region are similar to those in other regions of the eastern seaboard, we conclude that although there is no recent or historical evidence that other regions have experienced strong earthquakes, the historical record is not of itself sufficient ground for ruling out the occurrence in these other regions of strong seismic ground motions similar to those experienced near Charleston in 1886. Although the probability of strong ground motion due to an earthquake in any given year at a particular location in the eastern seaboard may be very low, deterministic and probabilistic evaluations of the seismic hazard should be made for individual sites in the eastern seaboard to establish the seismic engineering parameters for critical facilities.[7]

It may therefore be necessary to consider that earthquakes of up to Richter magnitude 6.5 or 7 could occur in NYC, though with very low probabilities (on the order of 0.01 percent APE or a 1 percent probability of being exceeded in 100 years) when evaluating construction standards along the eastern seaboard. The peak accelerations generated by earthquakes of this magnitude would be on the order of 0.40 to 0.50 g or up to seven times the design basis acceleration of 0.07 g. Thus, while the worst New York City hurricane or the worst Los Angeles earth-

quake may exceed the design basis values by up to a factor of 1.75, the worst
conceivable NYC earthquake could exceed the design basis proposed (though
not at present applied) by a factor of about 7.

It is important in considering this comparison to recall the difference in the
design philosophies applied to seismic and wind design. The aim of wind design is
to achieve occupant comfort in common windstorms and an essentially elastic
structural response in extreme winds. In the words of the Structural Engineers'
Association of California, the aims of the building code's seismic provisions are:

1. To resist minor levels of earthquake ground motion without damage;
2. To resist moderate levels of earthquake ground motion without structural
damage, but possibly experience some nonstructural damage;
3. To resist major levels of earthquake ground motion having an intensity
equal to the strongest either experienced or forecast for the building site, without
collapse, but possibly with some structural as well as nonstructural damage.[8]

The ability of a structure to withstand without collapse a major earthquake
depends on the "ductility" or energy-dissipating capacity that is provided to the
lateral-load-resisting system through detailing. Earthquake-resistant structures
are designed using typical elastic analysis methods with allowable material
stresses around 80 percent of tensile yield and lateral loads 1/10th to 1/5th the
expected loads to recognize the structure's ductility. This ductility, or, more
accurately, toughness, is simply the capacity of the system to dissipate energy
through many large displacement cycles of plastic rotation, shearing or extension
in the members. Moment frames yield at beam ends, braced frames "yield"
through inelastic extension and buckling of the braces, shear walls in chord
flexure or web shear, and coupled shear walls or eccentric braced frames in
specially designed "fuses." To insure that the predicted mechanism is obtained,
the members away from the yielding element and all the connections are sized and
detailed to remain essentially elastic up to the stage where a side-sway mechanism
is formed. Throughout, as individual elements yield and distort with each cycle,
energy is effectively dissipated. Tests on large-scale subassemblies of these structures have demonstrated that overall ductilities (i.e., maximum lateral displacement divided by that at yield) of up to 4–6 can be achieved and repeated through
several cycles. The difference between these values and the larger ratios used in
the code is the subject of some controversy, particularly since the 1985 Mexico
City earthquake.[9]

A structure designed in NYC by conventional code-specified seismic design
methods would be sized for lateral loads of roughly one-fifth those expected in an
earthquake of a magnitude equal to the largest yet experienced. In the extreme
case of a magnitude 6.5 event the structure would be subjected to inertial loads in
excess of 30 times those for which it had been designed, well beyond the effective
capacity of the structure to dissipate energy. It should be clear from this that the
nature of "moderate" seismicity suggests an even greater need for structural
"ductility" or energy-dissipating capacity than is the case in the areas of high
seismicity.

PARAMETRIC STUDIES

Direct comparisons of wind and seismic demands on building structures is
complicated by the essential difference in their action. Whereas wind pressures
act on the exposed surface area of a building, the effect of earthquake strong
ground motions is to accelerate the masses of the structure in increasing propor-

tion to their height above ground. For example, buildings with long, narrow plan dimensions, as are common in NYC, are designed for relatively small wind loads parallel to the long side. In many cases there is no direct lateral-load-resisting system provided other than the façade construction. Since the earthquake effects are essentially independent of plan configuration, such buildings would be potentially vulnerable in an earthquake.

As a means of scaling in general terms the relative magnitude and parameter sensitivity of the wind and seismic demands on structures in NYC, a simple equation has been developed for the unit seismic base shear as a function of the building geometry and density. The unit base shear is the total horizontal load acting on the structure divided by the cross-width of the building. The relationship has been derived on the basis of the approximate equations contained in the 1988 Uniform Building Code (UBC 88) for the building period and the proposed seismic zone factor of 0.15.[1] Referring to FIGURE 1, consider a simple building of height H and base dimensions B (width) and αB (depth). The wind or seismic effects are considered to act perpendicular to the width B. A height-to-width ratio β for the structure is defined as

$$\beta = \frac{H}{\alpha B}$$

The building weight can be calculated as

$$W = \rho(\alpha B^2 H)$$

where ρ is the average density of the building. The building weight per unit width is then given by

$$\frac{W}{B} = \rho\,\frac{H^2}{\beta}$$

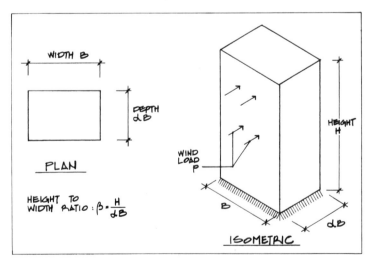

FIGURE 1.

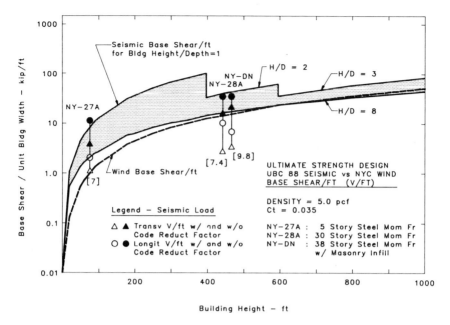

FIGURE 2. UBC 88 seismic versus NYC wind unit base shears (5 pcf).

The UBC 88 equation for the seismic base shear is

$$V = \frac{ZIC}{R_w} W$$

where R_w is the reduction factor relating to the structure's energy-dissipating capacity and, for NYC, Z = seismic zone factor = 0.15 (Zone 2A); I = importance factor = 1.0; C = response factor = $\frac{1.25S}{T^{2/3}}$; and S = soils factor = 1.0.

The structural period T is calculated using the approximate formula

$$T = C_t H^{3/4}$$

where C_t is an empirical factor depending on the type of structure and varying between 0.020 and 0.035.

From these the unit seismic base shear for NYC can be expressed as

$$V_s = 0.1875 \frac{\rho H^{1.5}}{C_t^{2/3} \beta}$$

The equivalent expression for the unit base wind shear is

$$V_w = p(Z)H$$

allowing for the variation of pressure p with height z.

The variation in unit base shear is plotted (semi-log) in FIGURES 2 through 4 for building height-to-width ratios β of 1, 2, 3, 5, and 8. The curves are cut off at

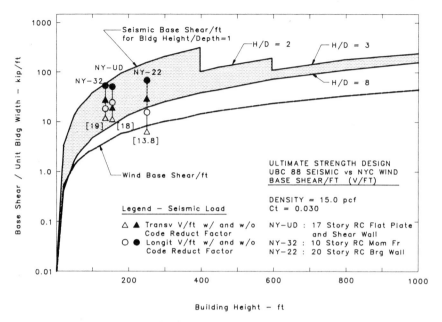

FIGURE 3. UBC 88 seismic versus NYC wind unit base shears (15 pcf).

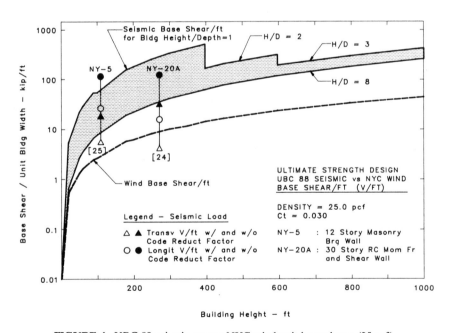

FIGURE 4. UBC 88 seismic versus NYC wind unit base shears (25 pcf).

heights considered excessive for a given β ratio. The curve of unit wind shears is shown, as are bars indicating prototype buildings reviewed in a trial design program conducted to assess the proposed ATC-3 Tentative Provisions for the Development of Seismic Regulations for Buildings.[10,11] The average densities for the trial design buildings are shown in brackets below the bars. As noted on the figures, values are shown for unit base shears with and without the code reduction factors that account for the structure's energy-dissipating capacity. The solid circle and triangle represent the demand on the structure, across both its axes, were it to remain elastic. Among the structures considered are

- 5-, 30- and 38-story steel moment frames with an average density of 8.5 pounds per cubic foot (pcf);
- 10-, 17- and 20-story reinforced concrete (RC) shear wall and frame structures with an average density of 16.5 pcf; and
- a 12-story brick masonry bearing wall structure and a 30-story RC moment frame and shear wall structures with an average density of 24 pcf.

The plots have been grouped into densities of 5, 15 and 25 pcf as representative of the range of NYC construction.

It is clear from the figures that the elastic seismic demand (for $R_w = 1.0$) on buildings calculated by the 88 UBC will in all cases exceed the design unit wind base shears.

THE INFLUENCE OF THE EARTHQUAKE FREQUENCY CONTENT

Studies of the shape of response spectra characteristic of earthquakes of various magnitudes have demonstrated that at periods greater than 1 second there is a substantial decrease in spectral accelerations with decreasing magnitude.[5,12] FIGURE 5, developed from Ref. 12 illustrates the trend. A comparison of response spectra applicable to NYC from various sources with one derived specifically for the city (from Ref. 6, and shown as NYC and NYC-mod, representing alternative proposals) reflects the implications of this trend. The normalized versions of these spectra shown on FIGURE 7 further demonstrate the basic difference. In FIGURES 8 through 9 the curves for unit seismic base shears have been modified to reflect the influence of these proposed NYC-specific spectra on the wind/seismic comparison. FIGURE 10 further reduces, for the very light buildings of a 5 pcf density, the seismic unit base shear by allowing for a ductility factor of 2. It is apparent that the shape of the response spectrum will have a substantial impact on the demand on structures over 200 feet in height.

CONCLUSIONS

Structures designed to modern seismic codes are provided with sufficient strength and stiffness to withstand frequent earthquakes with minor damage and sufficient toughness and energy-dissipating capacity to survive major events without life-threatening damage or collapse. Careful design and detailing of the structure's connections are essential to ensure their ability to withstand large inelastic cycles of deformation in excess of 4 to even 6 times the yield displacement. The number of structural system types with demonstrated energy-dissipating capacity is limited as well, so that consideration of building configuration and overall

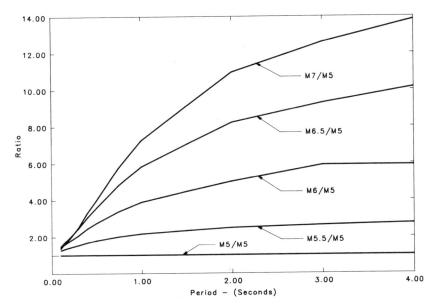

FIGURE 5. Spectral ordinate ratios normalized to Magnitude 5. (After Joyner and Boore.[12])

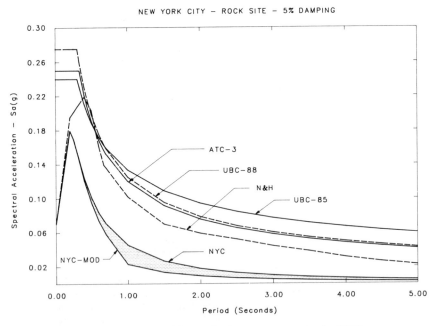

FIGURE 6. Proposed linear elastic response spectra for NYC.

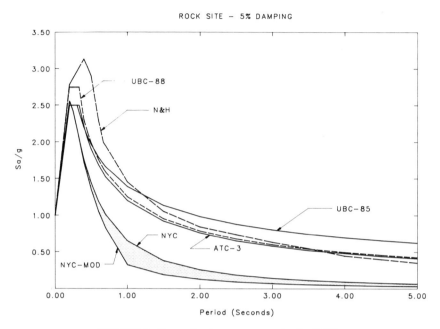

FIGURE 7. Normalized response spectra for NYC.

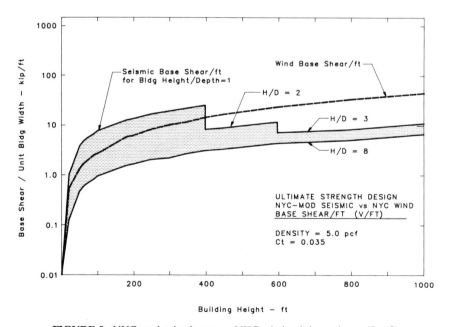

FIGURE 8. NYC-mod seismic versus NYC wind unit base shears (5 pcf).

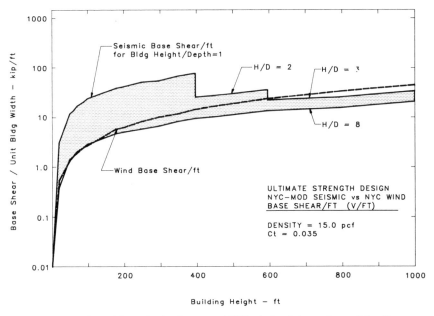

FIGURE 9. NYC-mod seismic versus NYC wind unit base shears (15 pcf).

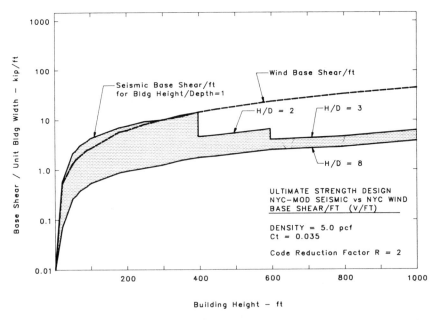

FIGURE 10. NYC-mod seismic versus NYC wind unit base shears (5 pcf and R = 2).

structural strategy are fundamental to effective seismic design. The design of tall structures for wind effects is at times a complex task, especially where aerodynamic effects are important. In general, however, wind-resistant design is quite straightforward and allows for considerable diversity in design strategies. The principal issues raised in this paper can be summarized as follows:

- The ratio of values representing the effects of a "maximum credible event" to those on which the building code is based is on the order of 1.5 to 2 for wind along the eastern seaboard and for earthquakes in the western United States. The equivalent value (or "marginal hazard") is closer to 5 to 7 for earthquakes in the East. This reflects the rare occurrence of large earthquakes in the East, particularly in relation to the return period on which the currently proposed seismic codes are based. This suggests that structures designed for earthquake resistance in the East should be provided with even greater energy-dissipation than is the practice in the West (if this is possible) or that the average return period chosen as the basis for the code-prescribed loads should be increased to reduce the magnitude of the marginal hazard.
- A comparison of unit base shears calculated by the code procedures for wind and seismic design demonstrates that for a representative range of building height, slenderness, and density the elastic seismic demands (allowing no load reduction for "ductility") in all cases exceeds the prescribed wind loading.
- If the probable frequency content for the type of moderate earthquake used as the design basis event for NYC is incorporated in the comparison (e.g., by modifying the response spectrum shape) the relative seismic demands are substantially reduced. This is especially true for structures over 200 feet in height.
- Clearly a further reduction in the difference is possible if even a nominal degree of energy-dissipating capacity is allowed for.

It should be evident that it is impossible to establish a direct relationship between code-prescribed wind forces and earthquake resistance. Both the physical nature of the phenomena and the design philosophy underlying the respective code procedures are fundamentally different.

Appropriate seismic design provisions are needed for adoption in New York City.

ACKNOWLEDGMENT

Mr. Tian Fang Jing played a central role in all aspects of the work.

REFERENCES

1. UNIFORM BUILDING CODE. 1988. International Conference of Building Officials. Whittier, CA.
2. BSSC (BUILDING SEISMIC SAFETY COUNCIL). 1985. Recommended Provisions for the Development of Seismic Regulations for New Buildings. Washington, D.C.
3. SIMIU, E. & R. H. SCANLAN. 1988. Wind Effects on Structures, 2nd ed. Wiley, New York, NY.
4. AMERICAN NATIONAL STANDARD INSTITUTE. 1982. ANSI Standard Building Code Requirements for Minimum Design Loads in Buildings and Other Structures, ANSI A58.1-1982. New York, NY.

5. IDRISS, I. M. 1985. Evaluating seismic risk in engineering practice. *In* Proceedings of the Eleventh International Conference on Soil, Mechanics and Foundation Engineering.: 255–320. A. A. Balkema. Accord, MA.

6. STATTON, C. T., R. C. QUITTMEYER & I. M. IDRISS. 1987. Probabilistic Seismic Hazard Evaluation for George Washington Bridge, New York–New Jersey. Woodward-Clyde Consultants. Wayne, NJ.

7. NUCLEAR REGULATORY COMMISSION. Division of engineering geosciences plan to address USGS clarification relating to seismic design earthquakes in the eastern seaboard of the United States. Memorandum from Richard H. Vollmer to Harold R. Denton, March 2, 1983.

8. SEAOC (STRUCTURAL ENGINEERS ASSOCIATION OF CALIFORNIA). 1988. Tentative Lateral Force Requirements. Seismology Committee, Sacramento/San Francisco/Los Angeles, CA.

9. BERTERO, V. V. 1987. The Mexico Earthquake of September 19, 1985: Performance of building structures. Presented at seminar of ASCE Met Section.

10. WEIDLINGER ASSOCIATES. 1984. BSSC Trial Design Program. National Institute of Building Sciences. ATC-3-06 Trial Design Program. New York, NY.

11. ROBERTSON FOWLER AND ASSOCIATES. 1984. BSSC Trial Design Program. National Institute of Building Sciences. ATC-3-06 Trial Design Program. New York, NY.

12. JOYNER, W. B. & D. M. BOORE. 1982. Prediction of earthquake response spectra. Presented at the 51st Annual Convention of the Structural Engineers Association of California. Sacramento, CA.

13. APPLIED TECHNOLOGY COUNCIL. 1978. Tentative Provisions for the Development of Seismic Regulations for Buildings, ATC 3-06, NBS-SP 510, NSF Pub. 78-8. Palo Alto, CA.

Concrete Structures

JACK P. MOEHLE

Department of Civil Engineering
College of Engineering
University of California, Berkeley
Berkeley, California 94720

INTRODUCTION

Central to the successful design of earthquake-resistant reinforced concrete structures is the adequate provision of three characteristics: stiffness, strength, and toughness. Awareness and judicious implementation by the structural engineer of these three first-order concepts will ensure a high degree of reliability and economy in design of earthquake-safe structures in the eastern United States. It is the objective of this paper to discuss these concepts and to suggest how they may be successfully implemented in design of reinforced concrete structures in the eastern United States.

STIFFNESS

In conventional practice, structures are designed to have strength below the force level that would develop if the structure were to remain elastic. The survivability of such a structure may well rely on capacity to deform beyond the elastic range. Although this current practice implicitly anticipates damage during the design earthquake,[1] the occurrence of damage after an earthquake is not so easily accepted. As noted by Sozen,[2] the most relevant performance criterion for a building structure that has survived an earthquake is the total cost of damage. Given this perspective, clearly attention to damage control should be a central concept in the design of structures to resist earthquakes.

Damage in buildings can often be related directly to the level of distortion occurring during the earthquake.[4] In the language of the earthquake engineer, this distortion is usually measured in terms of interstory drift, that is, the lateral displacement of one floor relative to an adjacent floor. Interstory drift is usually expressed as a fraction or percentage of the interstory height.

Drift can cause damage either to the building structure itself or to the building contents. The former is of importance because structural damage reduces the load-carrying capacity of the structure. The latter is of importance because damaged nonstructural elements may be a threat to life safety if not properly anchored and because the nonstructural content typically accounts for a significant majority of the total value of the building.

Interstory drift on the order of $0.005H$, where H is the interstory height, is known to result in damage to building partitions[3] if they are rigidly attached to the structure. Larger distortions can be accommodated if appropriate connections are made between structure and partition. Drifts on the order of $0.01H$ have been observed to result in minor damage to the building structure, and more severe damage to the building content. Drifts approaching $0.02H$ are generally associated with severe damage in the structure, and may require expensive repair or demolition.

Current building codes[1,4] seemingly treat the problem of drift control as a matter of secondary importance. Drifts are computed for a set of static loads that are lower than those likely to occur during the design earthquake and using a structural model that is stiffer than is likely. The result is a set of computed drifts well below those likely to occur.

To compensate partially for the underestimated drifts, code-computed drift is limited to the relatively low value of $0.005H$. Higher values are apparently anticipated by the code writers. For example, the codes require that vertical-load-carrying capacity of elements should be verified for lateral drifts equal to the code-computed drift factored by $3/K$, where K is factor ranging from 0.67 (for frames) to 1.0. Thus, drifts ranging from 3.0 to 4.5 times the computed drifts are contemplated.

Although the code procedure for computing drift does recognize the value of structural stiffness in controlling drift, it falls short of anticipating the magnitude of drift that will occur during the design earthquake. To demonstrate this shortcoming, an analytical study was recently conducted. In the study, reinforced concrete frame, frame-wall, and wall buildings, having heights equal to 5, 10, 15, and 20 stories, were designed according to the UBC.[4] Although designed for regions of high seismicity, the results of the study should be equally applicable to regions of lower seismicity. The floor plans had three bays in the direction of lateral loading. The frame-wall and wall buildings had areas of wall cross sections (along each axis) to floor areas equal to approximately 0.005 and 0.03, respectively. Fundamental periods, computed using gross cross-sectional member stiffnesses, were approximately N/10, N/15, and N/20 for the frame, frame-wall, and wall buildings. Lateral drifts for the gross-section models were computed according to the UBC.

Analytical responses of these generic building models to various earthquake records were computed using the computer program DRAIN-2D, as described by Kanaan and Powell.[5] Effects of inelastic flexure were considered. Effects of bond slip, inelastic shear, and foundation movements were not included. Thus, the analytical models tended toward the stiff side. Ground-motion records considered were the 1940 El Centro NS, 1985 Vina del Mar, Chile S20W, and 1985 SCT, Mexico S60E records. The El Centro and Vina records are considered representative of records on firm ground in California, although intensities may be below the maximum design intensity envisaged by the UBC. The Mexico record was measured on a deep bed of soft clay, and may not typify records in the United States.

FIGURE 1 plots the ratios between maximum drift obtained during the response analyses and drift obtained by the UBC analyses. As discussed previously, response drifts consistently exceed the UBC drift. However, ratios between response and UBC drift, instead of being limited to values between 3 and 4.5 (which, as discussed previously, is implied by the UBC), range between 4 and 25. The higher ratios obtained for the low-rise structures may not be relevant because drifts for these structures are often below the damaging threshold. Likewise, those obtained from the Mexico City record may not be relevant because of the unusual nature of the motion. Nevertheless, the ratios as high as 15 computed for buildings of ten stories and more are still noteworthy.

The possibility of lateral drifts on the order of ten times the code-computed drift is a sobering thought. If drifts equal to $0.005H$ are permitted in the code

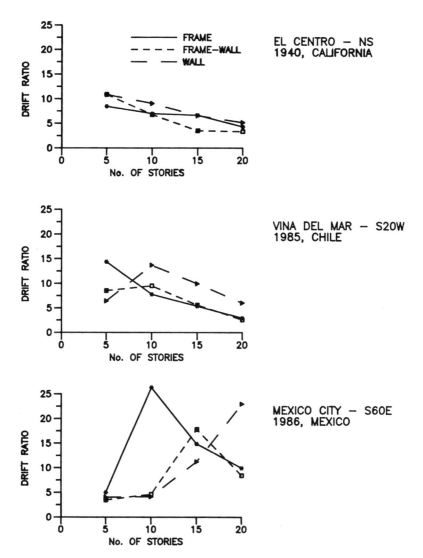

FIGURE 1. Ratios between drifts from dynamic response analysis and the UBC code procedure.

analysis, actual displacements equal to $0.05H$ can be visualized. Clearly, such large drifts cannot be tolerated from structural as well as nonstructural perspectives. The clear message is that the engineer should strive in earthquake-resistant design to obtain a relatively stiff structural system rather than one that just barely satisfies the code drift limits.

STRENGTH

As stated by Cross in an early paper on design of reinforced concrete structures, strength is essential, but otherwise unimportant. Given current seismic design philosophies wherein strength is traded for redundancy and ductility, this prescient view of the relevance of strength stands true. A minimum strength must be provided to ensure that corresponding ductility demands will not exceed available member ductilities. An excessively high strength, while certainly acceptable, may be unattainable economically. Any strength in between can be acceptable, provided attention is paid to the issues of stiffness and ductility.

The earthquake design base shear for a building in the eastern United States is approximately one-quarter of that specified for the highest risk areas in the western United States. A strong earthquake in the eastern U.S. may well impose motions on that structure that are approximately one-quarter of the intensity anticipated in the western U.S. Given the similarities in the ratios between earthquake intensity and required building strength in eastern and western regions, it is reasonable to conclude that for similar performances the amount of ductility required in an eastern U.S. building may be similar to that required in a western U.S. building. Of course, this ignores the complicating factors that arise because of the differences in return periods in the two regions. Although further research is clearly needed on this subject, it will be sufficient to conclude at this point that a structure in the eastern U.S. must possess a reasonable degree of toughness in order to survive an earthquake. The issue of toughness aside, the magnitude of lateral load strength can be considered, as Cross stated, unimportant.

Probably of greater importance than magnitude of strength is the distribution of strength. Buildings having irregular distributions of strength in plan are generally subject to torsional actions. Torsion leads to increased ductility demands in portions of the structure.[6] Buildings with nonsymmetric plan distributions of lateral-resisting elements should be avoided (FIG. 2).

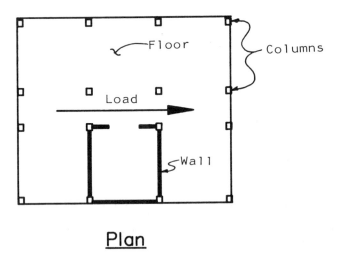

FIGURE 2. Structure having nonsymmetric strength distribution in plan.

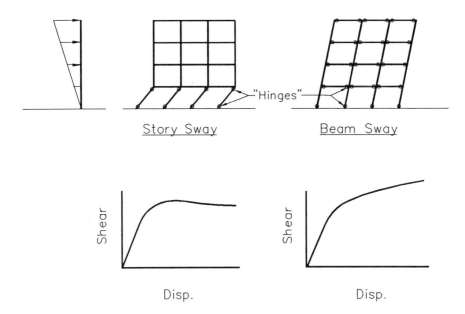

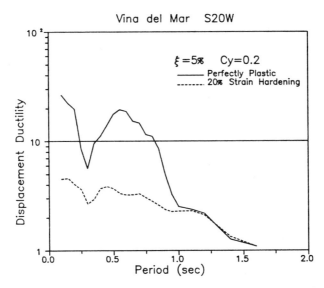

FIGURE 3. Effect of lateral-resistance mechanism on ductility demands. (*Top*) soft-story and distributed mechanisms; (*middle*) elastoplastic and strain-hardening load-deformation relations; and (*bottom*) displacement ductility demands for elastoplastic and strain-hardening systems.

Likewise, vertical strength discontinuities may result in increased ductility demands in the stories adjacent to the discontinuity.[7] Current codes emphasize the importance of providing a continuous path for seismic loads from upper stories of a building to the foundation.[1,8] It is important to note that current detailing recommendations are intended to be sufficient for reasonably regular buildings. They may be inadequate to satisfy the relatively large local ductilities required in irregular buildings.

Soft stories can be a particularly vexing problem in building design. Because of architectural requirements they may be difficult to avoid, especially in the ground floor, yet they are even more difficult to design with success. If a building should develop a soft-story mechanism (FIG. 3, *top*), two problems arise. First, a disproportional amount of the lateral building drift will be concentrated in the soft story. Second, because few elements contribute to the strength of the soft story, an effectively elastoplastic load-deformation relation may result for the entire building (FIG. 3, *middle*). Analytical studies suggest that ductility demands can be significantly higher for systems having elastoplastic response functions than for those that develop some degree of strain-hardening after yield (FIG. 3, *bottom*). In

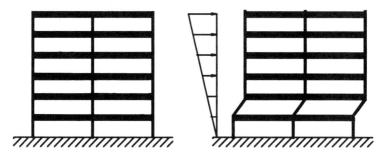

FIGURE 4. Story sway mechanism for structural system having strong girders and weak columns.

general, buildings that have a more continuous distribution of strength over height, and, in particular, dual systems such as frame-wall structures, exhibit a more gradual yield in the load-displacement relation. Such buildings are likely to have significantly lower ductility demands than buildings with discontinuous strength distributions.

Buildings having relatively long girder spans (FIG. 4) are likely to have beam strengths exceeding column strengths. Inelastic response of such a structure is likely to result in more damage to the relatively weak columns than to the girders, resulting possibly in very demanding story-sway mechanisms (FIG. 4). The possibility of such failures can be reduced significantly by providing column flexural strengths exceeding beam flexural strengths. If this is not possible, provision of shear walls over the height of the building are recommended to prevent formation of the story-sway mechanism.

The experience with the 1985 Chile earthquake[9] supports a commonly held view that strength should be distributed among several elements in the structural system. In the Chilean practice, inspection and ductile detailing have been relaxed in favor of stiffness, redundancy, and continuity in the lateral-load-resisting system. In the typical Chilean building, lateral resistance is provided by several structural walls that usually are well distributed in plan and elevation. Given the redundancy and continuity of this structural system, localized construction or detailing errors do not surface to cause extensive damage in the entire system. The exceptionally good performance of the majority of buildings during the 1985 Chile earthquake shows this to be an effective form of reinforced concrete construction. In eastern U.S. construction, attention to the continuity and redundancy in the structural system may also obviate the need for excessive and expensive ductile detailing.

TOUGHNESS

Toughness in reinforced concrete structures is achieved by providing continuity, redundancy, and strength in the structural system, and ductility in the structural elements. Given the longer return periods for intense shaking in the eastern U.S. as compared with the western U.S., it would not be economical to require that eastern structures possess toughness equivalent to that in western structures. Current design recommendations[8] stipulated for structural details in regions of low and moderate seismicity are less stringent than those required in regions of high seismicity. The less stringent detailing requirements are likely to result in structures having satisfactory performance if, in addition, a continuous and redundant structural system is provided.

Suggested frame details for buildings in nonseismic and moderate seismic risk regions are shown in FIGURE 5. Two major differences in the suggested details are evident. The first pertains to provision of continuous longitudinal reinforcement; the second pertains to provision of transverse reinforcement. The details are discussed in the following paragraphs.

In frames that do not resist earthquake motions, the direction and magnitude of design loads are fairly well established. The relatively predictable nature of moment patterns that develop from gravity loads enable use of relatively simple patterns of longitudinal reinforcement. Top longitudinal reinforcement is not generally required along the midspan of beams in such frames, because moment is predictably positive. Similarly, bottom reinforcement is required for flexure only at the midspan, and is not likely to be well anchored within the supports.

In frames resisting earthquake and gravity loads, magnitudes and directions of loads cannot be well established. Given the uncertainty of load effects, more continuity of longitudinal reinforcement is required. At supports, some continuous, well-anchored bottom reinforcement is required to cover the possibility of moment reversals due to severe lateral loading. Some of the top reinforcement should continue across the entire span to enhance continuity of the span.

Requirements for column longitudinal reinforcement are similar in both seismic and nonseismic designs. However, the possibility of tension in the reinforcement due to applied moments should be considered when designing reinforcement splices. Although splices are better placed away from the column ends (because of concentrations of action at the ends), recent research[10] has shown splices at column ends to function adequately if confined by column transverse reinforcement. Also, it is advisable to proportion the structural system to avoid excessive yielding in columns.

In design of frames to resist nonseismic loads, transverse reinforcement is provided to ensure that shear strengths exceed the shears that develop because of

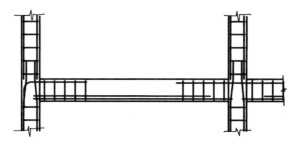

Nonseismic

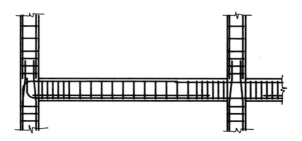

Moderate Seismic

FIGURE 5. Recommended frame details in nonseismic (*top*) and moderately seismic (*bottom*) regions.

the applied loads. In seismic design, because the design earthquake may induce yield in the structural system, it is common practice to provide shear strength so that flexural yielding at member ends can occur under lateral loads without shear failure. As illustrated in FIGURE 6,[8] a frame member is assumed to develop the flexural strength at both ends, and gravity loads are simultaneously applied to beam spans. Required shear strengths are computed for this load condition using conventional statics. Transverse reinforcement should be provided in both beams and columns to resist the computed shears.

Considering the uncertainty in direction and magnitude of seismic loads, it is generally accepted practice that transverse reinforcement should be provided throughout the length of all beam and column spans, regardless of the magnitude of computed shears. Although not shown in FIGURE 5, some codes[8] recommend placement of transverse reinforcement within the beam-column joint.

In design of reinforced concrete structures in the western United States, great import is placed on achieving member ductility by confining concrete through placement of a tight net of longitudinal and transverse reinforcement. This reinforcement is complicated (requiring extensive inspection) and expensive. It is not likely to be cost-effective in the eastern United States. If a structural system is selected that precludes the possibility of local failure modes such as shown in FIGURES 3 and 4, the requirements for member ductility, and hence for confinement, are greatly reduced.

Slab-Column Construction

Slab-column construction is widely used in the eastern United States. Behavior of this system under gravity loads is fairly well understood, and many years of service of millions of square feet of this floor system have proven its reliability and economy. Relatively little is known about its lateral-load response. Several recent studies shed light on some important aspects of the flat plate under lateral loads.

Lateral stiffness of slab-column frames is low in comparison with conventional beam-column frames because of the relative thinness of the slab. For slab-column frames of typical proportions, elastic analyses show that approximately half of the slab width can be considered effective as a beam-framing between columns. However, cracking under service-load conditions results in further reduction in stiffness to approximately one-half to one-third of the elastic stiffness, that is, one-half to one-sixth of the slab width can be considered effective. The result is a quite low stiffness of the slab-column frame.

FIGURE 7 compares computed stiffnesses for beam-column and slab-column frames. The beam-column frame has 24- by 24-inch columns, with 12- by 18-inch beams. Stiffness of the beams is taken as half the gross-section stiffness to account for cracking, and no slab is assumed to act to stiffen the floor beams. The slab-column frame has the same columns with 8-inch-thick slab, with one-sixth of the slab taken as an effective width (as described above). Even though the beam-

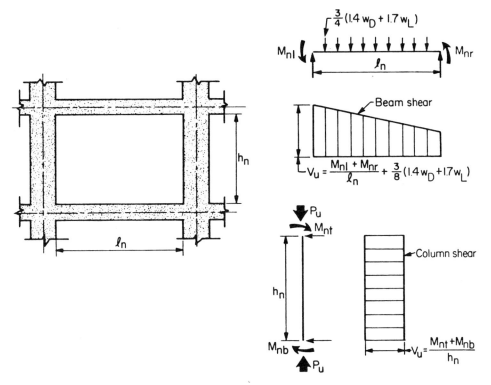

FIGURE 6. Load condition for determination of design shears.[8]

column frame has been selected and analyzed to produce a lower bound stiffness, the slab-column frame is computed to be much more flexible.

The large flexibility of the typical slab-column frame is apparent also in experimental measurements. FIGURE 8[11] is a plot of the hysteretic response of a well-proportioned and detailed multi-bay slab-column frame. Yield in the load-dis-

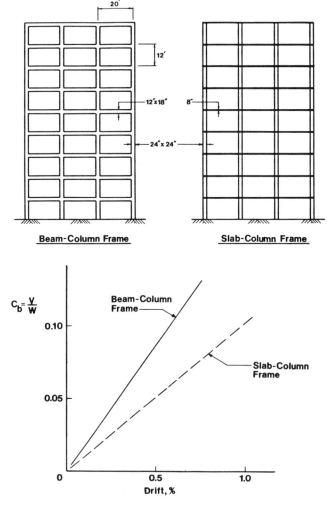

FIGURE 7. Relative stiffness of beam-column and slab-column frames.

placement response is not apparent until lateral drift reaches approximately 1 percent of interstory height. The relatively large drift at yield has two important implications: (1) If drifts during the design earthquake are to be controlled to values not to exceed between one and two percent, as has been suggested,[12] then only limited ductility can be utilized, and (2) design lateral forces should be

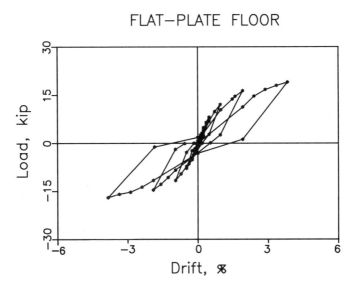

FIGURE 8. Measured load-deformation relation for well-proportioned slab-column frame 3 subjected to reversed cyclic lateral loads.

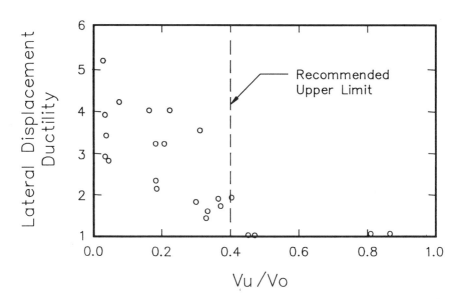

FIGURE 9. Effect of gravity load on the lateral displacement ductility of interior flat-plate connections.

established with due consideration to the limited ductility. The large reductions in design forces that are embodied in current codes,[4] and that are partially attributable to inelastic behavior, are not appropriate for structures in which the flat plate frame is the primary lateral-load-resisting system.

Given the relative flexibility of the flat plate, design for seismic forces may require use of shear walls or conventional frames for added stiffness. Even if such systems are used (except in cases where extremely stiff systems are provided), the possibility of yield in the slab-column connection should be considered. Recent data on behavior of interior slab-column connections reveals that lateral-displacement ductility is directly related to the level of gravity load carried on the slab (FIG. 9).[11] (In FIGURE 9 the quantity V_u/V_o is the ratio of the gravity direct shear to the punching shear strength in the absence of moment transfer.) In order to ensure that some ductility is available, it is recommended that the shear force acting on the slab-critical section not exceed 40 percent of the pure punching shear strength.

Because of the propensity toward shear punching failures around connections, and toward progressive collapse if the punching at a connection is not contained,

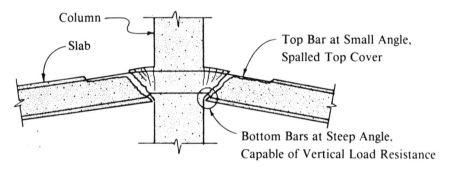

Column

Slab

Top Bar at Small Angle, Spalled Top Cover

Bottom Bars at Steep Angle, Capable of Vertical Load Resistance

FIGURE 10. The role of bottom reinforcement to arrest a punching failure.

provisions for structural integrity in slab-column connections are of great importance. As illustrated in FIGURE 10, bottom reinforcement is required over the column to suspend the slab in the event of a punching failure; top bars are inadequate to suspend the slab. Two of the main bottom bars in each direction placed within the column cage are usually adequate. Additional requirements for anchorage and continuity of reinforcement for slab-column frames in regions of moderate seismicity are provided in Ref. 8.

SUMMARY

Stiffness, strength, and toughness are identified as three issues that are central to the successful planning of any structure that may resist earthquake forces. Stiffness is predominantly a concern for control of damage during an earthquake. Strength and toughness are considered central to life safety, and are perhaps of greater concern in design in the eastern United States, where earthquakes of significant magnitude have long return periods.

Considering the longer return periods for strong shaking in the eastern United States, and the significant expense of comprehensive design for earthquake resistance, design procedures are necessary that differ from those applied in the western United States. It is recommended that a primary consideration for design in the eastern part of the country is appropriate selection of the structural system. The importance of continuity, redundancy, and symmetry cannot be overstated. Adequate provision of these three characteristics can avoid the need for excesses in ductile detailing. Nonetheless, detailing for anchorage and shear resistance are not to be overlooked. The special problems of slab-column construction are also noteworthy.

ACKNOWLEDGMENTS

I am thankful to the National Science Foundation, which has generously supported earthquake engineering research by me and others throughout the United States. The breadth of developments in this field precludes consideration and individual acknowledgement of all contributions. Nonetheless, the labors of the many are greatly appreciated. Special thanks for contributions to this paper are expressed to S.-J. Hwang, A. Pan, X. Qi, and J. Wallace, graduate research assistants at the University of California at Berkeley, and to S. A. Mahin, Professor of Civil Engineering at the University of California at Berkeley.

REFERENCES

1. 1980. Recommended Lateral Force Requirements and Commentary. Structural Engineers Association of California. San Francisco, CA.
2. Sozen, M. A. 1987. A frame of reference. Presented at the Peck Symposium, University of Illinois, Urbana, IL.
3. Freeman, S. A. 1985. Drift limits: Are they realistic? Earthquake Spectra 1: 355–362.
4. 1985. Uniform Building Code. International Conference of Building Officials, Whittier, CA.
5. Kanaan, A. E. & G. H. Powell. 1973. DRAIN-2D, a General Purpose Computer Program for Dynamic Analysis of Inelastic Plane Structures, Reports No. 73-6 and No. 73-22, Earthquake Engineering Research Center. University of California. Berkeley, CA.
6. Rosenblueth, E. & R. Meli. 1985. The 1985 Earthquake: Causes and Effects in Mexico City. Concrete International: Design and Construction 8: 14–34.
7. Aranda, G. R. 1984. Ductility demands for R/C frames irregular in elevation. In Proceedings of 8th World Conference on Earthquake Engineering. San Francisco, August 1984. 4: 559–566.
8. 1983. Building code requirements for reinforced concrete (ACI 318-83). American Concrete Institute. Detroit, MI.
9. Wyllie, L., et al. 1986. The Chile earthquake of March 3, 1985. Earthquake Spectra 2: .
10. Suvikumar, B., et al. 1983. Suggestions for the design of R/C/ lapped splices for seismic loading. In Concrete International. 5: 46–50. American Concrete Institute.
11. Pan, A. & J. P. Moehle. Ductility of R/C flat plates under lateral loading. SEMM Report, Department of Civil Engineering, University of California, Berkeley, CA, July 1987. Also accepted for publication in JACI.
12. Sozen, M. A. 1981. Review of earthquake response of R. C. buildings with a view to drift control. State-of-the-Art in Earthquake Engineering. In Proceedings of the 7th World Conference on Earthquake Engineering: 383–418. Turkish National Committee on Earthquake Engineering. Istanbul.

Seismic Design of Masonry

V. C. FENTON[a]

Adjeleian Allen Rubeli
Ottawa, Ontario KIP 5E7, Canada

INTRODUCTION

Conventional masonry consists of a stack of heavy, brittle stone, concrete or ceramic units insecurely interconnected by cementitious mortar having uncertain bond and tensile strength characteristics. This type of construction is especially hazardous in an earthquake zone because it simply falls apart if shaken or severely distorted and comes down, causing death and destruction to anything in its path. We have seen many examples of this catastrophic performance in the Middle East and Central America, where plain masonry construction is prevalent. These buildings have performed so badly under seismic loading because they have not been designed to resist the loads and distortions without overstress and do not have sufficient ductility to survive the overstress without losing structural integrity, particularly when subjected to cycles of reversing loads.

SEISMIC DESIGN OF MASONRY

The problem of designing masonry to resist seismic loading has a two-part solution:

1. *Reinforcement* can be provided in the joints and cores to improve the tensile and shear strengths to acceptable levels of reliability.

2. *Boundary conditions* can be provided for this modified material that will be compatible with its strengths and limitations.

REINFORCEMENT

Masonry can be reinforced by placing bonded reinforcement horizontally in the mortar joints between courses and vertically in grouted cores or vertical joints (FIG. 1). This reinforcement clamps the joints, mobilizing shear-friction by providing large forces normal to the joints. These forces prevent one side of the joint from riding up over the mechanical interlocking of the mortar into the roughnesses and edges of the other side of the joint (FIG. 2). The same clamping forces

[a] Address for correspondence: Adjeleian Allen Rubeli, Consulting Structural Engineers, 75 Albert Street, Suite 1005, Ottawa, Ontario KIP 5E7, Canada.

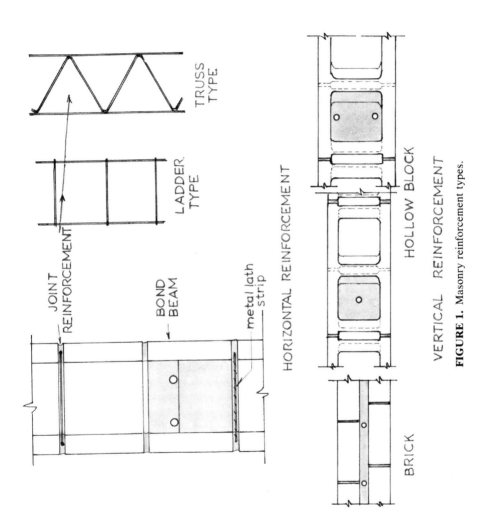

FIGURE 1. Masonry reinforcement types.

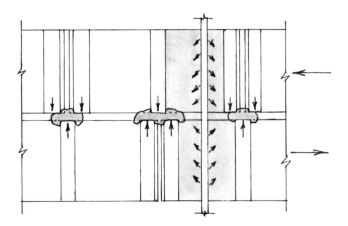

FIGURE 2. Reinforcement joint clamping action.

also provide predictable flexural resistance by eliminating the need to rely on mortar bond or tensile strength to resist flexural tension (FIG. 3).

Vertical reinforcement (FIG. 4) can be placed into the vertical cores as short pieces lapped to other short pieces above and below in a block core or vertical space between wythes. The bars are grouted or embedded in mortar as the wall is built. This permits the mason to fill the reinforced cores as he goes and limits the

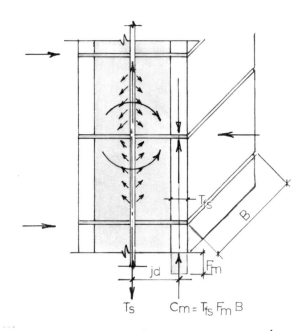

FIGURE 3. Reinforcement flexural resistance.

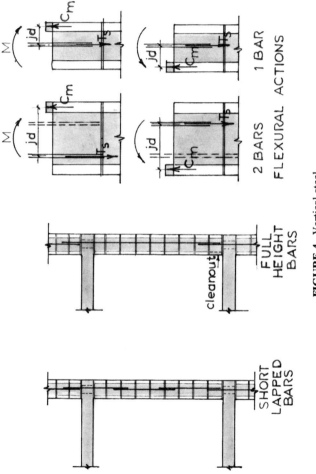

FIGURE 4. Vertical steel.

height of bar that he must lift the blocks over when he places them in the wall. This method has two major flaws:

1. The mason can use the same short piece of steel for the full height of wall, raising it a little when nobody is looking, and nobody is the wiser until the wall is called upon to resist a heavy wind or earthquake.

2. The bars are not usually well aligned with those above and below, the lap lengths are not easily controlled, and the grout or mortar may not be well compacted around the bars. A survey of tornado damage at London, Ontario revealed very poor performance from several walls that had been reinforced in this manner. They had simply fallen apart at lap points which in some cases had the lapped bars in opposite sides of the grouted cores.

A more reliable method is to construct the full storey lift of wall leaving a "clean-out" port at the bottom of each reinforced core so that the mortar droppings can be removed and the bar placement inspected before the "clean-out" port is covered with a board and the core filled with grout. The mason must be more careful with this method. The grouted cores must be sealed by mortar on the webs at either end to contain the grout, the mortar projecting into the cores must be trimmed to maintain working clearance for placing the steel and grout, and the droppings accumulated in the bottom of the core must be cleaned out through the clean-out port. This method may appear to be more costly, but will usually be found to be efficient and economical if properly organized to provide a labor crew to place the steel and grout, freeing the mason from these tasks and permitting him to concentrate on the block-laying.

Walls less than 10 inches thick are usually reinforced with one central bar. Walls 10 inches and thicker can be reinforced with two bars. This increases the flexural resistance of the masonry compression zone by increasing the effective depth of the moment couple. Note, however, that only one bar is effective in each direction since it is usually impractical to tie the bars as required to make them act reliably in compression.

Horizontal reinforcement can be placed in the mortar joints or in bond beams (FIG. 1). Joint reinforcement must be constructed of special welded wire mesh fabric since it must be accommodated in a 3/8-inch mortar joint without disrupting the work or lifting the unit as it is bedded into the mortar. The type of masonry joint reinforcement usually preferred by masons is 2- or 4-wire 9-gauge ladder or truss type. This may be used with full-height vertical reinforcing since it is almost entirely contained on the face shells, leaving the cores clear except for the occasional web wire, which causes little interference. Bond beam reinforcement, on the other hand, complicates full-height vertical reinforcement because it runs through the same core spaces as the vertical bars and the grouting of the horizontal bond beam must be interrupted at each grouted core if the full-height method is to be used. As you may have concluded from all this, joint reinforcement and full-height vertical reinforcement are considered to be the most practical methods of reinforcing masonry in most Canadian design practice.

BOUNDARY CONDITIONS

The design of the boundary conditions also breaks down into two primary considerations:

1. The first is the obvious requirement that the connections or anchors between a masonry element and its supports must have adequate strength to resist

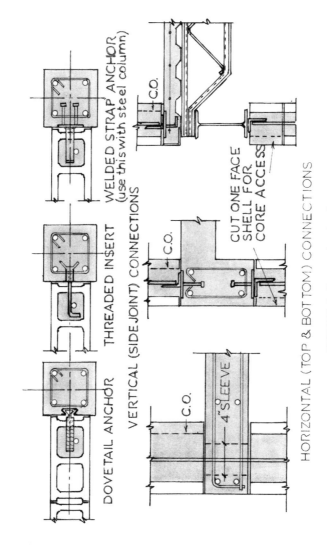

FIGURE 5. Connection details.

the inertial forces applied as the ground motion accelerates the building, which, in turn, accelerates the masonry element by exerting connection forces on it. There should be toughness and ductility in the connections if we are to have confidence in the structural integrity of the assembly. The best methods we have seen are illustrated in FIGURE 5.

Note that it is not easy to get vertical reinforcement into grouted cores in masonry built into most structural systems that are constructed ahead of the masonry. An exception to this is the flat plate, which can accommodate large sleeves for placing reinforcement and grout. In areas with low seismicity, the most economical combination is horizontal reinforcement, either in the joints or in bond beams, and vertical lateral supports at each end of a panel. In areas having high seismicity it is generally considered good design practice and is required by most of the codes to have steel in both directions and supports all around.

2. The second consideration is the less obvious requirement that we must protect the masonry from large distortions because, even though we have managed to obtain some reliable tensile and flexural strength by introducing reinforcement, the ductility or ability to resist reversing cycles of overstress is very low. The confining ties and other details required to produce ductility in reinforced concrete are usually impractical in masonry construction. So we might just as well accept this fact and design our buildings in such a way that the masonry elements are not overstressed during an earthquake. This can be accomplished in low buildings by designing masonry shearwalls with plenty of overstrength and redundancy or by providing strategically located concrete shearwalls to protect the masonry. Note that masonry panels are very stiff and that any protective element must be at least as stiff or the masonry walls will become "accidental shearwalls," which will be overstressed and damaged before the protective elements can resist the loads.

In high buildings the same concept applies. If the primary lateral resistance system is a frame, the storey drift may be substantial and any stiff masonry, finish or cladding elements may be overstressed and damaged before the frame acts to resist the loads (FIG. 6). High-rise buildings should therefore be provided with stiff lateral resistance systems that limit the storey drift and protect the masonry and nonstructural elements from damage during an earthquake. Concrete shearwalls or structural steel K-braced frames are the usual choice in our practice (FIG. 6)

BRICK VENEER

Brick veneer is designed in the same way. However it is difficult to provide vertical reinforcement unless giant hollow brick is used or the brick is laid-up stack bond and the steel is placed in the head joints (FIG. 7). It will usually be more practical to provide horizontal reinforcement at frequent intervals, say at every sixth brick course, which corresponds with every second block course, and ductile ties in the same joints, also at frequent intervals, securely anchored into the structural backing. The mortar bond and flexural strength of the masonry is called upon, in this case, to span the masonry vertically the six courses between reinforced joints. There are excellent wire masonry joint fabrics available which provide the ties and the joint reinforcement in both the backing and the veneer (FIG. 7).

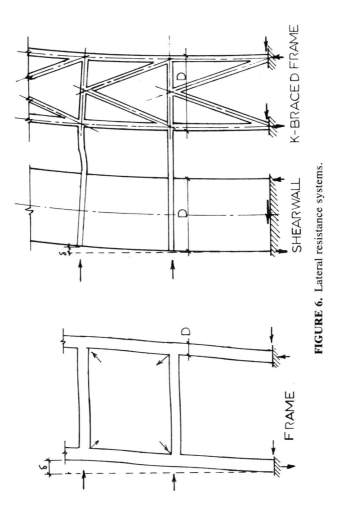

FIGURE 6. Lateral resistance systems.

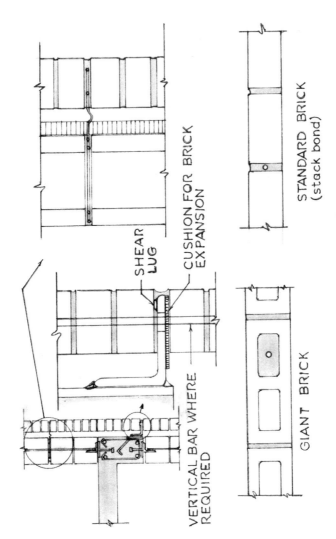

FIGURE 7. Veneer details

TABLE 1. Minimum Masonry Reinforcement Specified in Building Codes: 1965 to 1985

Date	Code	Masonry Type	Minimum Steel Pg-As/Ag and Maximum Spacing	E.Q. Zone
1965	CNBC[a]	All	0	Any
1966	DANAF[b]	Loadbearing walls (LBW)	.001 Vertical (V) @ 24	1 & 2
			.0014 V @ 24"	3 & 4
		Nonloadbearing walls (NLBW)	.0007 V @ 48"	1 & 2
			.001 V @ 32"	3 & 4
		All	.0007 Horizontal (H) @ 48"	all
1970 (& 1982)	CNBC	LBW and [d]Critical NLBW	0	0 & 1
			.002 H + V and not less than .0007 H or V @ 6t or 48"	2 & 3
1979	UBC[c]	LBW	.002 H + V and not less than .0007 H or V @ 24"	2, 3 & 4
1980 (& 1985)	CNBC	LBW	0	0 & 1
			.002a V and .002 (1 − a) H where .33 < a < .66 @ 6t or 48"	2 & 3
		[d]Critical NLBW	0	0 & 1
			.0005 H or V @ 16"	2
			.001 H or V	3

NOTE: The zones correspond to the following ratios:

Za or Zv	v(m/s) or a/g
	1965 to 1980 1985
1	.05
2	.1
3	.15
4	.2
5	.3
6	.4

[a] National Building Code of Canada.
[b] This refers to the manual *Seismic Design for Buildings*, published by the U.S. Dept. of the Army, Navy and the Air Force.
[c] U.S. Uniform Building Code.
[d] Masonry walls weighing more than 40 PSF, more than 10 feet high, or around elevator shafts or stairwells.

TABLE 2. Minimum Reinforcement Limit States Design of Masonry for Buildings

Wall Type	Velocity or Acceleration Zone						
	0	1	2	3	4	5	6
Shearwalls	C	C	A	A	A	A	A
Loadbearing walls	C	C	A	A	A	A	A
Nonloadbearing exterior and special partitions[a]	C	C	B	B	B	B	B

Reinforcement shall be continuous between lateral supports and shall be spaced at not more than 8 feet.

Reinforcement equivalent to 1–10 M bar shall be provided around each masonry panel and opening therein, lapped and anchored to develop the bars at corners and splices.

CLASS A

Minimum reinforcement:
$Av = .002$ ag.a
$Ah = .002$ Ag$(1 - a)$
where $.33 < a < .66$ and maximum spacing is 6 t or 4 feet

CLASS B

Minimum reinforcement:
Av or Ah $= .0005$Ag in zone 2 and
$.001$Ag in zones 3 to 6
and maximum spacing is 6 t or 4 feet

CLASS C

Minimum reinforcement:

Walls required to have Class C reinforcement shall be reinforced vertically or horizontally, or both, as required to resist the effects of factored loads.

[a] Partitions weighing more than 40 psf, more than 10 feet high, or around stairwells or elevator shafts.

MINIMUM REINFORCEMENT

The minimum reinforcement required to hold a masonry panel together so that it will respond reliably when it is subjected to seismic loading has been the subject of much discussion among Canadian designers and code writers. TABLE 1 reviews various codes dating back to 1965. Note that the requirements specified in the design manual published by the Departments of the Army, the Navy, and the Air Force in 1966 are, with minor variations, identical with the modern requirements of the Canadian National Building Code and the American Uniform Building Code. It looks to us in Canada as though we have all accepted the original judgment call of that pioneer Code writer without checking to see whether it was based on more than intuition. If any reader is able to recall the origin and basis of these requirements, we would be very pleased to hear about it.

In 1979 the Canadian Standards Association Technical Committee S304 requested the University of British Columbia, which has a good structures laboratory equipped with a seismic shaking table, to investigate this problem of minimum reinforcement and maximum spacing. However, after some preliminary study and preparation, the work was stopped for lack of funds and we are still wondering whether the numbers in the Code are appropriate. The next edition of the Canadian masonry standard CSA S304-LSD will have the minimum masonry reinforcement requirements specified more or less as shown in TABLE 2. The matrix approach has been used here in an attempt to simplify the selection of appropriate values for the many zones in the new seismic requirements and to permit the insertion of additional classes of reinforcement requirements if thought appropriate at some future date when the aforementioned research project has been completed.

A Review of Current Practices for Seismic Design of Building Cladding

GLENN R. BELL AND THOMAS A. SCHWARTZ

Simpson Gumpertz & Heger
297 Broadway
Arlington, Massachusetts 02174

In spite of the considerable risk to life safety represented by falling building cladding during seismic shaking, relatively little research has gone into improvements in analyses and design practices for seismic design of cladding. Design forces for out-of-plane loading for cladding are based on simplistic methods that employ C_p factors that the profession generally assumes to be conservative. However, the application of these factors is subject to various interpretations, leading to inconsistent and sometimes inadequate design practices. Design forces assume various levels of ductility that may or may not be present in actual cladding components.

Adequate provision in the design for isolation of the cladding from the building frame is frequently not provided, typically in regions of low to moderate seismic activity.

This paper discusses the fundamentals of seismic design for building cladding, reviews examples of the performance of cladding in past seismic events, compares various code provisions for seismic design of various cladding types, and identifies areas of practice that require further research.

INTRODUCTION

Exterior cladding has become one of the most complex building components to properly design and construct. Cladding must be designed and detailed to provide a weather enclosure for the building in resisting air and water infiltration, while being capable of resisting seismic, wind, dead, and thermal loads, independent of the building frame. Cladding is required to accommodate thermal movements, resist moisture penetration, have adequate strength, and look pleasing.

Some of these requirements suggest contradictory solutions: through-wall flashing in a masonry wall to control water entry may introduce a line of weakness, rendering the wall structurally insufficient if proper anchors are not provided; a wide isolation joint required for seismic movements may be difficult to waterproof or be aesthetically unacceptable; a metal-and-glass curtain wall system with good thermal isolators may not have adequate strength and ductility in its connections to resist seismic loads.

Given such a complex set of design criteria, one might expect that a few materials and systems might emerge for use in suitable cladding designs. In fact, the reverse is true. Since the 1950s we have seen an explosion of innovation in cladding materials and systems. Traditionally thick, massive masonry-clad walls have been replaced with thinner, more materially efficient wall systems. While these new types of cladding have generally become lighter and are therefore subject to lower lateral loads due to seismic shaking, they have become less forgiving in performance in many other respects. Building frames have become more flexible. Today buildings are commonly clad in masonry, stone, concrete, aluminum, glass, steel, wood, stucco, and various new composites of these materials. The forms of framing systems used for cladding are far more varied than building frames. These trends demand that much more engineering effort be placed in cladding design.

Frequently, architects have relied on material suppliers for advice on exterior wall design, rather than structural engineers and curtain wall specialists. But material suppliers are in business to sell products, not technical advice. Even if they are knowledgeable about the design of their system and convey that knowledge skillfully, they are generally not qualified to evaluate building frame movement and the interaction of the cladding and building frame. Proper curtain wall design is a complex task requiring the efforts of the architect and the structural engineer or curtain wall consultant, who are capable of evaluating wind loads, seismic effects, waterproofing, and energy considerations. Some aspects of the system design, such as structural adequacy of components, and air and water infiltration may be evaluated by laboratory or field testing.

The demand for more sophisticated engineering design coupled with the complex process by which we design and build curtain walls presents many opportunities for unsatisfactory results.

This paper is concerned with assessment of current cladding seismic design practices from the point of view of a design engineer. The following topics are presented:

- fundamentals of seismic design of cladding;
- performance of cladding in past events;
- comparison of specific code provisions; and
- conclusions and recommendations

FUNDAMENTALS OF SEISMIC DESIGN OF CLADDING

Seismic Design Objectives

The primary objective of seismic design for building cladding is life safety. The most obvious threat from failed cladding is the direct falling hazard of cladding panels or components. Some seismic design criteria, such as ATC-03,[1] recognize that the consequence of the direct falling hazard depends on the likelihood of occupancy in a falling zone. Fallen wall components that block building exits, streets, or fire access lanes also represent life safety hazards. Unintentional interaction of cladding components with structural frames may also have an adverse effect on building structural performance.

Damage control and mitigation of economic loss may also be objectives of a seismic cladding design. It is a direct objective of the ATC-03 regulations to reduce damage to essential facilities, which must remain functional after a major seismic event, through adjustment of design forces for seismic hazard exposure groups. More commonly used regulations like the Uniform Building Code (UBC)[2] account for damage control by Importance Factor I. Presently, adequate data do not exist on which a definitive cost-benefit analysis could be performed to design against economic loss, and a seismic design to reduce economic loss remains a nebulous objective.

The consequences of cladding failure vary with the cladding type; contact of adjacent precast concrete wall panels during seismic activity caused by inadequate separation of adjacent panels may result in falling pieces of panel, a serious life safety hazard; the same deficiency in the design of a metal cladding may result only in some crushing of the panel edges. At present, seismic design provisions do not rationally address these differences; our ability to predict damage is better than our capacity to estimate hazards to life.

General Structural Behavior

The commonly employed structural design model assumes that cladding does not participate structurally with the primary structural system in responding to seismic movements. The principal structural design requirement for the cladding is resistance to the inertia forces generated by the acceleration of the cladding component, and accommodation of the displacements imposed on the cladding attachment points to the structural frame; generally the cladding is isolated from the structural frame by movement joints.

Even in the most carefully designed building, some interaction between the cladding and the frame occurs. Bolted joints with slotted holes develop friction resistance; elastomeric sealants in movement joints develop force and damping. Generally, the interaction of a well-designed cladding with the structural frame can have beneficial damping and stiffening effects. Reference 2 demonstrates the beneficial effect of limited-slip bolted connections for large panel concrete construction. However, incomplete isolation of cladding from the structural frame can seriously degrade seismic performance of the building by shortening its natural period, altering the distribution of story shears from that assumed in the design, introducing torsional components to the structural response or concentrated forces on structural elements for which they were not designed.[3,4]

Reference 13 documents an analytical study done for seismic response of a precast-concrete-clad building; it shows that even when provisions of the Uniform Building Code and the Prestressed Concrete Institute were followed in the design of the panel system, complete isolation was not achieved.

A small fraction of the research in earthquake engineering has been applied to the study of building cladding or the interaction of cladding with the primary structural system. A summary evaluation of state-of-the-art knowledge and practice of seismic response of cladding from the proceedings of a recent Earthquake Engineering Research Institute (EERI) workshop[5] is given in TABLE 1. Further research in interaction of cladding and building frame is essential.

The simplified concept that cladding is designed to be isolated from the building frame and is expected to bear seismic loads due only to its own inertia forces is discussed in more detail in Sections 2.3 through 2.5.

Design Loads on Cladding

Theoretical Considerations

The inertia loads induced in a properly isolated cladding panel will depend on the response of the structural frame, the panel, and the attachment of the panel to the frame, which generally have excursions into the inelastic range during severe

TABLE 1. Summary Evaluation of State of the Art Knowledge and Practice in a Seismic Design of Building Cladding: EERI Nonstructural Workshop[5]

Nonstructural Elements Architectural	Potential Interaction with Structure	Damage Source		Understanding Behavior		Reliability of Analytical Model	Availability of Design Guidelines	Design Practice	Information Needed
		Distortion	Vibration	Field Data	Laboratory Data				
Masonry infill walls	Yes	In plane	Out of plane	Much	Some	Poor	Insufficient	Poor	Full-scale assembly testing both reinforced and unreinforced
Heavy-weight claddings	Yes	In plane	Out of plane	Little	Little	Poor	Yes	Inconsistent	Full-scale component and connection testing
Light-weight claddings	Unlikely	In plane	Out of plane	Little	Little	Poor	Insufficient	Inconsistent	Full-scale connection testing
Glazing	Unlikely	In plane	Out of plane	Much—needs compilation	Little	Poor	Inadequate	Inconsistent	Field damage documentation; coherent program of in-plane testing and design guideline development

earthquakes. Typically, vertical accelerations are ignored in cladding design, and only the out-of-plane component of horizontal accelerations and interstory drift may be important.

If the cladding itself and its attachments are very rigid with respect to the building frame, which is often the case, the inertia loads are practically determined by the accelerations of the cladding attachment points to the steel frame. It is important to understand that the inertia loads on the cladding element are a function of the absolute acceleration of the element and not the acceleration relative to ground motion.

Rigorous analysis of loads on nonstructural elements are performed for nuclear power plants, but generally not for buildings. Building codes employ simplified relationships as described below.

Simplified Code Relationships for Forces on Nonstructural Elements

The general simplified code relationship for forces on nonstructural elements is:

$$F_p = AC_p W_p P M_e M_x$$

where

F_p = design horizontal load on the nonstructural element (part);
A = spectral acceleration associated with the structural fundamental period;
C_p = element seismic coefficient;
W_p = weight of the element;
P = performance factor;
M_e = attachment magnification factor; and
M_x = height magnification factor.

The acceleration magnification effects due to attachment response or vertical location on the building cladding in most code provisions are assumed to be relatively unimportant, so that the factors M_e and M_x are taken as 1.0. The spectral acceleration is simplified to be a rough function of only the expected level of ground-shaking, and does not directly include consideration of the building's response. The performance factor, P, allows the design loads to be augmented to achieve a relative performance characteristic level, which recognizes the effects of seismic hazard and the consequences of failure. The element seismic coefficient, C_p, varies with the type of element. It is chosen on the basis of engineering judgement and prior experience with the type of element in past events, rather than theoretical analysis.[6]

This procedure is a gross simplification of actual loads on cladding. The resulting loads are quite conservative; generous conservatism for the sake of simplicity is justified in light of the relative economy for which out-of-plane seismic resistance of cladding can be achieved. Because C_p factors are based on semiempirical rather than rational analysis, problems of interpretations of code provisions arise when a novel wall system or detail is encountered that was not envisioned by the code developers. Another serious shortcoming of this procedure is that the prior performance data on which the C_p factors are based on do not distinguish between performance of elements properly designed and constructed for seismic effects and those that are not.[6] While code provisions require well-anchored, ductile elements, clearly the poor performance of some cladding in past events was due to some elements that did not have good ductility or anchorage.

Isolation of Cladding from Building Frame

Code philosophy for isolation of cladding from building frames requires that joints and connections be capable of absorbing deflections under extreme seismic movements. Deflections calculated under code-equivalent lateral forces are factored to obtain realistic estimates under maximum seismic loads. In the case of the UBC Code,[2] for example, isolation must be provided for a story drift of $3(R_w/8)$ times the calculated elastic story displacement but not less than ½ in. Under this requirement a building frame with a 12'–0" story height, $R_w = 8$, and designed for the maximum drift limitation of 0.005 h_s will require a movement capability of ±2 in. Vertical frame movements under seismic motions are generally negligible, although as ATC-03 points out, special consideration must be given to vertical movements of cantilever beams.

Connection and joint design must consider other sources of movement or adjustability, which may add to the requirement for seismic movement. These include fabrication and erection tolerances, thermal and moisture movements of the cladding, short-term live load deflections, long-term deformations of the building frame under sustained dead load, and creep of the (concrete or masonry) building frame.

Designers face a tradeoff decision to achieve damage control through appropriate measures of building frame drift limitation and movement provision through isolation joints and connections.

Detailing and Ductility

A prerequisite of code cladding design is that the cladding and its attachments be designed and detailed for ductile behavior or the design forces must be scaled up to ensure elastic or limited-ductility response. This is inconsistent with empirically derived C_p factors, which are based, in part, on nonductile behavior in past events. Unfortunately, detailed performance information on building cladding to develop a fully rational design criteria is not available.

Some codes contain detailed provisions for design and connection detailing. For example, the UBC code requires that all masonry and concrete elements in seismic Zone 2 and higher be reinforced to qualify as reinforced masonry or concrete. Fasteners embedded in concrete must be attached to or hooked around reinforcing steel. These requirements do not represent the myriad of cladding materials and systems in use. Designers often face a difficult decision when new or unusual wall configurations are developed for which no detailed code provisions for detailing and ductility are specified. These considerations are discussed in more detail for specific cladding types in Section 5.

Other Design Considerations for Cladding

In addition to resisting in-plane and out-of-plane forces generated from seismic loads, without loss of structural integrity, building cladding must perform the following functions:

□ *structural performance*—transfer gravity loads (from its dead weight) and wind loads to the structure
□ *water penetration*—resist uncontrolled water entry into the wall

- □ *air infiltration*—resist excessive air infiltration or exfiltration
- □ *moisture migration*—control the flow of moisture vapor to avoid condensation within the wall
- □ *movement capability*—accommodate thermal and moisture movement without damage or distortion
- □ *thermal capacity*—resist the flow of heat or cold into or out of the building
- □ *sound transmission*—attenuate noise transmission through the wall
- □ *fire safety*—resist fire and smoke spread from one area of the building to another
- □ *aesthetics*—maintain color, reflectivity, and shape
- □ *durability*—resist deteriorating exposures, such as freeze-thaw cycles, corrosion, and ultraviolet degradation
- □ *maintenance*—provide means to preserve above-listed performance characteristics and to replace components that fail to perform properly

PERFORMANCE OF BUILDING CLADDING IN PAST SEISMIC EVENTS

Of all nonstructural building elements, the exterior cladding is the most prone to damage during seismic activity due to the common use of brittle, heavy facing materials (e.g., stone and masonry) with complex anchorages. Although catastrophic failures of building cladding have been rare, the threat to life safety, when even limited failures occur, is great. In addition, the cost of repairing noncatastrophic cladding damage can be a significant fraction of the overall building value.

Our review of the literature shows that the following types of cladding damage, resulting in fracture and/or loss of attachment, have occurred during earthquakes:

- excessive in-plane distortion leading to brittle failure;
- impact from pounding of adjacent structures;
- fracture or pull-out of attachment anchors; and
- fracture from stress concentration at points of hard contact between cladding segments or between cladding and subframes

In addition to these primary forms of damage, some breakage of brittle cladding has resulted from the impact of airborne debris, resulting from the loss of cladding at higher elevations.

These types of damage are discussed in greater detail below.

Damage from Excessive In-Plane Distortion

The most common type of cladding damage resulting from seismic motion results from the inability of cladding, spanning floor to floor, to accommodate the interstory drift of the structure. The problem is especially acute for masonry in-fill panels. During the 1964 Alaska earthquake[7] this type of wall design caused both cladding and structural frame failures.

The masonry in-fill cladding failures are caused by the racking of the structural frame and the subsequent shear deformation of the masonry, often resulting in diagonal racking of the masonry emanating from the corners of the window openings. The interaction of the structural frame and the masonry in-fill also causes damage to the structural frame. The racking of the relatively flexible structural frame is resisted by the stiff masonry in-fill panel, resulting in concentrated stresses in the frame corners that can fail the structural members.

Most glass breakage during earthquakes occurs in large lites of glass in fixed frames near the base of the building, where interstory drift and frame-racking are severe. Damage to glass set in operable sash is rare, because the clearance between the edges of the sash and the frame allows the frame to rack substantially before the sash is distorted, thus limiting stress and breakage in the glass. Although breakage is usually limited, the racking of frames frequently requires adjustment of sash to restore operability. The problems of large lites set in fixed frames is discussed in Section 3.4.

Damage from Pounding

Curtain wall damage due to pounding or hammering between adjacent structures and between separate segments of cladding within a single structure is common during major earthquakes.[7-9]

Inadequate clearance between structures with different dynamic characteristics causes impact damage to the cladding at the points of contact. Damage of this type has occurred in building complexes composed of buildings of different height that are not structurally integrated, i.e., where low-rise portions of a tower complex abut the high-rise portions.

In addition to lack of clearance between individual structures, damage has occurred where joints between individual segments of the cladding are inadequate to prevent contact between the segments during seismic shaking. Such joints are typically provided to limit the effect of thermal and moisture movement or differential foundation settlement on the cladding. While the joints may be adequate for these purposes, they may not be adequate to accommodate lateral drift and distortion of cladding segments during earthquakes.

Loss of Cladding Attachment by Fracture or Pull-Out of Anchors

Catastrophic cladding failures have occurred as a result of the failure of anchors used with rigid heavy cladding systems. For example, the collapse of the precast concrete panels on the facade of the J.C. Penney Building during the 1964 Alaska earthquake resulted in two fatalities.[6] During that incident, the relatively flexible building frame deflected and transferred part of the lateral load to the cladding. Since the cladding was rigid, it resisted deformation, resulting in the concentration of stress in the connection between the structure and the exterior panels. The anchors, which were embedded in the edge of the concrete slab, pulled out of the concrete.

In general, lightweight flexible cladding systems have not suffered from anchorage failure during earthquakes, because the cladding is able to deform with the structure and reduce stress concentration in the anchors.

Fracture of Cladding by Lateral Movement and Contact of Cladding Components

Glass breakage during earthquakes has occurred typically in large fixed lites of glass near the base of the building. The probable principal causes of this damage are:

(1) large interstory drift, metal frames, and glass rack to the point where glass and frame come into contact; and

(2) lower amplitude, higher frequency vibration, which causes glass to shift laterally into contact with the metal frame.

Such lateral movement or "walking" of the glass within its frame occurs in buildings subjected to vibration from HVAC operations, and external and internal wind loads.

When glass moves in contact with its metal frame, the resulting high-contact stress crushes the glass edge and severely weakens the glass plate. Once the glass edge is damaged, the glass can fracture when subjected to relatively small applied loads. Despite the fact that glass manufacturers have recognized the need to provide perimeter cushioning and adequate space between the glass and its framing since the early 1970s and have incorporated such requirements in glazing standards, many installations today do not have such design features. Some wall designers and manufacturers rely on sealants applied between the glass and frame to secure the glass and prevent "walking" damage. However, such elements are unlikely to prevent glass-to-metal contact and consequential glass fracture during a major earthquake for the following reasons:

- experience with past events shows that the large displacements between glazing and frames during seismic shaking causes these seals to "de-bond" and roll out of the gap between the glass and metal frame;
- large racking displacements of the frame will cause glass-to-metal contact even if the glass does not move laterally.

SPECIFIC CODE PROVISIONS FOR SEISMIC DESIGN OF CLADDING

General

The seismic design provisions for cladding prescribed by most U.S. building codes or model codes contain the following fundamental provisions:

- design for out-of-plane lateral forces, based on simplified "C_p" relationships as described in Section 2.3;
- special augmented forces for design of the connection of the cladding to the frame;
- special ductility requirements for design of the connection of the cladding to the frame;
- other special detailing requirements for connection of the cladding to the frame;
- limitations on drift of the structural frame, intended, in part, to preclude frame/cladding interaction;
- special reinforcement for masonry and concrete walls;
- required isolation or movement joints.

Summary of Code Provisions

TABLE 2 summarizes the cladding design provisions of several U.S. codes.

- The 1988 Uniform Building Code[2] contains progressive seismic provisions commonly used in western and southwestern United States. Its provisions are largely based on the SEAOC "Blue Book."[10]
- The 1987 Building Officials and Code Administrators[11] requirements are commonly used in eastern United States in zones of low to moderate seismicity. Its requirements are more traditional than the other codes referred to in TABLE 2.
- The Applied Technology Council ATC-03-06[1] provisions are model code requirements that, while now ten years old, represent the most progressive and comprehensive United States seismic regulations.
- The Applied Technology Council ATC-14 provisions[12] are new provisions for rapid screening of the seismic vulnerability of existing buildings.

Discussion

Lateral Force on Elements

All four codes express the lateral force F_p as a fixed fraction of the weight of the element, all contain a factor (I or C_c) that allows performance to be enhanced based on "importance" or desired level of reliability, all contain a factor representing the zone or expected intensity of ground motion, and all contain a lateral force coefficient that depends on the type of element under consideration. As mentioned in Section 2.3, none of these criteria rationally addresses the actual dynamic response of the structural frame, member, or its connections.

The C_p factors do not distinguish between material or system types, even though the expected performance level can vary greatly. Also, the definitions of the C_p factors are unclear, and do not represent the myriad of wall systems in use. For example, in application to UBC 1988, should a stone cornice, which is part of an exterior masonry wall, be designed for a C_p factor of 0.75, as part of an exterior wall, or 2.0 as an exterior ornamentation and appendage?

Special Force Requirements for Connector Design

Most codes prescribe an increased level of design load for connector design in order to preclude failure of the usually more brittle connection. However, the definition of the "body of the connector" is vague, making this requirement difficult to apply. ACT-3 has special force requirements for connections of concrete and masonry walls to diaphragms, but no similar requirements are given for other cladding types.

Ductility Requirements for Connectors

UBC-1988 requires all elements of the connecting system to be designed for 4.0 F_p. This is done in an attempt to preclude brittle fracture of nonductile elements, but this is not explicitly stated, as it is in BOCA 1987. When a component of a connection should be designed for $1\frac{1}{3} F_p$ or 4.0 F_p is unclear.

TABLE 2. Comparison of Some Provisions for Seismic Design of Building Cladding

	UBC—1988[2]	BOCA—1987[11]	ATC 3-06[1]	ATC14[12]
Lateral force on elements	$F_p = ZIC_pW_p$ where Z = zone factor $\quad 0.075 \leq Z \leq 0.40$ I = importance factor $\quad 1.0 \leq I \leq 1.25$ $\quad$ except 1.0 for the entire connector. C_p = coefficient for a part of the building (see Table No. 23-P) W_p = weight of part under consideration	$F_p = ZIC_pW_p$ where Z = zone factor $\quad \frac{1}{8} \leq Z \leq 1$ I = importance factor $\quad 1.0 \leq I \leq 1.5$ $\quad$ except 1.0 for the entire connector assembly. C_p = coefficient for a part of the building (see Table No. 113.10) W_p = weight of part under consideration	$F_p = A_vC_cPW_c$ A_v = effective peak velocity $\quad 0.05\,g \leq A_v \leq 0.40\,g$ C_c = coefficient for a component of the building (see Table No. 8B) P = performance criteria factor $\quad 0.5 \leq P \leq 1.5$ W_c = weight of component under consideration	$F_p = A_aIC_pW_p$ A_a = effective peak acceleration $\quad 0.05\,g \leq A_a \leq 0.40\,g$ I = importance factor $\quad 1.0 \leq I \leq 1.25$ $\quad$ except 1.0 for the entire connection C_p = coefficient for a part of the building (see Table No. 4.8) W_p = weight of the part under consideration
Special force requirements for connector design	Body of connector shall be designed for $1\frac{1}{3} F_p$.	Body of connector shall be designed for $1\frac{1}{3} F_p$.	Connections of concrete and masonry walls to floor and roof diaphragms shall be designed for not less than 1,000 A_v lb/ft of wall.	Body of connector shall be designed for $1\frac{1}{3} F_p$.
Ductility requirements for connectors	All elements of the connecting system shall be designed for $4.0 F_p$. Bodies of connectors shall have sufficient ductility and rotation capacity so as to preclude fracture of the concrete or brittle failures at or near welds.	Fasteners attaching the connector to the panel or structure shall be designed to ensure ductile behavior of connector or shall be designed for $4.0 F_p$ Bodies of connectors shall have sufficient ductility and rotation capacity so as to preclude fracture of the concrete or brittle failures at or near welds.		Fasteners attaching connector to panels or structure should have a capacity of $4.0 F_p$. Bodies of connections should have sufficient ductility and rotation capacity so as to preclude fracture of the concrete or brittle failure at or near welds.

Detailing requirements for connectors	Fasteners embedded in concrete shall be attached to, or hooked around, reinforcing steel or otherwise terminated so as to effectively transfer forces to the reinforcing steel.	Fasteners embedded in concrete shall be attached to, or hooked around, reinforcing steel or otherwise terminated so as to effectively transfer forces to the reinforcing steel.	None	None
Drift limitations	Story drift shall not exceed: • story height <65 ft (.04/R_w) nor 0.005 times story height • story height >65 ft (.03/R_w) nor 0.004 times story height Drift limits may be exceeded when it is demonstrated that greater drift can be tolerated.	Story drift limited to 0.005 times the story height, unless demonstrated that greater drift can be tolerated. The horizontal displacement calculated from the application of lateral force shall be multiplied by $1/K$ to obtain the drift ($1/K \geq 1$). *Exception:* story drift for buildings with unreinforced masonry limited to 0.0025.	Story drift limited to 0.10 to 0.15, depending on seismic exposure hazard. Multiply computed displacements by C_d, where C_d = deflection amplification factor $1/4 \leq C_d \leq 6\frac{1}{2}$	Story drift limited to 0.005 times the story height unless demonstrated that greater drift can be tolerated. Displacements computed under equivalent lateral force procedure loads.
Reinforcement for masonry or concrete	For Zones 2, 3, and 4, all concrete or masonry elements must qualify as reinforced.	Not specifically required	Not specifically required	None
Accommodation of story drift	Connections and panel joints must accommodate story drift of not less than two times story drift caused by wind, $3(R_w/8)$ times calculated elastic story drift caused by design seismic forces, or ½ in., whichever is greater.	Connections and panel joints must accommodate story drift of $3.0/K$ times calculated elastic story displacement caused by seismic forces or ½ in., whichever is greater.	Attachments to building frame shall have sufficient ductility and rotational capacity to accommodate story drift of C_d times calculated D based on elastic analysis. The architectural components must accommodate the story drift described above, except components assigned a performance factor of L (low requirement) may provide for ½ the design story drift.	Connections and panel joints must accommodate story drift of not less than (⅜) R_w times the calculated elastic story drift caused by seismic forces, or ½ in., whichever is greater (see Table 4.4 for R_w).

Detailing Requirements for Connectors

UBC 1988 and BOCA 1987 have explicit requirements for anchorage of fasteners embedded in concrete. These would commonly apply to threaded inserts or cast-in weldments used in heavy masonry or precast concrete cladding. However, they do not apply to other concrete anchors, such as power-driven or expansion anchors.

CONCLUSIONS AND RECOMMENDATIONS

General

Current design provisions for exterior cladding oversimplify actual behavior. While such provisions are generally considered to be conservative, they are vague, contradictory, and difficult to apply in design practice, leaving the opportunity for unsatisfactory results. The provisions do not distinguish between the actual performance characteristics of different cladding materials and systems.

Research and Code-Development Recommendations

Research and improvements in building codes are needed in the following areas:

(1) Development of lateral force levels for components that directly account for the dynamic response of the structure and the cladding.

(2) Better understanding of curtain wall/frame interaction, including beneficial effects of damping provided by "isolated" cladding and damaging effects of interstory drift imposed on cladding.

(3) Development of typical design details for cladding components that incorporate all necessary design considerations.

(4) Study of seismic response of typical cladding materials and systems, including those more commonly used in regions of low to moderate seismicity.

(5) Development of code-prescribed performance characteristics for cladding that recognize the differing response of various cladding systems and materials.

(6) After major and moderate seismic events, careful documentation of the cladding performance, including study of whether unsatisfactory performance resulted from improper design, improper construction, or deficiencies in state-of-the-art design practices.

Design Recommendations

Premise: Curtain wall should sustain little damage in modest earthquake.

Curtain wall should accommodate severe quake without severe damage and loss of attachment.

(1) Separate buildings and parts of buildings for seismic drift to avoid pounding.

(2) Design control/expansion joints to allow for seismic distortion of adjacent components without hammering.

(3) Use lightweight, flexible components with flexible structures.

(4) Use rigid structures to support rigid, floor-to-floor-spanning cladding.

(5) Use well-thought-out connection details that are ductile and positively anchor the cladding to the structure.

(6) Avoid connection details with eccentricity or indirect load paths. Isolate cladding for frame.

(7) Edge cushion brittle materials or provide separation and prevent lateral walking.

(8) Back or laminate opaque brittle materials to prevent loss of fragments.

(9) Design walls for seismic movement magnitude based on the maximum possible interstory drift as opposed to deflections computed for code seismic forces.

(10) Develop continuity of horizontal wall reinforcing at corners.

(11) Design anchors and joints to accommodate simultaneous earthquake motion in all three directions, while allowing for all other forms of movement and tolerance required by design.

(12) For glass:
- Design to limit risk once failure occurs, i.e., use tempered or laminated glass for first floor storefronts to either completely shatter into small fragments or be retained in frame. Laminated glass maintains level of security after breakage.
- Use framing/gasket system that locks gasket in to prevent roll-out during earthquake distortion.
- Follow FGMA for lateral cushioning of glass but increase glass-to-cushion clearance per earthquake analysis interstory drift requirements.

(13) When in doubt, test for drift damage in mock-ups.

(14) Do not use light gauge metal ties to anchor stone and masonry veneers.

(15) Avoid brittle inserts as thermal isolators that take tension in metal curtain walls.

(16) Design for damage control.

REFERENCES

1. APPLIED TECHNOLOGY COUNCIL. 1984. Tentative Provisions for the Development of Seismic Regulations for Buildings, ATC 3-06 Amended, 2nd Printing. Applied Technology Council, Palo Alto, CA.
2. INTERNATIONAL CONFERENCE OF BUILDING OFFICIALS. 1985. Uniform Building Code. International Conference of Building Officials. Whittier, CA.
3. WAKABAYASHI, M. 1986. Design of Earthquake-Resistant Structures. McGraw-Hill. New York, NY.
4. POP I., D. VERDES & L. TUTU. 1985. Nonstructural elements in interaction or without interaction with structure. In Proceedings of the 8th European Conference on Earthquake Engineering, Lisbon, Portugal.
5. EARTHQUAKE ENGINEERING RESEARCH INSTITUTE. 1984. Nonstructural Issues of Seismic Design and Construction, EERI Publication No. 84-04, Earthquake Engineering Research Institute, Berkeley, CA.
6. STRUCTURAL ENGINEERS' ASSOCIATION OF CALIFORNIA. Tentative Lateral Force Requirements 1985. Sacramento, CA.
7. THE GREAT ALASKA EARTHQUAKE OF 1964. National Academy of Sciences. Washington, DC.
8. MURPHY, L. M. 1973. San Fernando, California Earthquake of February 9, 1971. Washington, DC.

9. ELSESSER, E. 1984. Life hazards created by nonstructural elements. *In* Nonstructural Issues of Seismic Design and Construction, EERI Publication No. 84-04. Earthquake Engineering Research Institute, Berkeley, CA.

10. STRUCTURAL ENGINEERS' ASSOCIATION OF CALIFORNIA. Tentative Lateral Force Requirements 1971. Sacramento, CA.

11. BUILDING OFFICIALS AND CODE ADMINISTRATORS INTERNATIONAL, INC. 1986. The BOCA National Building Code/1987, 10th ed. Building Officials and Code Administrators International, Inc. Country Club Hills, IL.

12. APPLIED TECHNOLOGY COUNCIL. 1987. Evaluating the Seismic Resistance of Existing Buildings, ATC-14, Applied Technology Council, Palo Alto, CA.

13. PALL, A. S. & C. MARGH. 1979. Seismic response of large-panel structures using limited-step bolted joints. *In* Proceedings, Third Canadian Conference, Earthquake Engineering, Montreal, Canada.

Earthquake-Resistant Bridge Design

JAMES H. GATES

California Department of Transportation
Division of Structures
Sacramento, California 94274-0001

INTRODUCTION

The design of earthquake-resistant bridges in the United States took a dramatic turn in February 1971. The heavy damage observed as a result of the 1971 San Fernando earthquake was unprecedented in the history of bridge design.[1] Before 1971 California had seen less than $100,000 in earthquake damage to bridges. The 1971 earthquake initiated an effort by the California Department of Transportation (CALTRANS) and the Federal Highway Administration (FHWA) to improve the seismic resistance of highway bridges. A comprehensive program was started to evaluate and improve both the design specifications and details.

In 1973 new criteria were implemented by CALTRANS that considered fault activity and the soils at a bridge site as well as the vibrational properties of the bridge itself.[2,3] These criteria departed from traditional seismic design criteria and presented several new items including site-specific elastic response spectra, specified reductions for ductility and risk, and a modular arrangement of variables to permit future adjustment.

In 1975, the CALTRANS criteria were modified and adopted by the American Association of State Highway and Transportation Officials (AASHTO) for national use.[4]

In 1981, the Applied Technology Council (ATC) completed development of the *Seismic Design Guidelines for Highway Bridges*.[5] This report (sometimes referred to as ATC-6) was funded by the Federal Highway Administration and the National Science Foundation and presented guide specifications for national use. The ATC-6 document was adopted in 1983 by AASHTO as a guide specification for the design of highway bridges.[6]

During the short time since 1971, all of these changes have had a dramatic influence on the analysis, design, and detailing of earthquake-resistant bridge structures throughout the United States. This paper reviews the development of earthquake-resistant bridge design criteria in California and assesses significant criteria, analysis, construction, and retrofit procedures.

THE CALTRANS SEISMIC DESIGN CRITERIA

A primary requirement in the development of the CALTRANS criteria was to permit the criteria to be easily modifiable to reflect new developments in earthquake engineering. Each component of the criteria was designed to represent the independent influence of a different discipline. The following factors affecting the response of a structure to seismic forces are included in the criteria:

- The location of the bridge relative to active faults.
- The effect of a maximum credible earthquake on the fault.

- The dynamic responses of the bridge to the strong motion.
- The dynamic properties of the underlying soils at the site.
- The reduction in design forces for ductility and risk.

Four components were developed from these factors to describe the member forces in the bridge, (A, R, S and Z). These components are briefly defined as follows:

A—The peak rock acceleration, determined from seismologic studies of fault activity and attenuation data gathered from historic events.[2,7-9]

R—An elastic acceleration spectra for rock based on actual recorded data.[2,10]

S—A spectral oriented soil amplification factor, based on both computer studies and actual recorded data.[2,10]

Z—A component-oriented reduction factor for risk and ductility.[2,10]

The product of the first three factors (ARS) results in an elastic response spectra that would result from a maximum credible event on the closest fault. These spectra are presented in the criteria as 28 curves.[2,11]

Division of the elastic forces by the factor Z is the final step. This then results in the design force for the particular component of the structure under consideration. By requiring this reduction to be component-dependent, the design force depends not only on the seismicity and site conditions, but also on the ductile capacity of the structural component itself, permitting members of varying toughness to be introduced into the structure in a rational manner.

RECENT CRITERIA DEVELOPMENTS

Recent national criteria developments have been dominated by the ATC-6 project.[5] This project, funded by the Federal Highway Administration and the National Science Foundation (NSF), was started in 1977 and was completed in 1981. These guidelines were developed to be applied on a national basis and are the composite recommendations of a team of experts composed of consulting engineers, university personnel, state bridge engineers, and federal representatives. The guidelines contain several new concepts that are significant departures from previous design provisions:

- Four seismic performance categories are defined on the basis of expected acceleration levels at the bridge site.[12]
- Different degrees of complexity and sophistication of analysis and design are specified for each of the four seismic performance categories.
- Uncertainty in the direction of loading is accounted for by combining loads in two orthogonal directions.
- Column reinforcement in plastic hinge zones is based on work in New Zealand by the University of Canterbury.[13]
- Bridge bent analysis is performed by the distribution of overstrength plastic moments in the bent.
- Minimum seat widths are specified to prevent span drop.

The ATC-6 guidelines were adopted by AASHTO as a guide specification in 1983, to be utilized in lieu of the regular AASHTO Design Specifications for Highway Bridges in all parts of the United States and Puerto Rico.

The current CALTRANS seismic bridge criteria[11] generally are a modification of the 1983 AASHTO guidelines, incorporating improved reinforced concrete specifications but retaining their own seismic force level and reduction values from previous versions.[2,3]

SIGNIFICANT ASPECTS OF CURRENT SEISMIC DESIGN CRITERIA

Site-Specific Design Spectra

Both the CALTRANS criteria and the AASHTO guide specification permit the development of special site-specific design spectra.

At CALTRANS, special spectra have been developed for several large projects including the new Century Freeway in Los Angeles.[14,15]

The current method used at CALTRANS to determine site-specific spectra consists of estimating surface motions by computing the propagation of rock motions to the surface of a one-dimensional soil column. The SHAKE computer program[16] is used for this analysis.

The computation of special ARS spectra is performed using a procedure similar to that used to develop the original CALTRANS standard elastic design curves.[2,3,10] Recent investigators[17,18] have described the procedure as capable of obtaining a reasonable approximation of the motions at a site.

Analysis Procedures

Current criteria permit the use of several analytic methods ranging from a simple static analysis to a complex time history dynamic analysis.

The AASHTO guidelines[6] permit the use of simpler analysis procedures in lower seismic regions and for less complex bridges; however, a single-mode procedure is presented in the guidelines. This procedure requires a minimum of computer analysis to be used efficiently and may also be applied with hand methods.

After several unsuccessful attempts to develop simplified procedures at CALTRANS, it was concluded that most of the decisions required for an effective simplified procedure were automatically made by a response spectrum modal analysis. A modal analysis currently is used on bridge designs at CALTRANS.

The primary analytic tool at CALTRANS is the STRUDL computer program.[19] This large, general purpose analytic program with a preprocessor program called STRUBAG provides an efficient analysis that requires a minimum of user input. The generated structural model also is used to evaluate the effects of wind, temperature, and other lateral loads.

The computer program SEISAB[20] is also available for use nationwide. This program was specifically developed for the seismic analysis of bridges. The program uses both modal analysis and the AASHTO guide single-mode procedure and features a free format input utilizing bridge engineering terms.

Having a good analytic tool only solves the problem of performing the analysis of the given structural model. It is probably more important that the model accurately represent the structure's dynamic characteristics. The STRUBAG and SEISAB programs were the first to build in several bridge-modeling assumptions.

Verification of Modeling Assumptions

In order to verify the modeling assumptions used in the dynamic analysis of bridges, the Federal Highway Administration funded a 3.5-year field measurement project,[21,22] which was completed by CALTRANS. The project recorded vibrational data on 57 typical bridges and compared these results with current elastic analysis procedures.

System identification techniques were utilized to determine superstructure properties and boundary conditions for three of the bridges instrumented for strong-motions.

Significant results from this project include charts which may be used to estimate the approximate fundamental frequency of an average bridge. Moments of inertia of various concrete superstructure members were found to vary as follows:

For reinforced concrete bridges:
• IX (Torsion) Equal to the gross value.
• IZ (Normal bending) 40–60% of the gross value.
• IY (Transverse bending) 60–80% of the gross value.

For prestressed concrete bridges:
• IX (Torsion) 200% of the gross value.
• IZ (Normal bending) 120–140% of the gross value.
• IY (Transverse bending) 100–120% of the gross value.

Column and Foundation Design Requirements

Both the AASHTO guide specification and the current CALTRANS criteria require that any connections expected to undergo plastic deformation be able to develop the plastic moment and resultant shear as a limiting value. In regions of high seismic activity, this means that the plastic forces usually become the limiting design values to be used in the design of the connections. At CALTRANS, the design of the column member is highly automated. An interactive computer program, YIELD,[23] is used to evaluate the column member for factored, service, and plastic loads and to determine vertical, shear, and confinement reinforcement.

The current CALTRANS design specification requires the extension of the confining spiral reinforcement into the adjacent cap or footing. Shear and confinement reinforcement in the column consists of continuous closely spaced spirals (usually No. 6 or 19-mm bars). These spirals are spliced by welding or approved couplers. As an alternative, individual circular welded hoops may be substituted for spirals at the option of the contractor. The intent of these specifications is to ensure that the spiral will remain effective when the column is subjected to large excursions into the ductile range. Rectangular columns are reinforced for shear and confinement by using a series of interlocking spirals or hoops spaced not greater than 0.75 times the spiral diameter.

The design of the footing is also automated at CALTRANS using the FOOT program.[24] The footings must be evaluated for plastic loads from the column in addition to service and factored loads. Plastic moments are applied to the footing at 15-degree intervals to consider the effect of plastic action in all directions.

Top footing and shear reinforcement are designed for all imposed loads including uplift on piles and the weight of footing and overburden. Allowable soil stresses or pile loads vary depending on the class of loading. Current

CALTRANS criteria require an allowable stress on soil to vary from 0.33 times ultimate for service loads to 0.50 for factored to 1.00 for seismic or plastic hinge loads. These factors vary from 0.50 to 0.75 to 1.00 for loads on piles.

Pile Shaft Foundations

Drilled pile shaft foundations are being used at an increasing number of locations on California bridge projects. The drilled shaft foundation generally is cheaper than other foundation types and permits columns to be located in tight places with a minimum of disturbance to existing facilities.

The use of this foundation generally is limited to areas where soil conditions permit economical excavation for the shaft and where ground water is not encountered. The presence of ground water does not prohibit the use of the drilled shaft; however, the cost becomes considerably higher in this case and since it is more expensive to maintain the shaft walls underwater (usually through the use of bentonite slurry) and underwater concrete placement is required. Vertical load tests by CALTRANS on carefully installed slurry seal piles have shown that the resulting pile is equal in strength to a normal dry constructed shaft.[25] Drilled shaft dimensions vary from 6 to 8 feet (1.8 to 2.4 m) in diameter and the shaft extends 30 to 60 feet (9 to 18 m) into the ground. Construction of the foundation consists of drilling of the shaft excavation, installation of the reinforcement cage, and placement of the concrete.

Column reinforcement is continuous but may be spliced by welding or approved connectors; however, staggered lap splices are permitted for No. 11 (36-mm) bars or smaller in columns more than 34 feet (10 m) high. Lap splices are not permitted near areas where plastic hinges are expected to form.

The design problems involved with the use of the pile shaft are slightly different from those of ordinary pile and spread footing foundations. The pile shaft has a softer lateral stiffness and therefore requires more refined foundation data at an earlier stage in the design process. This softer lateral stiffness also must be considered in the design and analysis of the superstructure as well as the substructure components. CALTRAN[26] has adopted the following procedure to deal with this problem:

1. Determine an equivalent column length for each column support which is to be a pile shaft. This determination has been standardized at CALTRANS through the application of the PILE computer program[27,28] from the University of Texas.

2. Analyze the bridge with the equivalent column lengths and determine column loads to be used in the YIELD and PILE computer programs.

3. Design the top of column reinforcement for all AASHTO load groups I through VII using the YIELD program and the equivalent column lengths.

4. Using the PILE program, determine the maximum moments in the pile shaft for the components of group loads I through VII. Detailed soil data from the foundation engineer is required for use of the PILE program.

5. Using the YIELD program, determine the amount of vertical reinforcement required to resist the maximum moment in the pile shaft. The plastic moment capacity of the column and shaft also is determined at this time.

6. Using the PILE program, analyze the pile shaft for the plastic condition. The plastic moment, the associated axial load, and the assumed plastic shear are applied at the top of the column. The shear is increased until a plastic hinge forms in the pile shaft. The shear reinforcement is then designed for the lesser of the

shears resulting from seismic plastic hinging or group loads I to VI and unreduced elastic group load VII.

7. Perform a final check of the overall stability using the PILE program. The pile shaft is considered stable when a substantial decrease in pile shaft length does not result in excessive deflection.

SEISMIC RETROFIT GUIDELINES

Guidelines for the retrofit of bridges have been developed in the Applied Technology Council project ATC-6-2.[29] This project, funded by the Federal Highway Administration, was completed in 1983. The guidelines are intended to be applied in the United States on a national basis and are the composite recommendations of a team of nationally recognized experts, including consulting engineers, university personnel, state bridge engineers, and Federal representatives from throughout the United States. The guidelines are basically an extension of the AASHTO Guide Specification.[5] Much of the material covering analysis procedures is utilized and the guidelines are compatible with the AASHTO Guide Specification wherever possible.

In the ATC-6-2 Guidelines, the seismic retrofit process is divided into three major steps:

1. A preliminary screening procedure, which is used to rank or prioritize a large number of bridges by vulnerability.

2. A detailed evaluation procedure, which is used to identify the particular components in the bridge that require retrofitting.

3. A presentation of various retrofit measures along with guidelines on their selection and use.

RETROFIT OF BRIDGES IN CALIFORNIA

The California retrofit program began almost immediately after the 1971 San Fernando Earthquake. Projects that were under construction were modified where possible by contract change order to meet increased seismic requirements. This effort was extended to include bridges in the design phase. The California Department of Transportation identified about 1300 bridges (among about 13,000) that were primarily deficient in seat width. These unrestrained joints represent the initial focus of the Department's retrofit program. To date, almost all these bridges have been retrofit at a cost of nearly $55 million.

The average California retrofit project consisted of the addition of steel restrainer cables at hinge and expansion joints to prevent spans from collapsing. Force levels were identical to those used on new structures. It is expected that the potential for collapse can be minimized even though extensive damage is experienced.

The California design details for restrainer units have evolved to the point where reliable and economical systems are performing satisfactorily under service conditions in the field.

Selection of structures for retrofit in California was based on a priority system which takes into account the expected accelerations at the site, the estimated cost to retrofit the structure, the cost of replacement in the event of loss, the length and availability of detours, and the average daily traffic as well as other factors that reflect the importance of the bridge in the transportation system. California developed a simplified restrainer analysis procedure[8,9] which considers many of the nonlinear features of multi-hinged bridges, such as expansion joint gaps and the impacting of the superstructure against the abutment fill. This procedure is now used to compute restrainer forces in restrainers in both old and new construction.

THE OCTOBER 1987 WHITTIER EARTHQUAKE

As soon as reports of damage from the early-morning M5.9 Whittier earthquake were received from CALTRANS Structures Maintenance personnel, the Sacramento-based Post Earthquake Investigation Teams (PEQIT) from the Division of Structures was dispatched to the area. In addition, CALTRANS Structures Maintenance personnel performed routine inspection and survey of all of the structures in the region of strong ground-shaking. Bridges and sound walls along all the state highways in the area were inspected in detail.[30]

A total of 24 bridges showed evidence of movement or damage. One bridge suffered moderate damage, sixteen bridges received minor damage, and seven bridges showed evidence of movement without visible damage. At least eleven retrofitted bridges were located in the region of highest shaking.

The one bridge that suffered moderate damage was the Route 605/5 separation bridge, a 9-span structure with a total length of 567.5 feet and a width of about 124 feet. The bridge is supported by a series of 5-column bents and seat-type abutments. The two main spans over Route 5 are simply supported prestressed I girders. The remainder of the superstructure consists of continuous cast-in-place reinforced concrete box girders. The structure was built in 1964 and retrofitted for seismic forces in 1981. The retrofit consisted of the addition of longitudinal cable restrainers at bents 5, 6, and 7, plus added pedestals at the abutments.

The heavy damage occurred at bent 6, the center support for the two main spans. All five columns were damaged, with the three interior columns receiving more damage than the exterior columns. Major cracking occurred on the east and west faces (wide sides) of all columns.

All keeper plate bolts sheared at the rocker bearings at abutment 10. The bearings remained plumb, but most came to rest with a transverse displacement of 2 inches. At abutment 1, only some of the rocker bearings showed damage similar to that of abutment 10, indicating only a small movement.

Bent 6, with its short columns and continuous footing is considerably stiffer than the other bents in this structure. The shear capacity of these short columns was limited due to the low amount of shear reinforcement (#4 @ 12 inches). The bent 6 columns drew a large portion of the seismic force and failed in shear.

The earthquake restrainer cables at bent 5 and 7 showed no sign of damage. In fact, there was no evidence that they had been engaged during the earthquake.

Because of the extent of the column damage and the potential for aftershocks, both traffic routes were closed and temporary shoring was installed. Within 22 hours after the quake occurred, the 605/5 separation bridge was reopened to traffic. The contract to rebuild the five columns of bent 6 was advertised on November 3, 1987 and the bids were open on November 17, 1987. The low bid was $376,608. The project was completed before the end of January 1988.

FUTURE RETROFIT DIRECTIONS IN CALIFORNIA

The installation of hinge restrainers added a considerable amount of seismic reliability to the California bridge system. There remain, however, several weaknesses in various bridges in the network, especially those located in areas of very high seismicity. The task of strengthening columns and substructures will occupy much of our time in the next several years. By applying nonlinear analysis techniques it is possible to better predict failure modes and to analyze the effects on overall stability when various components fail. Preliminary indications are that by increasing the amount of longitudinal restrainers at joints and strengthening select columns, we can maintain the stability of long, tall ramp structures, even though considerable damage will occur. Another area for potential strengthening is the addition of infill walls between very weak two-column bents.

The task of selection, design, and construction of this second phase of retrofit is just beginning.

REFERENCES

1. FUNG, G., R. S. LEBEAU, E. D. KLEIN, J. BELVEDERE & A. F. GOLDSCHMIDT. 1971. Field Investigation of Bridge Damage in the San Fernando Earthquake. CALTRANS. Sacramento, CA.
2. GATES, J. H. 1976. California's seismic design criteria for bridges. J. Struct. Div. ASCE 102(ST12): 2301–2313.
3. GATES, J. H. 1979. Factors considered in the development of the California seismic Design Criteria for Bridges. In Proceedings of a Workshop on Earthquake Resistance of Highway Bridges. :141–162. Applied Technology Council. Palo Alto, CA.
4. AMERICAN ASSOCIATION OF STATE HIGHWAY AND TRANSPORTATION OFFICIALS, HIGHWAY SUBCOMMITTEE ON BRIDGES AND STRUCTURES. 1975. Interim Specifications, Bridges 1975. AASHTO. Washington, DC.
5. APPLIED TECHNOLOGY COUNCIL. 1981. Seismic Design Guidelines for Highway Bridges, Report ATC-6 and FHWA/RD-81/081. Federal Highway Administration. Washington, DC.
6. AMERICAN ASSOCIATION OF STATE HIGHWAY AND TRANSPORTATION OFFICIALS, HIGHWAY SUBCOMMITTEE ON BRIDGES AND STRUCTURES. 1983. Guide Specifications for Seismic Design of Highway Bridges 1983. AASHTO. Washington, DC.
7. GREENSFELDER, R. W. 1974. A Map of Maximum Expected Bedrock Acceleration from Earthquakes in California, Map Sheet 23. California Division of Mines and Geology. Sacramento, CA.
8. SEED, H. B. & P. B. SCHNABEL. 1972. Accelerations in Rock from Earthquakes in the Western United States, Report EERC 72-2. University of California, Earthquake Engineering Research Center. Berkeley, CA.
9. SEED, H. B., I. M. IDRISS & F. W. KIEFER. 1968. Characteristics of Rock Motions During Earthquakes, Report EERC 68-5. University of California, Earthquake Engineering Research Center. Berkeley, CA.
10. CALIFORNIA DEPARTMENT OF TRANSPORTATION, OFFICE OF STRUCTURES DESIGN. 1984 (July). Memos to Designers, 15–10 Earthquake Design Criteria (Commentary). CALTRANS. Sacramento, CA.
11. CALIFORNIA DEPARTMENT OF TRANSPORTATION, OFFICE OF STRUCTURES DESIGN. 1985. Bridge Design Specifications Manual. CALTRANS. Sacramento, CA.
12. APPLIED TECHNOLOGY COUNCIL. 1978. Tentative Provisions for the Development of Seismic Regulations for Buildings, Report ATC 3-06. NBS Special Publication 510. U.S. Department of Commerce, National Bureau of Standards. Washington, DC.
13. PRIESTLEY, M. J. N. & R. PARK. 1979. Seismic resistance of reinforced concrete bridge columns. In Proceedings of a Workshop on Earthquake Resistance of Highway Bridges. :253–283. Applied Technology Council. Palo Alto, CA.

14. CALIFORNIA DEPARTMENT OF TRANSPORTATION, OFFICE OF THE TRANSPORTATION LABORATORY. 1983 (August). Century Freeway Ground Response Study. CALTRANS. Sacramento, CA.

15. GATES, J. H. 1984. Seismic considerations for the Century Freeway. In Proceedings of the ASCE Symposium on Lifeline Earthquake Engineering. American Society of Civil Engineers. New York.

16. SCHNABEL, P. B., J. LYSMER & H. B. SEED. 1972. SHAKE: A Computer Program for Earthquake Response Analysis of Horizontally Layered Sites, Report EERC 72-12. University of California, Earthquake Engineering Research Center. Berkeley, CA.

17. BELL, J. M. & R. A. HOFFMAN. 1978. Design earthquake motions based on geologic evidence. Proceedings of the ASCE Geotechnical Engineering Division Specialty Conference, Earthquake Engineering and Soil Dynamics, 1: 231–271.

18. HAYS, W. W. 1980. Procedures for Estimating Ground Motions. U.S. Geol. Surv. Prof. Pap. 1114. Reston, VA.

19. LOGCHER, R. D., B. B. FLACHSBART, E. J. HALL, C. M. POWER & A. J. FERRANTE. 1968. ICES STRUDL II—The Structural Design Language, Engineering User's Manual, Vol. 1, Report R68-81. MIT Department of Civil Engineering. Cambridge, MA. (Proprietary versions also available as STRUDL, STRUDL DYNAL, STRUDL PLOTS from McAuto, MultiSystems, Inc., and as GTSTRUDL from Georgia Institute of Technology.)

20. IMBSEN, R. A., J. LEA, V. KALIAKIN, K. PERANO, J. GATES & S. PERANO. 1984. SEISAB-I—Seismic Analysis of Bridges: User's Manual and Example Problems. Engineering Computer Corp. Sacramento, CA.

21. GATES, J. H. & M. J. SMITH. 1982. Verification of Dynamic Modeling Methods by Prototype Excitation. CALTRANS (FHWA/CA/SD-82/07). Sacramento, CA.

22. GATES, J. H. & M. J. SMITH. 1984. Results of Ambient Vibration Testing of Bridges. Earthquake Engineering Research Institute, Proceedings of the Eighth World Conference on Earthquake Engineering, VI: 873–880. Prentice-Hall. Englewood, NJ.

23. CALIFORNIA DEPARTMENT OF TRANSPORTATION, OFFICE OF STRUCTURES DESIGN. 1983 (July). Column design by yield surface, 'YIELD.' In Bridge Computer Manual, Chapt. 5-3. CALTRANS. Sacramento, CA.

24. CALIFORNIA DEPARTMENT OF TRANSPORTATION, OFFICE OF STRUCTURES DESIGN. 1983 (July). Footing analysis/design program, 'FOOT.' In Bridge Computer Manual, Chapt. 8-2. CALTRANS. Sacramento, CA.

25. WILHELMS, R. C. 1975. Load Test—48 Inch Diameter Pile, Constructed by Slurry Displacement Method, Guadalupe River Viaduct, Bridge No. 37-308 R/L, 04-Scl-87-5.8. CALTRANS Engineering Geology Branch. Sacramento, CA.

26. CALIFORNIA DEPARTMENT OF TRANSPORTATION, DIVISION OF STRUCTURES. 1985 (August). Pile shaft procedure. In Bridge Design Aids. CALTRANS. Sacramento, CA.

27. REESE, L. C. & W. RANDALL SULLIVAN. 1980. Documentation of Computer Program COM624. The University of Texas, Geotechnical Engineering Center. Austin, TX.

28. CALIFORNIA DEPARTMENT OF TRANSPORTATION. 1983 (July). Shaft pile analysis—PILE. In Bridge Computer Manual, Chapt. 8-1. CALTRANS. Sacramento, CA.

29. APPLIED TECHNOLOGY COUNCIL. 1983. Seismic Retrofitting Guidelines for Highway Bridges, Report ATC-6-2 and FHWA/RD-83/007. Federal Highway Administration. Washington, DC.

30. MELLON, S. & G. KLEIN. Seismic Report, Whittier, CA Earthquakes of October 1 & 4, 1987. Division of Structures Post Earthquake Investigation Team Report. CALTRANS. Sacramento, CA.

Seismic Design Considerations for Buried Pipelines[a]

T. D. O'ROURKE

School of Civil and Environmental Engineering
Cornell University
Ithaca, New York 14853-3501

INTRODUCTION

Design and planning for the seismic performance of buried pipelines are influenced by uncertainties in material properties, loads, and the mechanics of how stresses and deformations are conveyed between pipe and adjacent soil. Pipelines are composed of various brittle and ductile materials. They are equipped with special joints, tees, tie-ins, connections, and fittings. Pipelines are affected by differences in installation techniques, standards of manufacture and construction, levels of corrosion, and stresses accumulated over time from temperature fluctuations, surface loads, and adjacent construction activity. Pipeline performance during an earthquake is subject to geologic setting, soil profile, soil properties, and groundwater levels. Variations in geotechnical site conditions add to variations in both pipeline composition and strong ground-motion characteristics to establish an environment for which the planner must assess over-all system performance, and the designer must evaluate parameters and failure mechanisms for component- or site-specific design.

Fortunately, simplifications in design can be made by considering a particular component and optimizing its performance over a range of potential seismic loads. Likewise, a specific site can be identified and a portion of the pipeline system can be analyzed and configured to accommodate various levels of ground deformation. Site-specific design usually is performed for locations in which large permanent differential movements may occur, such as at active faults and locations of potential liquefaction and soil settlement.

This paper starts by summarizing observations of pipeline performance during recent earthquakes. Lessons learned from the earthquakes are discussed according to the types of ground movements caused and pipelines affected. Simple models then are examined for buried pipeline response to permanent ground deformation. Sources of uncertainty in the simple models are reviewed, the results of parametric studies are presented, and recommendations for design and planning are proposed.

[a] This work was supported by the National Center for Earthquake Engineering Research at Buffalo, New York, and the National Science Foundation under Grant No. 873007.

PREVIOUS EARTHQUAKE PERFORMANCE

Observations of pipeline performance help to identify the most serious earthquake hazards and vulnerable system components. A summary of previous earthquake ground movements and pipeline performance has been presented by O'Rourke et al.,[1] and the format adopted in their work is used in this study. TABLE 1 summarizes observations for twelve recent earthquakes. The table is intended as an update and extension of the previous work, with Japanese, New Zealand, and Central and South American earthquakes added.

The summary begins with the 1971 San Fernando earthquake. This event, which caused more than 2400 breaks in mains and services of the water, sewer, and gas distribution systems,[2] stimulated considerable interest in lifeline earthquake engineering, and is regarded as being a major impetus for U.S. work in this field. The twelve earthquakes included in the table were selected because of their relatively high levels of damage and because reliable records of pipeline behavior were available for each event.

The table indicates that permanent ground movements were prominent in nearly all the earthquakes and that large soil displacements were associated with buried pipeline damage. Soil slumps at cut-and-fill boundaries and ground movements generated by soil liquefaction were especially troublesome during the 1978 Miyagiken-Oki and 1983 Nihonkai-Chubu earthquakes, respectively. Soil liquefaction played an important role in pipeline damage during the 1971 San Fernando, 1985 Chile, and 1987 Edgecumbe earthquakes.

Perhaps the most graphic example of pipeline damage from permanent ground movements is the destruction of the 660-mm-diameter TransEcuadorian Pipeline during the 1987 Ecuador earthquakes. Earthquake damage to approximately 40 km of the TransEcuadorian Pipeline represents the largest single pipeline loss in history. It deprived the country of 60% of its export revenues for six months, and cost roughly $850 million in lost sales and reconstruction.

The pipeline was composed of X-60-grade steel with a wall thickness of approximately 9.5 mm in the area of strong ground-shaking. Most of the damaged portion of the line was supported above ground, either on "H" pile support frames or concrete pedestal foundation saddles. Seismic shaking had only a limited effect on the line pipe, whereas permanent ground deformation had a severe and extensive influence. Landslides, debris flows, and flooding caused most of the pipeline damage, and contributed to all ruptures and lost sections of line.

The area of most severe disruption is shown by the hatchured zone in FIGURE 1. This general region includes the pipeline route from its crossing of the Aguarico River to its crossing of the Salado River. The two earthquakes of March 5, 1987 occurred after a month of heavy rains, during which 600 mm of precipitation were measured at a nearby rain gauge station. Strong ground-shaking triggered rock avalanches, land slips, and debris flows which drained northward into the Dúe River and south and eastward into the Salado and Coca Rivers. Flooding along the Dúe spread to the Aguarico River and destroyed the pipeline and road bridge across the Aguarico approximately 50 km northeast of the Salado Pump Station.

FIGURE 2 provides an expanded view of the area (hatchured zone in FIGURE 1) of greatest pipeline damage. Approximately 12 km of the pipeline were destroyed along the banks of the Coca River, from just east of the Salado Pump Station to a location roughly 2 km east of the confluence of the Malo and Salado Rivers. Along this section, damage was generated by landslides and debris flows in the residual soils and igneous parent rock of the hills flanking the northern side of the Salado

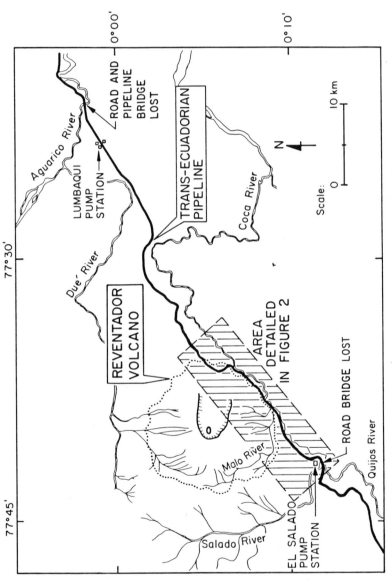

FIGURE 1.

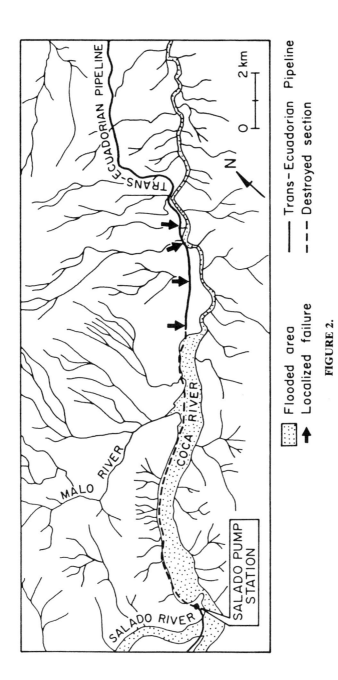

FIGURE 2.

TABLE 1. Summary of Ground Movements and Pipeline Performance for Twelve Recent Earthquakes: 1971–1987

Earthquake	Magnitude/ Intensity[a]	Surface Faults	Landslides	Liquefaction/ Earthquake Settlements	Reported Pipeline/Conduit Damage	Remarks	References
San Fernando, 1971	Magnitude = 6.4 Intensity = XI	Reverse oblique fault dipping 70° north. Most movement distributed across zone 100-m wide. Max. combined disp. = 1.9 m; max. throw = 1.4 m; total length of faulting = 15 km.	Many landslides including slumps and rockslides in nearby foothills of San Grabriel Mtns. and extensive slumping around Upper and Lower Van Norman Reservoirs. Sliding was predominantly in bedrock, alluvial soils, and fills. Deep slumping generally absent.	Liquefaction-induced lateral spreading near Upper Van Norman Reservoir. Sand boils observed at several locations. Liquefaction of hydraulic fill in Lower Van Normal Reservoir dam.	Extensive pipeline damage, especially in areas of active faulting. 113 gas-line breaks, 52 breaks in one 410-mm-diameter gas transmission line. Overall, 17,000 customers lost gas service because of the earthquake; 38 km of sewer pipe required replacement.	Earthquake caused extensive pipeline damage by faulting, landsliding, and liquefaction. 25 to 50% of pipeline damage occurred within the fault zones, which represented only 0.5% of the area affected by strong ground-shaking.	Stienbrugge et al.[2], O'Rourke and Tawfik[3]

	Magnitude/Intensity	Faulting	Landslides	Liquefaction	Pipeline damage	Remarks	Reference
Managua, Nicaragua, 1972	Magnitude = 5.6 Intensity not reported	Four main, strike slip surface faults; max. left lateral disp. = 381 mm; max. throw = 102 mm; max. length mapped 5.9 km; max. width of zone of faulting = 2.5 km	Landslides primarily along steep natural slopes and steep granular embankments	No obvious movement related to liquefaction. No significant settlements reported from consolidation of granular soils.	Extensive damage to water distribution system with many pipelines ruptured at fault crossings. Substantial damage to asbestos-cement pipe and cast iron pipelines with lead joints. Severe strain on electric transmission line caused by landslide.	Extensive destruction attributed to surface faulting. Lack of liquefaction effects, in part, due to low water table and short duration of shaking.	Cajina[4]
Guatemala, 1976	Magnitude = 7.5 Intensity = IX	Up to 3.5 m of left lateral strike slip on Motagua fault; length of fault = 240 km. Substantial movements on secondary faults.	More than 10,000 landslides. 11 large slides, each with more than 100,000 m³.	Numerous sand boils, liquefaction-induced subsidence, and lateral spreading.	Numerous breaks in water and petroleum pipelines at oil refinery at Puerto Cortes.	Girth-welded steel pipelines withstood soil displacements of 0.5 m without rupture.	Mahan et al.[5]

(TABLE 1. *Continued)*

Earthquake	Magnitude/ Intensity[a]	Surface Faults	Landslides	Liquefaction/ Earthquake Settlements	Reported Pipeline/Conduit Damage	Remarks	References
Miyagiken-Oki, 1978	Magnitude = 7.4 Intensity = VIII	No surface faulting observed.	Slumps and slides in loose fills.	Substantial settlements and lateral displacements near cut-and-fill boundaries.	More than 1600 pipeline failures, primarily in small-diameter lines. Damage concentrated in alluvial and cut-and-fill deposits.	Damage prominent at threaded small-diameter steel pipelines and as pull-out of mechanical joints.	Katayama[6]
Imperial Valley, 1979	Magnitude = 6.6 Intensity not reported	Up to 0.6 m of coseismic right lateral strike slip on Imperial fault; vertical disp. of 190 mm; vertical offsets of 150 and 200 mm on Brawley and Rico faults.	Relatively flat terrain. Widespread slumping of embankments.	Numerous sand boils and liquefaction-induced subsidence up to 1.2 m. One lateral spread with approx. 2.1 m max. disp.	Approx. 30 breaks in water pipelines. Most breaks in older cast iron pipelines in areas parallel to Imperial Fault.	Three welded steel gas pipelines were deformed at fault crossings, but not broken.	Waller and Ramanathan[7]
Coalinga, 1983	Magnitude = 6.6 Intensity = VIII	No surface rupture observed after main shock. Max. vertical and horizontal offsets of 600 and 200 mm, respectively, following an aftershock.	Numerous landslides and rockfalls. Up to 200,000 m³ of soil involved in largest landslide.	Minor settlement of road fills. Lateral spreading along river banks.	Approx. 13 leaks in water pipelines, mostly in cast and wrought iron, and asbestos cement pipe.	Many of the pipeline breaks were associated with corrosion or areas of known weaknesses.	Isenberg and Escalante[8]; Isenberg and Taylor[9]

Earthquake	Magnitude/Intensity	Surface Faulting	Ground Movement	Other Effects	Pipeline Distribution Damage	Pipeline Damage	Reference
Nihonkai-Chubu, 1983	Magnitude = 7.7 Intensity = VIII-IX	No surface faulting observed.	Downslope movements mostly related to soil liquefaction.	Lateral spreads with maximum 5-m displacement. Liquefaction-induced subsidence; conspicuous sand boils.	Substantial damage to water and gas distribution piping caused by liquefaction-induced soil movements.	Compressive and tensile failures of 80-mm-diameter welded steel gas pipelines.	Hamada et al.[10]
Chile, 1985	Magnitude = 7.8 Intensity = VIII-IX	No surface faulting observed.	Landslides in steep slopes and slides in hillsides of loose sand.	Tailings dam failures from liquefaction; lateral movement of bridge piers and liquefaction failures along sea walls.	Ruptures in major aqueducts and extensive failures in water distribution pipelines.	Pipeline damage due to permanent soil deformation with cast iron and asbestos cement pipeline damage reported.	Wyle et al.[11]

(TABLE 1. Continued)

Earthquake	Magnitude/ Intensity[a]	Surface Faults	Landslides	Liquefaction/ Earthquake Settlements	Reported Pipeline/Conduit Damage	Remarks	References
Michoacan, 1985	Magnitude = 8.1 Intensity = IX	No surface faulting observed.	Some landslides and rockfalls in mountainous terrain.	Settlements in fill and adjacent to pile-supported buildings in Mexico City.	More than 5300 repairs of trunk lines and mains in the water distribution system of Mexico City. More than one month required to reinstate the system.	Pipeline damage caused predominantly by traveling ground waves; substantial damage to asbestos cement and cast iron pipelines. Compressive failures at pipeline joints.	Ayala and O'Rourke[12]
San Salvador, 1986	Magnitude = 5.4 Intensity not reported.	No conclusive evidence of surface faulting, although several long linear surface cracks were observed.	Numerous landslides in man-made fills and natural soil and rock. Majority of landslides in poorly consolidated volcanic tuff along steep slopes.	Compaction and differential settlement of fills. No observation of soil liquefaction.	Approximately 2500 breaks in the water distribution system of San Salvador.	Damage conspicuous in ferro cement pipelines.	Morgan[13]

Edgecumbe, 1987	Magnitude = 6.3 Intensity = IX-X	Seven linear surface ruptures mapped. Principal normal fault 7 km long. Maximum vertical disp. of 1.5 m; maximum extension of 1.3 m across fault.	No major slides reported.	Extensive liquefaction and lateral spreading near rivers; sand boils reported.	More than 400 breaks in water and wastewater pipelines. Damage observed in pipelines at bridges because of adjacent soil movement.	A 114-mm-diameter gas main crossing the principal fault was undamaged. Extensive damage in asbestos cement pipelines, mainly as compressive failures at joints.	Pender and Robertson[14]
Ecuador, 1987	Magnitudes = 6.1 and 6.9 Intensity = XI	No surface faulting detected	Extensive rock falls, landslides, and debris flows destroying approx. 100 km² of forest and involving as much as 100 million m³ of debris.	Settlements of 50–100 mm in river terrace deposits at the Salado Pump Station. No liquefaction detected.	Approximately 40 km of the 660-mm-diameter Trans-Ecuadorian and 150-mm-diameter Poliducto pipelines destroyed or severely damaged by debris flows and flooding.	Loss of Trans-Ecuadorian pipeline was the largest single pipeline loss in history.	Crespo et al.[15]

[a] Maximum MMI reported in the literature or inferred from personal reconnaissance of the author.

River. Much damage was caused by flooding of the Coca River. The flood zone is shown by the stippled area in the figure.

In contrast with the earthquakes in Ecuador, the 1985 Michoacan earthquake caused extensive pipeline damage almost entirely from traveling ground-wave effects. More than 5300 repairs in trunk lines and distribution mains of the water supply of Mexico City were reported.[12] Most damage of both pipelines and buildings was confined to portions of the city underlain by soft lake-bed clays and silts. Before the earthquake, settlement caused by consolidation of these clays may have weakened pipeline components. Permanent ground movements, however, should be regarded only as an indirect factor affecting the condition of the system. Virtually all pipeline damage during this event was a result of ground-wave interactions.

There are several noteworthy trends in pipeline performance which emerge from a general review of the earthquake records, each of which is summarized and discussed according to the subheadings that follow.

Asbestos Cement Pipelines

Repairs evaluated after the 1978 Miyagiken-Oki, 1983 Coalinga, 1985 Chile, 1985 Michoacan, and 1987 Edgecumbe earthquakes show that substantial damage was sustained by asbestos cement pipelines as a consequence of both transient and permanent movements. Repairs per kilometer after the 1978 Miyagiken-Oki earthquake were higher for asbestos cement than any other type of pipeline by a factor of nearly two or greater.[6] In portions of Mexico City, where approximately 90% of the water distribution mains were made of asbestos cement, the damage was extensive. Asbestos cement is a brittle material which fails at relatively low strains. In addition, a typical joint consists of a cylindrical coupling within which two tapered pipe ends are inserted and sealed against leakage by circular gaskets. The tapered ends are fragile and susceptible to battering and compressive failure. This weakness has affected performance as is evidenced by the many compressive failures observed at the joints of asbestos cement pipelines after the 1985 Michoacan and 1987 Edgecumbe earthquakes.

Locations of Corrosion and Reduced Wall Thickness

It has been well established that leaks in water and gas piping develop after earthquakes at corrosion pits in steel, "graphitized" sections of cast iron, and threaded connections in steel pipe.[1,16] Sections of pipeline systems most susceptible to corrosion are therefore likely candidates for seismic damage. Repairs and replacements, which aim to reduce corrosion-related damage may also be regarded as measures for strengthening systems against earthquakes.

Pipeline Connections

Locally high stress will develop at pipeline connections with larger structures if sufficient flexibility is not provided to accommodate differential movement. Damage of pipelines and telecommunication conduits at penetrations of manholes and subsurface vaults was observed after the 1983 Nihonkai-Chubu and 1985 Michoacan earthquakes. Damage at pipeline penetrations of bridge abutments was observed after the 1978 Miyagiken-Oki and 1987 Edgecumbe earthquakes. Differential settlement of fill adjacent to bridges highlights a potentially weak location. Many large-diameter, high-pressure pipelines cross rail and highway routes at bridges, and it is prudent to provide flexibility where these facilities penetrate the more rigid bridge structures.

Plastic Pipelines

Plastic piping is being used in increasingly greater quantities as distribution mains and services, with the great majority composed of either polyvinyl chloride (PVC) or polyethylene. Damage per kilometer to PVC pipelines, 50 to 100 mm in diameter, was relatively high after the 1978 Miyagiken-Oki earthquake. More than two-thirds of the damage in Sendai City was confined to cut-and-fill areas in which large permanent ground movements occurred, and less than 6% was found in the relatively stable terrace deposits located nearby.[6] Although damage per kilometer of asbestos cement and cast iron water mains was relatively high after the 1983 Coalinga earthquake, there was no damage observed in the 0.32 km of PVC pipelines in operation at the time.[9] Observations after the 1987 Edgecumbe earthquake indicate that 100-mm-diameter PVC water mains performed well.[14] As a result of the 1985 Michoacan earthquake, damage at water services and service connections was extremely high in areas of the Federal District, which did not use PVC in services. Where accurate repair records were available in the State of Mexico, not a single break at a PVC service or service connection was reported.[12]

Because of its flexibility, it appears that PVC piping has been able to accommodate traveling ground waves without high repair rates. In areas of permanent ground movement during the 1987 Miyagiken-Oki earthquake, there is evidence that repair rates were comparable to those of other types of pipeline.

Welded Steel Pipelines

There is little evidence of damage to continuous girth-welded steel pipelines when construction of the pipeline was performed according to modern specifications and quality-control procedures. There is virtually no record of damage to this type of pipeline from traveling ground-wave effects, with the exception of a 660-mm-diameter water trunk line in which compressive failures were observed after the 1985 Michoacan earthquake.[12] Ruptures of a 150-mm-diameter liquid fuel as well as 80-mm-diameter gas pipelines were caused by liquefaction-induced soil movements during the 1971 San Fernando[3] and 1983 Nihonkai-Chubu[10] earthquakes, respectively.

Evidence from the 1933 Long Beach, 1952 Kern County, and 1971 San Fernando earthquakes[1] indicates that some steel pipelines constructed before and during the early 1930s do not perform well in response to both permanent and traveling ground-wave movements. Pipelines of this vintage did not benefit from the same quality controls during construction that apply today. As a result, failures at a large number of low-quality girth welds have occurred during several earthquakes. These pipelines should be regarded as potential "bad actors."

SOURCES OF SEISMIC DEFORMATION

Because pipelines are affected by both permanent movements and traveling ground waves, the question often arises as to which source of disturbance is most important. The question can be misleading because it implies an order of importance without considering the context of operation, state of repair, and system characteristics. Traveling ground waves cover a larger area than do localized permanent ground movements, and thus the opportunities for disturbing weakened or constrained portions of a piping system are correspondingly greater. Permanent ground movements, however, are likely to exceed the peak ground displacements from seismic waves. Accordingly, they represent the conditions of maximum distortion for buried pipelines and conduits, and generally can be taken as the most severe deformations possible during an earthquake.

When considering the effects of permanent and traveling ground-wave movements, it is best to think first of the type of piping that will be affected. Jointed pipelines, pipelines composed of brittle materials, such as asbestos cement and cast iron, and pipelines subject to corrosion and reduced wall thickness are vulnerable to ground-wave effects.

Permanent ground movements generally are critical when working with large, high-pressure pipelines, for which there is concern related to system supply and safety. Many high-pressure pipelines, and virtually all modern gas and liquid fuel transmission lines, are made of girth-welded steel. These facilities are sufficiently strong in general to resist traveling ground-wave effects, but are vulnerable to large permanent soil and rock displacements.

PERMANENT GROUND-MOVEMENT EFFECTS

Permanent ground movements during earthquakes can be caused by soil liquefaction, landslides, surface faulting, and the consolidation of relatively cohesionless fills and loose natural deposits. All these sources of permanent deformation involve some distribution of the ground movement, for which the extreme case is an abrupt or knife-edge displacement in which all relative movement is concentrated along a single plane.

Two simplified models for buried steel pipeline response to abrupt surface faulting and similar types of ground rupture have been proposed by Newmark and Hall[17] and Kennedy et al.[18], and these are discussed with the aid of the illustrations in FIGURE 3. The figure provides plan views of a pipeline that crosses a right lateral strike-slip fault at an angle, β. The pipeline has been oriented to sustain elongation, or tension, as a result of differential movement. Thin-walled pipelines are especially vulnerable to compression, which may induce local wrinkling. A tensile orientation, however, is able to avoid buckling and wrinkling and take advantage of the inherent ductility of the steel.

As shown in FIGURE 3b, Newmark and Hall[17] assumed that the pipeline is subjected to constant frictional resistance and that the pipeline deforms in an antisymmetric pattern of two broad circular arcs, each extending from the fault intersection to an anchor point at a distance, L_a, from the fault. Anchors may be caused by bends, tie-ins, or other constraints which resist axial movement. Alternatively, the anchor point may represent an effective anchor length, beyond which there is no axial stress imposed in the pipeline from fault movement.

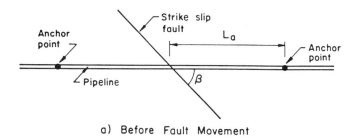

a) Before Fault Movement

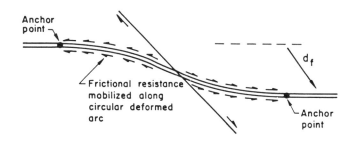

b) After Fault Movement
Newmark and Hall[17]

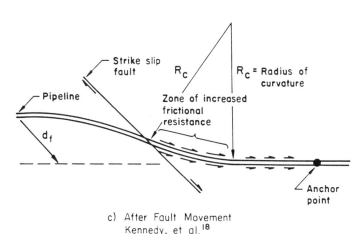

c) After Fault Movement
Kennedy, et al.[18]

FIGURE 3.

Newmark and Hall[17] used bi- and trilinear representations of the stress-strain response of steel, and did not consider bending strains.

As shown in FIGURE 3c, Kennedy et al.[18] assumed that the pipeline is subjected to increased axial friction in the zone of curvature near the fault. The radius of curvature is related to both the axial tension in the pipe and the lateral earth pressure mobilized against transverse movement of the pipe in the curved zone.

Kennedy *et al.*[18] assumed that the pipeline behaves as a cable under large deformation. Their model accounts for both axial and bending strains, and uses a Ramberg-Osgood formulation[19] to represent the nonlinear stress-strain behavior of the steel.

The maximum frictional resistance per unit pipe area, f, is evaluated by means of

$$f = 1/2(1 + K_o)\gamma H \tan \delta \qquad (1)$$

in which γ is the unit soil weight, H is the depth to center of pipe, K_o is the coefficient of earth pressure at rest, and δ is the angle of shearing resistance mobilized at the interface between the pipeline and adjacent soil. In the zone of pipeline curvature near the fault, Kennedy *et al.*[18] assume that the frictional resistance is increased to values of $2.4f$ and $3.3f$ for depths of cover equivalent to one and three times the pipe diameter, respectively.

To evaluate the sensitivity of the predicted pipeline response to modeling assumptions and variations in the steel and soil properties, a reference condition was studied which is illustrated in FIGURE 4. It was assumed that the pipeline crosses a right lateral strike-slip fault at $\beta = 65°$; it has a depth to centerline, diameter, and wall thickness of 1.45 m, 1.07 m, and 14 mm, respectively. The pipeline is surrounded by sand with a unit weight, $\gamma = 16.7 \, kN/m^3$, and a friction angle, $\phi = 35°$. Unless otherwise noted, the angle of interface shearing resistance was taken as $\delta = 28°$. The condition was analyzed in which there are no bends or tie-ins near the fault, and for which L_a is determined by the total frictional resistance needed for equilibrium against the axial force in the pipe at the fault line.

FIGURE 5 shows the uniaxial tensile stress-strain plots used to characterize the X-60-grade steel of the pipeline. FIGURE 5a shows three bilinear approximations, each with a different plastic modulus, E_2, superimposed on the measured stress-strain plot of Steel A. FIGURE 5b shows the measured stress-strain plots for two specimens of X-60 steel. Ramberg-Osgood formulations[19] have been used to represent the nonlinear characteristics of Steels A and B, in the form

$$\epsilon = \frac{\sigma}{E_1} \left[1 + \frac{\alpha}{(r + 1)} \left(\frac{\sigma}{\sigma_0} \right)^r \right] \qquad (2)$$

in which E_1 in Young's modulus, α and r are Ramberg-Osgood coefficients, and σ_0 is the effective yield stress of the steel. The parameters used for each steel are

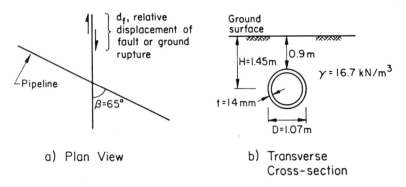

a) Plan View   b) Transverse Cross-section

FIGURE 4.

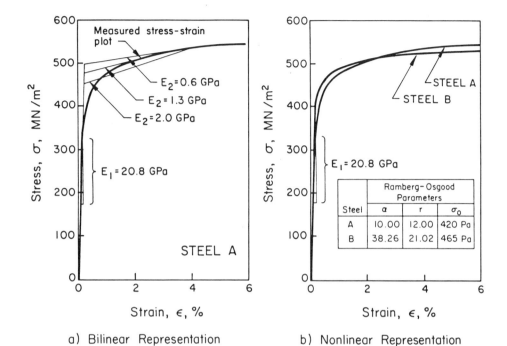

a) Bilinear Representation   b) Nonlinear Representation

FIGURE 5.

listed in FIGURE 5b. Steels A and B correspond to the X-60 steels used by Kennedy *et al.*[18] and the Committee on Gas and Liquid Fuel Lifelines,[20] respectively, in previous work explaining the use of simplified fault crossing models.

FIGURE 6 shows the analytical results obtained from the Newmark and Hall model[17] for the three bilinear representations of Steel A depicted in FIGURE 5a. The maximum pipe strain is plotted as a function of total fault displacement. Pipeline deformation is elastic until a fault movement between 3.4 and 4.3 m, after which there is substantial increase in strain with additional displacement. The sharp change in response is related to the bilinear characterization of the steel. It can be seen that the plastic modulus, E_2, has a profound influence on the predicted strain, provided that movements are sufficiently large to cause plastic elongation. Depending on the choice of E_2 and magnitude of fault displacement, there is a five- to six-fold difference in the maximum strain predicted.

FIGURE 7 shows the maximum pipeline strain plotted as a function of the ratio of the pipe interface to soil friction angles, δ/ϕ. The analytical results were obtained with the Newmark and Hall[17] model in which the maximum fault displacement was 4.6 m and the plastic modulus of the steel, E_2, was taken as 2.0 GPa. Two pipe-wall thicknesses, 9.5 and 14 mm, were studied. The predicted pipeline response is extremely sensitive to the level of frictional resistance along the pipeline, as referenced by δ/ϕ. Depending on the wall thickness and δ/ϕ, the predicted maximum strain for the same amount of fault displacement can differ by an order of magnitude. Part of this discrepancy arises from the choice of a bilinear model for stress-strain behavior of the steel. As the pipeline is deformed from the

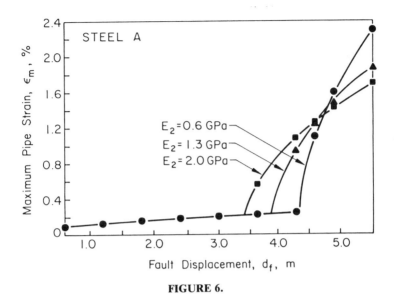

FIGURE 6.

elastic range, characterized by E_1, to the plastic range, characterized by E_2, there is an abrupt transition in the amount of strain for given level of fault movement. For example, as the interface friction angle for a pipe with 14-mm wall thickness increases by 40% from $\delta/\phi = 0.6$ to 1.0, the maximum predicted strain increases from 0.2 to 1.9%.

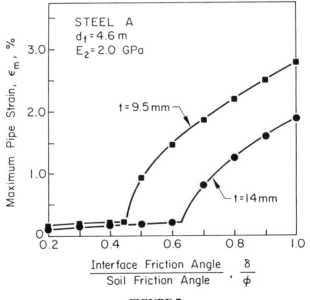

FIGURE 7.

By convention, the interface friction is expressed by geotechnical engineers as a fraction of the shear strength capacity of the adjacent soil in the form δ/ϕ. In the limit the interface angle of friction, δ, equals ϕ, when the pipe surface is rough and oxidized such that a bond forms between the pipe and adjoining soil. Under these conditions, shear failure occurs entirely through the soil next to the pipe surface, with the resistance to shear controlled by ϕ.

TABLE 2 summarizes the range in δ/ϕ for various types of pipe surfaces, including bare and coated steel. As part of corrosion protection measures, many steel pipelines are coated with thermosetting resin epoxies or polyethylenes. These hard, polymer surfaces will generally be characterized by $0.5 \leq \delta/\phi \leq 0.6$ when new. Polyethylene coatings are chemically resistant and stable. Epoxies, on the other hand, may soften when they age so that, in the long term, the surface hardness may decrease. To account for this possibility, resin epoxy coatings are accorded a range of δ/ϕ between 0.6 and 0.8 in TABLE 2. Coatings, such as coal tar epoxies, provide a surface in which adjacent sand and gravel can become embedded such that the shear surface forms through the soil immediately adjacent to the pipe. Under these conditions the angle of interface shear resistance approaches ϕ.

The value of δ/ϕ is not likely to be less than 0.5. A prudent lower bound choice of δ/ϕ is 0.6, which is applicable for polyethylene or polyolefin coatings. It is

TABLE 2. Pipeline Interface Angles of Friction for Contact with Sand

Pipeline Interface Material	Ratio of Interface to Soil Angle of Friction, δ/ϕ
Rusted and pitted steel, partially cemented and bonded to adjacent soil	1.0
Soft coatings and wrappings, such as coal tar enamels, hot or cold applied mastics, and coal tar epoxies	0.8–1.0
Rough steel, some oxidation and rusting of surface with minor pitting	0.7–0.9
Resin epoxy coating[a]	0.6–0.8
Polyolefin or polyethylene coating	0.5–0.6

[a] Assumes some aging and softening.

unlikely that δ/ϕ for smooth steel will apply because steel oxidizes, thereby roughening the surface. Moreover, it is mandated by U.S. Federal regulations that new gas and liquid fuel pipelines be coated. Values for bare steel would apply for pipelines that have been operating for some time, in which case δ/ϕ should be taken for an oxidized and roughened surface.

In the Kennedy et al.[18] model the radius of curvature, R_c, of the pipeline immediately adjacent to the fault is estimated as

$$R_c = \frac{\pi \sigma D t}{p_u} \qquad (3)$$

in which σ is the axial stress in the pipeline at the fault, D and t are the pipe diameter and wall thickness, respectively, and p_u is the lateral soil resistance per unit length of pipe caused by relative movement between soil and pipe near the fault.

There is considerable uncertainty in estimating p_u. Different methods of evaluation presented by the Committee on Gas and Liquid Fuel Lifelines[20] can result in

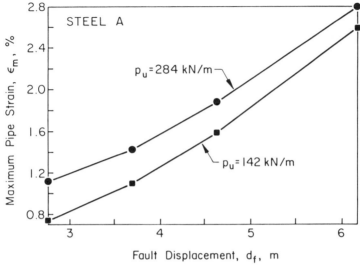

FIGURE 8.

values with a two-fold difference for identical soil, depth, and pipe diameter. There is a good mechanical basis for this uncertainty. The experimental backup for evaluating p_u consists of full-scale tests in dry coarse to medium sand. Relatively dense, partially saturated fine and silty fine sands may develop additional lateral resistance through soil suction as the soil dilates in the response to shear strain.

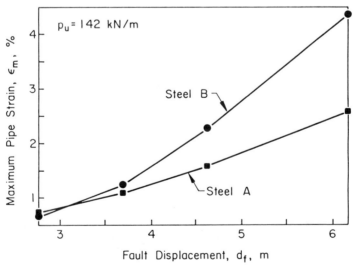

FIGURE 9.

For the reference conditions illustrated in FIGURE 3, analyses were performed with the Kennedy *et al.* model in which p_u was estimated as 142 kN/m by the methods proposed by Trautmann and O'Rourke.[21] To account for uncertainties on the unconservative side, the analyses were performed again for values of p_u that were twice as high. The results are shown in FIGURE 8, in which the maximum pipe strain is plotted as a function of fault displacement. For fault displacements between 4 and 5 m, there is about a 20% increase in the predicted maximum strain as p_u doubles from 142 to 284 kN/m.

FIGURE 9 shows the maximum pipe strain predicted from the Kennedy *et al.* model as a function of fault displacement for the two different stress-strain plots, corresponding to Steels A and B. Although there is little apparent difference in the stress-strain properties shown in FIGURE 4b, there is a significant difference in the analytical results as fault movements increase from 4 to 6 m. At 5 m of fault

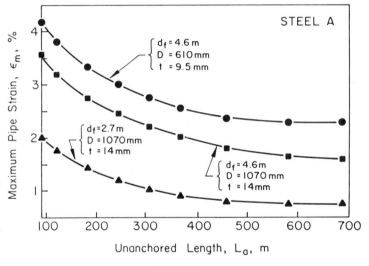

FIGURE 10.

displacement, there is more than a 50% increase in predicted strain for Steel B. In uniaxial tension, Steel B is characterized by a flatter, lower modulus relationship between stress and strain than Steel A for strains exceeding 1%.

It has been well established by previous studies[17,18] that the location of anchors, or points of restraint, will have a strong influence on pipeline strains. In FIGURE 10, the maximum pipeline strain predicted by the Kennedy *et al.* approach is plotted as a function of the effective anchor length. This anchor length can be defined conservatively as the distance from the fault centerline to a sharp bend or tie-in. Analytical results for the reference conditions in FIGURE 4 are plotted for fault movements of 2.7 and 4.6 m, respectively. In addition, results are shown for a 610-mm-diameter pipeline with 9.5-mm wall thickness, subjected to 4.6 m of fault movement with the same crossing angle, cover depth, and soil properties as the reference conditions. These analytical results corroborate the findings of previous studies[18,22] that pipeline strains are particularly sensitive to anchor locations within about 200 m of the fault. It has been asserted in previous work that pipeline performance is not sensitive to unanchored length so long as the length exceeds about 200 m[22]. Caution should be exercised when implementing this observation. As can be seen in FIGURE 10, there is a 20 to 40% reduction in maximum pipe strain as the unanchored length increases from 200 to approximately 300 m, after which there is only a small incremental change with distance. This reduction in deformation can be beneficial. Extending the unanchored length from 200 to 300 m, for example, would provide an ample margin of safety against uncertainties in the lateral resistance of the soil, p_u.

CONCLUSIONS AND RECOMMENDATIONS

Observations after previous earthquakes show that permanent ground movements and traveling ground waves are both major causes of pipeline damage. Traveling ground waves affect a broader area than do more localized permanent movements so that opportunities for disturbing weakened or constrained portions of a piping system are correspondingly greater. System components most susceptible to ground-wave effects include jointed pipelines, pipelines composed of brittle materials, and pipelines subject to corrosion and reduced wall thickness. Permanent ground movement represents the condition of most severe distortion for buried piping during an earthquake. This type of movement is caused by soil liquefaction, landslides, surface faulting, and the dynamic consolidation of cohesionless soils. Permanent ground movements generally are critical when working with high-pressure pipelines, for which there is concern related to system supply and safety.

Analytical models of pipeline response to fault movements are influenced strongly by the type of stress-strain representation used for the steel. It is recommended that continuous nonlinear representations, such as Ramberg-Osgood formulations,[19] be used. Bi- or trilinear representations can be useful as an approximation, if care is taken to check that the strain predicted by a simplified bi- or trilinear plot is very close to that of the measured stress-strain curve for the level of stress compatible with the design ground movement.

Buried pipeline response to faulting is very sensitive to frictional force mobilized between the pipeline and surrounding soil. For granular soils, the interface frictional resistance between pipe and soil often is estimated from a ratio of interface to soil friction angle, δ/ϕ. Ranges of δ/ϕ for various types of pipe surfaces and coatings are recommended in this paper. As a practical matter, δ/ϕ should not be less than 0.5, and should be selected on the basis of the long-term pipe surface characteristics.

ACKNOWLEDGMENTS

Sincere thanks are extended to W. D. Meyersohn for his careful analyses of pipeline response to fault movement. The efforts of Laurie Mayes, who typed the manuscript, and Ali Avcisoy, who drafted the figures, are gratefully acknowledged.

REFERENCES

1. O'ROURKE, T. D., M. D. GRIGORIU & M. M. KHATER. 1985. Seismic response of buried pipes. *In* Pressure Vessel and Piping Technology—A Decade of Progress. C. Sundararajan, Ed.: 281–323. ASME. New York.

2. STEINBRUGGE, K. V., E. E. SCHADER, H. C. BIGGLESTON & C. A. WEERS. 1971. San Fernando Earthquake February 9, 1971. Pacific Fire Rating Bureau. San Francisco, CA.

3. O'ROURKE, T. D. & M. S. TAWFIK. 1983. Effects of lateral spreading on buried pipelines during the 1971 San Fernando earthquake. *In* Earthquake Behavior and Safety of Oil and Gas Storage Facilities, Buried Pipelines and Equipment. Vol. PVP77: 124–132. ASME. New York.

4. CAJINA, A. 1972. The Managua earthquake and its effects on the water supply system. *In* Proceedings of the Earthquake Engineering Research Institute Conference. Vol. 2: 768–790. Oakland, CA.

5. MAHAN, J. F., H. J. DEKENKOLB, D. F. MARON & K. V. STEINBRUGGE. 1973. Engineering aspects. Managua, Nicaragua Earthquake of December 23, 1972. Reconnaissance Report: 27–214. EERI. Berkeley, CA.

6. KATAYAMA, T. 1980. Seismic behaviors of lifeline utility systems. J. Natural Disaster Sci. 2: 1–25.

7. WALLER, R. & M. RAMANATHAN. 1980. Site visit report on earthquake damages to water and sewage facilities El Centro, California, November 15, 1979. Imperial Valley, California Earthquake, October 15, 1979. Reconnaissance Report: 97–106. EERI. Berkeley, CA.

8. ISENBERG, J. & L. E. ESCALANTE. 1984. Damage to lifelines. Coalinga, CA, Earthquake of May 2, 1982. Reconnaissance Report: 227–247. EERI. Berkeley, CA.

9. ISENBERG, J. & C. TAYLOR. 1984. Performance of water and sewer lifelines in the May 2, 1983 Coalinga, California earthquake. *In* Proceedings of the Symposium of Lifeline Earthquake Engineering: Performance, Design and Construction. ASCE. New York.

10. HAMADA, M., K. KUBO & K. SAITO. 1985. Large ground displacement and buried pipe failure by soil liquefaction during 1983 Nihonkai-Chubu earthquake. *In* Seismic Performance of Pipelines and Storage Tanks. Vol. PVP98-4: 11–18. ASME. New York.

11. WYLLIE, L. A. *et al.* 1986. The Chile Earthquake of March 3, 1985. Earthquake Spectra. Vol. 2(2). 249–513.

12. AYALA, G. & M. J. O'ROURKE. 1988. Effects of the 1985 Michoacan Earthquake on the Water Systems in Metropolitan Mexico City. Tech. Report NCEER-88-0023. National Center for Earthquake Engineering Research. Buffalo, NY.

13. MORGAN, J. R. 1987. Performance of buried lifelines during the 10 October 1986 San Salvador earthquake. *In* Structures and Stochastic Methods. A. S. Cakmak, Ed.: 263–267. Elsevier. New York.

14. PENDER, M. J. & T. W. ROBERTSON. 1987. Edgecumbe earthquake: Reconnaissance report. Bull. New Zealand Natl. Soc. Earthquake Eng. 20: 201–249.

15. CRESPO, E., K. J. NYMAN & T. D. O'ROURKE. 1987. 1987 Ecuador earthquakes of March 5, 1987. EERI Newsl. 21: 1–4.

16. ISENBERG, J. 1979. Role of corrosion in water pipeline performance in three U.S. earthquakes. *In* Proceedings of the 2nd U.S. National Conference on Earthquake Engineering.: 683–692. Stanford, CA.

17. NEWMARK, N. M. & W. J. HALL. 1975. Pipeline design to resist large fault displacement. *In* Proceedings of the U.S. National Conference on Earthquake Engineering.: 416–425. Ann Arbor, MI.

18. KENNEDY, R. P., A. W. CHOW & R. A. WILLIAMSON. 1977. Fault movement effects on buried oil pipeline. J. Transport. Eng. Div. 103(TE5): 617–633. ASCE. New York.

19. RAMBERG, W. & W. R. OSGOOD. 1943. Description of stress-strain curves by three parameters. Technical Note TN902. National Advisory Committee for Aeronautics. Washington, DC.

20. COMMITTEE ON GAS AND LIQUID FUEL LIFELINES. 1984. Guidelines for the seismic design of oil and gas pipeline systems. ASCE. New York.
21. TRAUTMANN, C. H. & T. D. O'ROURKE. 1985. Lateral force-displacement response of buried pipe. J. Geotech. Eng. **111:** 1068–1084. ASCE. New York, NY.
22. KENNEDY, R. P. & R. H. KINCAID. 1985. Fault crossing design for buried gas and oil pipelines. *In* Seismic Performance of Pipelines and Storage Tanks. Vol. PVP98-4: 3–10. ASME. New York.

General Discussion: Part VI

IRWIN CANTOR (*The Office of Irwin Cantor, New York, N.Y.*): I'm a practicing engineer in New York. Mr. Nordenson, I will refer to studies you've done on steel buildings and concrete buildings—and I recognize that perhaps you changed the concrete frame completely in a high-rise concrete building. Assuming that the frame in the concrete building could be made similar to the current concrete-type building flat plates we're using, what would the increment in construction costs be for a steel or concrete building following the UBC-2A code?

GUY NORDENSON (*Ove, Arup & Partners, New York, N.Y.*): The only work that I know of that looked at that carefully is the ATC trial design program, and there were weaknesses in that study because it didn't really work itself into the actual construction practice that we have here. For the few buildings that were similar to the ones here, the percentage increase was probably on the order of about 1 and maybe 2 percent in the total cost of construction; but that's a very rough number. I did a lot of the trial designs and I know that it was very crudely estimated. We need to do a lot of work on that kind of problem.

UNIDENTIFIED SPEAKER: The comparison of spectra and other things from different codes could be truly misleading because it doesn't include other parameters of a code such as load factors.

Also, I have a real problem with this wind and unit weight density issue. For the building, the wind is acting on one surface. Where do the depth of the building or the dimension of the building perpendicular to the plane that you're considering come in? Is it all for one square building of known base in each direction? Is it unique to the one building? I don't understand it.

NORDENSON: The analysis is crude, but I think it's a correct comparison. Remember that we showed height to depth ratio of the structure, so the actual proportion of the structure is included and the density is the range of density that one finds. There's no claim that this is absolutely scientific.

What we are trying to do here is to elaborate a point that is often made, namely: by providing wind resistance, we automatically provide seismic resistance in certain types of buildings. We just compare the load on the building, without recognizing the different inherent energy-dissipating capacities of different types of systems. We assume that we are to design structures for earthquakes the way we design structures for wind, expecting elastic response in the worst-case event. We need to look at the problem of earthquake resistance in greater detail and not simply say that the wind resistance will provide.

Summary Panel Discussion II

ROBERT WHITMAN (*Massachusetts Institute of Technology, Cambridge, Mass.*): I approach this topic from the kind of schizophrenic vision of both researcher and engineer. As a researcher, I focus on the things we're not doing well now and how we might do them better in the future. The engineer part of me, on the other hand, tends to focus on what can we do now with the knowledge that is already available to us.

A very important point has been made already: There is a lot more to earthquake engineering than knowing what forces to put on buildings and how to analyze for those forces. Simply understanding how buildings behave and how to design a building that will perform well during an earthquake are very important.

Let me just focus on two topics: building code provisions, which govern the design of most buildings and structures, and ground motions, which form the basis for the analysis and design of structures not covered by codes.

With regard to both of those topics there are three possible approaches. We might decide to "go for it" and adopt an already-available code, or a recommended set of ground motions, for analysis and design. Alternatively, we might decide to pause briefly and review recent developments, and then modify some existing procedures. Or, we might decide to do nothing until research is further advanced. Although one might conceivably argue that we know so little about some aspects of the topic that the risk of doing something is worse than the risk of doing nothing, I do not believe one can really argue to do nothing at this time. Hence, I'm going to discard that last possibility, and come back to the others.

With regard to building codes, one strategy is to adopt one of the existing model codes. It would be interesting to discuss which one, but I don't want to get into that right now. Or, we might decide to modify some of the provisions in an existing code. One kind of modification would be to change either the seismic zone and/or the lateral force equation; in other words, to influence the force level to be used in the design of structures. Another possible change would be to modify in some way the detailing or ductility provisions.

We require ground-motion specifications for analysis and design of buildings that don't meet the ductility and detailing requirements of the code; for example, existing buildings. Such specifications are needed for structures that are not covered by codes. We would be interested in ground-motion specifications if we want to try to make sure that the contents of a structure—critical equipment and so forth—continue to function, or the structures involved require special treatment because they're deemed essential to the community, are especially hazardous, or are owned by someone who wishes to reduce risk of damage to the buildings. The discussion so far has emphasized some of the issues concerning ground motions, in terms of frequency content, duration, site-specific effects, and the probability of exceedance.

To come back to options for action, any community might decide to simply adopt one of the existing model code provisions, which is not a bad strategy for a community within eastern North America. For ground-motion specifications, we might, as Klaus Jacob has suggested, decide to go with the ATC-14 spectra for the purpose of examining existing buildings and their adequacy.

Under the second option, we might pause and review recent developments and modify available standards. In that regard, I'll simply leave you with a story that some of you may have heard. I attribute it to George Housener, who at a meeting once observed that it's obviously impossible to construct a seismic zoning map. Scientifically, we simply don't know enough to do it. But, he went on, if you take five good, knowledgeable people and put them in a room and tell them they can't come out until they've done it, they will in fact do it. And they will come out with a product that is really quite good and useful. If a decision is made to pause and review recent knowledge and then to come up with some modification to existing available procedures, that's the approach I would recommend. Get five good people to sit down and do it, and present those specific recommendations to the community for further discussion and eventually for adoption.

CHARLES SCAWTHORN (*EQE Inc., San Francisco, Calif.*): I would like to discuss the notion of realistic policies, and I suspect that in this panel we're going to be hearing seven or eight versions of reality; certainly, the Whittier earthquake was reality. And so I thought I'd share a little of that experience with you. I want to talk about a couple of buildings in Rosemead, which is in an area with Modified Mercalli Intensity 7.

Walnut Street is on an alluvial outwash. Adjacent areas are urbanized, but they're on firmer soils. The alluvial outwash was not built up until very recently. On the firm soils there was absolutely no damage. Chimneys and typical California housing had no cracks or damage.

On the outwash Rutgers recorded about .4 g as the peak ground acceleration. There was obvious amplification. On the 5 percent response spectra, there were accelerations on the order of 1 g at the fundamental period of about eight-tenths of a second.

Let's talk about buildings. One was built in the late 1960s and, in round numbers, sustained about 10 million dollars worth of damage. It was next to the recording device. Another building of 300,000 square feet was built in 1983. It was a four-story braced frame with K braces and wide flange structures that buckled. The building sustained about 5 million dollars worth of damage, but the "down time" for this building, which is the corporate center and computer center for a major corporation, was only a couple of days. Most of the damage was actually water damage due to sprinklers.

The building right next door, built in 1979, is another corporate headquarters. It was a two-story tilt-up, the second story of which was a light metal framing. Structurally, the building didn't make sense when I walked through it. Damage included major cracks in the floors, and the building had to be braced throughout. This corporation just moved out of the building and lost the use of it for an estimated eight months. So, here we see three buildings with 10 million dollars worth of damage, 5 million dollars worth of damage, and untold damage in the third case. All of this was for a magnitude 5.9 earthquake—that's structural damage.

For past earthquakes, fire damage has been compared to shaking damage. In some cases the damages are comparable.

The Office of Emergency Services has estimated the total losses in the Whittier earthquake at 400 million dollars. That's a very crude number and it's climbing. Some persons have put the damage at 500 million dollars. The California Department of Insurance issued a call to all insurance companies involved. To date more than 8000 claims have been settled by the insurance industry, with a total pay-out of more than 72 million dollars.

About 38 percent of the 72 million dollars was due to earthquake shaking policies, and 19 percent to homeowner's policies, generally to pay to house homeowners who had no earthquake insurance, but who had left their houses because they were afraid to live in them. Fire claims represented 10 percent of that 72 million dollars, commercial loss 13 percent, workman's compensation 5 percent, and auto claims 1 percent. These are very preliminary numbers. The NSF has recently funded a study which will go into this in greater detail.

I want to remind you that even with a modest earthquake of magnitude 5.9, modern buildings built in the 1960s, in 1979, and in 1983 sustained millions of dollars worth of damage and business for just one corporation interrupted for eight months. Earthquake shaking damage represents only a fraction of the total damage.

The best way to get a handle on reality is to do a cost/benefit analysis such as the ground-breaking work that was done at M.I.T. in the 1970s, which led to the adoption of a seismic building code in Massachusetts. Cost/benefit analyses must be done throughout all the various communities.

IRWIN G. CANTOR (*The Office of Irwin G. Cantor, PC, New York, N.Y.*): I've come away from this conference thoroughly convinced that there are and will continue to be earthquakes in California. I'd like to quote from Dr. Jacob in an abstract from the folder we were given at the beginning of the conference: "The brevity of the historic seismic record, under four hundred years, makes it uncertain whether future earthquakes are restricted to regions of recent activity and whether maximum magnitudes will exceed those in the historic record in any given region. The uncertainty is most consequential for the largest events with magnitudes greater than 6 or 7, which infrequently can cause great damage. The virtual absence of strong ground-motion records from significant eastern earthquakes prevents high-confidence estimates of future ground motions as function of distance and earthquake magnitude. Dynamic models, computed motions and analogs from other parts of the world are now used instead to estimate the ground motions for the East. An option is to adopt conservatively high design ground-motion levels and thus allow for conservative safety margins. Economic constraints tend to prevent this option to be realized. Therefore, eastern building codes and eastern construction practices hardly mitigate the effects of moderate-sized earthquakes and are almost certainly insufficient to provide effective protection. This leaves the public exposed to infrequent, potentially catastrophic risks that decision-makers must start to take into account."

Professor Whitman has observed that first we must acknowledge that there is a problem before we can begin to try to solve it. Now I have been accused by my Community Planning Board as being a tool of real estate interests because, aside from being a practicing engineer, I am concerned about the costs of construction. And I am finding it very difficult to be convinced that the problem, as it is being posed, is one that we can handle in a logical and economic way.

The numbers that were presented at this conference are percentages varying between 1 and 2 percent of the cost of construction. From what I understand, in New York in the past year somewhere between 2 and 3 billion dollars worth of buildings were constructed. And that's only in the immediate metropolitan area. And 2 percent of 3 billion dollars is about 60 million dollars a year spent on possibly protecting the metropolitan region from earthquakes with MMIs of approximately 5, which, from my limited understanding, have occurred either once or twice within the past 350 years. Consequently I have a really difficult time trying to persuade the developer of a 30- or 40-million dollar project, which is not an uncommon sized venture, to spend another 120 to 150 thousand dollars to reinforce the structure when these same developers try to whittle the cost of construction by 10 thousand dollars here and 20 thousand dollars there. That poses a big problem right there.

CHARLES A. DEBENEDITTIS (*Real Estate Board of New York, New York,*

N.Y.): The objectives of this conference are to review the scientific basis for assessing earthquake hazards in the eastern United States, to develop a realistic estimate of the extent of such hazards, and to assess alternative policies for the engineering design community and related regulatory agencies in response to these risks.

Coincidentally in Los Angeles a conference is being held on the current trends in planning, design, and construction of tall buildings in seismic regions. There is an obvious difference in concern between these two parts of the country, but there is also another very important difference, and that is the extent to which structures existed before these issues were addressed and earthquake codes were created.

In Los Angeles, I was a construction superintendent in the late 1950s on a building that was, at 13 stories, among the highest in the city. The building code prohibited higher buildings because of concern for earthquakes. Subsequently, with the enactment of a code that deals with this concern, the height limit was rescinded and all buildings over 13 stories were designed and built to comply with several generations of seismic code requirements.

Here in the east we have the problem of contending with a considerable number of the tallest buildings in the world which are already in existence. What do we do about them? What is the relative importance of dealing with remote possibilities of major earthquake damage in new buildings only? We should examine the overall picture and if we conclude that some preventive seismic measures are to be taken, then we should find the weakest links in the chain of issues that comprise the risks and make sure that we address those with the highest risk/reward ratio. Furthermore, I believe that because of the remoteness of what we'll call a "100-year earthquake" (100 years is longer than the projected economic life of buildings), we should concern ourselves with risks to life rather than to property damage. And, although those risks cannot be entirely separated, it is safe to say that, with buildings, for example, the exterior cladding can be more threatening to life than the structural frame would be.

Fortunately in New York City we have in the past several years developed legislation requiring the periodic detailed review of the structural stability of exterior walls. This probably addresses in New York City what would be one of the weakest links, if not the weakest, in the chain of life safety concerns due to seismic activity. That is the direction that appears to be indicated until a much more serious threat of major and imminent earthquake is documented for this region of the country.

UNIDENTIFIED SPEAKER: I was involved in the ATC study, the Building Seismic Safety Council study, to try to assess the cost impact of adopting the ATC code for New York City. I believe that the numbers given for cost increases to include the ATC seismic provisions for buildings in New York City in the range of 1 to 2 percent are probably true. Thus, the cost is not very much relative to the total cost of the building. And the ATC study did not include very tall buildings. Of course, with taller buildings, and lower natural frequencies, the earth-shaking effect would be less, and, on the basis of our studies, it looked as if gross structural effect would probably be pretty close to nil for steel buildings higher than thirty stories and for concrete buildings of about forty stories. However, some additional considerations have to be given if a building is going to be made resistant to earthquakes.

The cost is not very much, but it's a cost that developers are not going to be willing to bear if it isn't shown to return some benefit. The perspective on this probably has to get much broader and persons who occupy and use the buildings will have to make their feelings known.

It is also true that while the additional cost of new construction would not be too significant, many problems will arise if you try to upgrade existing buildings in New York City. Many of the buildings are probably strong, but just the fact that buildings are built so close together is a problem.

RICHARD TOMASETTI (*Thornton-Tomasetti, PC, New York, N.Y.*): I'd like to tell a story. A couple of years ago, a client came to us saying "I'd like you to go and look at this hotel in Pasadena. The structural engineers out there say it's in a lot of danger. We're in the process of renovating it, and I'd like you to go and give a second opinion." I said "Well, we've done a reasonable amount of earthquake analysis and design, but you have a tremendous amount of talent out there on the West Coast." And the client says "I know that, but I must have someone advise me who has an East Coast mentality." So, first, on the East Coast we have to look at things with an "East Coast mentality." So I went out to Pasadena and, after reviewing everything, was convinced that everything that had been done was done properly, and I participated in condemning this most gorgeous, beautiful hotel—the Huntington Sheridan—which was built around the turn of the century. I had constant guilt feelings about this. And then the city hired a consulting engineer who argued that we, the "East Coasters," were crazy to make the judgments we had about this hotel. So, then I wondered just who it was who had the East Coast mentality. This consulting engineer was fighting to maintain this beautiful structure for the city of Pasadena. So even some of the people on the West Coast were saying the building shouldn't have been condemned.

About several months later there was a significant earthquake, and there was some significant damage to this hotel. It's still standing, but it is definitely going to be condemned.

What I learned from that experience, is that there is a real problem in dealing with existing structures. I'm not even sure that the policies in various communities in California have any consistency to them. That hotel would not have been condemned if the owners had not decided to do major renovations. If they just kept the old layouts and had not sought to upgrade the hotel, it would not have been condemned. There was no building code requirement to condemn it, or to study it or analyze it. It would still be used right now and totally occupied. It was only condemned because the owner decided that he wanted to do some renovations, and, once the structure was touched, an analysis had to be done. There is something crazy there.

Before we consider whether or not to upgrade our codes, we have to emphasize two things. First, although I think we have to leave the question of risk to the seismologists we should nevertheless get five seismologists into a room, as was suggested, and have them go to work on whether we have a risk. Second, I believe that the cost of new buildings will take care of itself. I believe that we'll be innovative enough to minimize the effect of seismic design on cost. But before we implement codes for the new buildings, we have to have a policy for existing structures.

Finally, keep in mind that, prior to 1968, many buildings in this city were built up to about 100 feet or so with no lateral resisting elements.

CORNELIUS DENNIS (*Deputy Commissioner of the Department of Buildings of the City of New York, New York, N.Y.*): This conference is an example of how a building code gets changed. The current New York City Building Code basically was enacted in 1968. The structural portions of it were written not by persons employed by the City, or by Building Department employees. It was written by private structural engineers on a committee basis. The committees got together and came up with the idea that codes such as the ACI and AISC codes were good.

They were adopted as national consensus-type codes. As mentioned previously, they radically changed the wind stress criteria of a building but applied it only to new buildings. Conceivably a hundred years from now, all buildings will comply with that wind stress situation. Currently, very few buildings do.

If the structural engineering fraternity obtains input from the owners, input from the tenants, and input from the insurance industries, who conceivably would pay huge amounts of money if an earthquake did strike, and establishes a consensus with all of these groups, recommendations might very well ensue that would lead to a change in the building code. They might even require retroactive provisions for earthquake design.

The Department of Buildings is in the position today of waiting for guidance from the industry, and we will be waiting for the civil engineers and the insurance companies to come forward with some plan that will cover New York City. It will take cognizance, for example, of the clay soils in Staten Island, which might become water-saturated. The Borough Superintendent of Manhattan might start worrying a little bit about the fault zone across 125th Street, and the fault zone that constitutes the East River and the Hudson River. The Borough Superintendent of the Bronx might worry about three rock strata all meeting at one place; in fact a New York City Board of Education school was built on a site that straddles all three rock zones.

ROBERT KETTER (*NCEER, SUNY at Buffalo, Buffalo, N.Y.*): In the literature, over the last twenty years, when earthquakes have been discussed, whether they be in Whittier, San Fernando, Mexico City, or El Salvador, the statement has been made that those buildings that met the California criteria (or the earthquake criteria of the particular area in question) performed exceedingly well under the subsequent earthquakes that took place. California, however, still has not solved the retrofit problem. Long Beach has enacted a requirement that all structures must be taken care of, but most people agree that we don't know how to do it for many types of existing structures. This did not keep California from requiring over the last twenty to thirty years that new building designs must take earthquakes into account. They did not put off that requirement until all problems had been solved. What I have heard so far from several on the panel is that until we have all of the problems solved so that everyone is in an economically stable situation, we shouldn't do anything to change the status quo. But I may be hearing incorrectly.

JOSEPH KELLY (*Port Authority of N.Y. and N.J., New York, N.Y.*): I'll address my remarks to new construction because retrofitting existing construction is a significant issue which requires a major policy position that we have not really confronted. But in new construction we face a paradox. We build structures both in New Jersey and New York, including Yonkers, and we also build bridges. We are already required to design for seismic effects in at least two of these areas. Then we ask ourselves, "Why not New York?" and we don't have an answer. So, we're concerned; we are looking into it and we are exploring the issue of why New York is not equally covered/protected. As far as insurance is concerned, I don't know whether a lot of facilities are insured against earthquake damage. Many people are essentially self-insured. So insurance is not the answer. The answer should be sound engineering and knowing what is going on in the world as far as seismic technology is concerned.

In New Jersey we're required to provide for seismic design. In AASHTO bridge design we're required to provide for it. In New York, we say: "why not?" for new building construction.

MICHAEL PERO (*General Reinsurance Corporation, Stamford, Conn.*): I'm representing the insurance industry; basically I'm in the risk business. Our job in insurance is to try and separate risk from uncertainty.

You've all used that phrase: risk and damage and dollars. But, for us, underwriting is the key in terms of risk assessment. A risk is something we know something about so that we can at least have some idea of how to predict results. And then we make a financial bet.

Now, we have trouble when you seismologists and geologists talk about a return period of maybe 100, 150, or 200 years. We, of course, are trying to satisfy a board of directors every quarter of every year with profitable results.

But by and large we're very much interested in the earthquake area. We are always involved in major catastrophes. But, as you've seen, wind, storm and floods are not of the same financial order of magnitude as earthquakes are, as exemplified by the business interruption mentioned just now in that one building that did not sustain much physical damage. This interruption was enormous relative to the level of physical damage.

We want to be able to price our product better, so we need better risk assessment. We're also interested in the total amount of liabilities we incur in any one particular area, and we have to take care of our accumulation of risk. How many buildings would we take on and for how much would the insurance be provided in a particular area? These are the kinds of issues that we have to deal with to provide you with some sort of financial security. In that regard, my original purpose for coming was to find out what's going on in the scientific and engineering community relative to risk and to determine substantive, discrete and quantitative information so that we can do a better job with our product.

KETTER: This is a topic that the insurance industry has been addressing for awhile. It was discussed at a meeting of the insurance industry that was held last December in St. Louis, and Aetna has financed a large study at Stanford University. And FEMA, through its Expert Committee, has indicated that the question of whether the federal government should provide backup coverage, as they do for major floods, should be seriously examined. The insurance question is on many peoples' agenda.

PERO: Yes. The point with insurance is that it is not necessarily a unified endeavor on our part. Several years ago the State of California required all insurance companies to report their earthquake liabilities in that state. This was a modest first step in making sure that the public interest is served to establish how much insurance is there and ensure that the companies providing this insurance have the financial resources to pay those claims if and when a catastrophe occurs. Certainly that is on the minds of my board of directors as well. Certainly federal regulation is anathema to us, but we have to be able to do a better job in order to prevent that from happening or serve the public needs ourselves.

TOMASETTI: About ten or twelve years ago, we were investigating seven sites for the Veterans Administration just outside the Boston area. As we walked through the buildings and investigated them for a potential upgrading for earthquake hazards as defined by the Veterans Administration, we were advised not to tell anybody what we were doing. "They'll think we're crazy," they said. Later some changes were made in the codes and the Veterans Administration has done a lot in that area. What I'm trying to say, relative to existing buildings, is not that we should do nothing until we solve all the problems, but that we had better be very sure about what we're saying when we state that, for example, in New York City, we should be starting to design buildings for earthquake hazards. A lot of people are going to question the validity of our statements so we have to be very sure that we have as good an answer as possible from those five people, alluded to earlier, in that room. I think that a little more time should be spent with the five people in the room so that the risk might be defined a little better.

CANTOR: I'm in total agreement with Richard Tomasetti, but I want to remind you that when we talk about money, we're not speaking about the landlord's money. We're speaking about the consumer's money. We're talking about the cost of apartments, the cost of office space, the cost of medical space, the cost of living in a city that is already claiming it is saturated with costs and has difficulties in providing proper and adequate housing for its middle class and its poor and working classes. We are struggling in this city to find ways to build low- and middle-income housing. And while we may be talking about figures of 1 or 2, or 3 percent added cost, we have to remember it's *another* 1, or 2, or 3 percent.

The issue still comes down to this: If there is a problem, then we in the engineering community have a responsibility to solve it in the most efficient and economic manner. And certainly, I, as a practicing engineer, subject to lawsuits every Tuesday and Thursday, can very easily accept the notion that "what the hell, I'll throw in extra tonnage; it doesn't mean anything and it's going to cover me and protect me against the attorney sitting to my left."

But there doesn't seem to be a consensus of opinion among you professionals as to the severity of the problem here in New York. And as long as you guys can't decide how severe this potential problem is, I find it very difficult to get enthusiastic about adding costs to my buildings.

UNIDENTIFIED SPEAKER: We have been hearing that while the possibility of major catastrophes such as an earthquake always exists, the time interval between them has been and predictably will continue to be of great duration. On that basis it seems to me that we should attempt to separate concerns between property damage and life safety. That is probably not an easy thing to do, but it is conceptually useful. I would say that as owners—and maybe it's only because we've got an insurance company standing behind us—we would be less concerned with the property damage potential over a 100- or 200-year interval when the economic life of a building is only 40 to 50 years. We might be willing to take the chance that the frame of a building would be wrecked, but perhaps we should focus more on the things that might fall off a building and cause pedestrian damage and life safety. I would suggest then that in dealing with less frequent catastrophic situations we might focus more on life safety than on property damage.

COMMENT: Several statements have been made to the effect that nothing should be done about requirements on new construction until something is done about the existing building problem. I'll accept the point that as soon as code requirements are put in for new construction, some people are going to ask questions about the existing construction, although I don't think it's going to be a major problem.

Certainly the argument that we should not go ahead and put in requirements for new construction until we've dealt with the problem of existing buildings has some logic to it; I recognize that. But we have to be careful that it's not simply being used as an argument to do nothing. The mere fact that earthquakes are infrequent in this part of the world is the argument for going about things in a piecemeal way. If we really thought we would have a major earthquake within the next five or ten years, we'd embark on a crash course to do something about all these buildings. But because we don't really think that it's likely to happen immediately, we think that we've got some time and that it is sensible to adopt a piecemeal approach of requiring new construction to be better.

I would also like to comment on the point about renovation triggering problems. If a building owner embarks upon a renovation project, it's because he thinks that by so doing, he is going to get some economic value: he's going to prolong the life of that structure and allow it to be used for purposes that are going to make more money for him. That is the logical time to require him to face up to the safety issues with regard to that building.

KETTER: Let me make one stray comment: PBS—public television—has indicated its desire to put on a series of programs having to do with earthquakes. The first of these is to deal with history. I was interested when they said that their preliminary studies, which cover time as far back as Ancient Greece, indicate that no government has survived where there was a major earthquake to which the government did not adequately respond.

RENÉ LUFT (*Simpson, Gumpertz and Heger, San Francisco, Calif.*): It has often been observed that the only way of looking into the future is to understand the past. We are now looking at alternatives for seismic requirements in the eastern United States and in New York City in particular. The Commonwealth of Massachusetts went through a very similar process about fifteen years ago and as a result adopted a State Building Code that includes earthquake regulations.

At the time when the earthquake regulations were discussed, questions on cost came up that were similar to the ones raised here. And by doing the cost analysis, the Seismic Advisory Committee concluded that, if cost were the issue, one could not justify having any earthquake regulations. In other words, the cost, taken as an insurance policy, of reinforcing buildings would not offset the anticipated reduction in damage from future earthquakes because of their long return period. As a result of that finding, the Seismic Advisory Committee decided on a policy of no damage control.

Today we have heard quite a number of arguments that perhaps damage control should, in fact, be considered. But the cost/benefit study in Massachusetts indicated that a policy of no damage control made sense for very long return period earthquakes. However, on the same basis and on the same data we found that engineers and society have an obligation to provide a minimum level of protection to people who occupy the buildings. Describing that minimum level of protection, the Massachusetts Code says "the objective of these provisions is to protect life safety by limiting structural failures."

If I were to look at the question Mr. Cantor has asked, I would have to agree with his implicit conclusion: if a client is building a building and asks whether he should spend an additional 120 thousand dollars to provide for some earthquake reinforcement the answer would probably be no. However, as a society, it seems that the answer is different. If society decides to adopt a certain code that does not change the competitive position of one person relative to another, and if we are looking at it from the point of view of providing a minimum level of safety, which society expects in its buildings, it seems to make sense to have earthquake provisions in the code.

IRWIN CANTOR: I don't believe that the thrust of my comments was to protect one individual against another individual in the economic sphere. I thought I made it quite clear that society has a way of paying for every dollar that's spent. My concern still is that I don't think there's a unified consensus among the professionals, among the experts, as to what should be done. And as long as there isn't a unity consensus, I don't think the solution should be: do something; let's spend money. If the professionals did have a cohesive conclusion that there really is a risk, then I would be among the first to say: let's go for it.

ROBERT WHITMAN: On this question of consensus, during all the discussions we've had over the last couple of days I have not heard anyone knowledgeable on the question of the earthquake hazards suggest that the design criteria in the model building codes are too stringent. I have heard suggestions to the effect that they may not be strong enough; that a place like New York ought to be requiring more. Most of these arguments have been over details, mostly concerned, for example, with the kind of response spectra that one should use for special structures.

UGO MORELLI (*FEMA, Washington, D.C.*): "Truth in advertising" forces me to admit to a couple of things. One, I work for the federal government. (But there's no check in the mail!) Second, I'm the project officer for those series of documents that were on the table yesterday. With that statement out of the way, I can now make a few comments.

About this question of consensus: the documents that were on the table represent just such a consensus of the building industry. They were put out by the Building Seismic Safety Council, which represents sixty member organizations, ranging from the trade unions to the professional societies. It took a number of years to develop it. They were based on ATC-3, which took an additional four or five years to be developed. That effort has been subsequently reviewed. The first edition of these provisions was put out in 1985 and will be updated by the end of the summer of 1988. These provisions are current; they represent a consensus; and they're also flexible and of national applicability. They offer a set of contour maps, which are easier to use, as well as maps showing county boundaries. Seismicity indices are given too.

I agree that individual communities have the privilege as well as the responsibility of deciding whether or not to do something in this area. But I am reiterating that a consensus set of seismic provisions does exist. I would like Bob Whitman to add the NEHRP provisions to his list of model codes. They're available, they're flexible, and they're a consensus document accompanied by a number of other documents that make it easy to apply them. They include a guide with several examples on how to use them as well as an explanation for the layman that can be used to generate interest among citizens and elected and appointed officials. I encourage you to use these documents. You've paid for them with public funds, so let's use them.

Second, while I love New York we must recognize that New York is not all of the eastern United States, just as California is not the whole of the United States. Each has different problems. So even if New York were to elect not to do anything in this area of earthquake design, there are many other communities across the eastern United States that might elect to do so, and I urge them to do something and to do it now.

Third, the provisions that I have spoken about are directed at life safety. There may be damage control that comes with it, but that is not the primary purpose of the provisions.

Fourth, on the question of whether we should do anything on new buildings before we're ready to do something about existing buildings I would say let's not perpetuate a problem if we can avoid it. Let's do something now on new buildings. We are also preparing a set of documents that will deal with existing buildings. Unfortunately, they're not as far along as those on new buildings. Some of them will be coming out in the next couple of years, and I hope they will be useful.

The final point is this: I was privileged to attend a meeting in Knoxville in 1981 dealing with the central part of the United States and with the Southeast. A few members of this audience were there also. We heard the same arguments. "There's no problem." "Why should we do anything about it?" At that meeting they were faced with the same kinds of uncertainties in that part of the world as we are in this, and, as a matter of fact, they were less well prepared to deal with the problem, because no materials were developed yet. Now, however, there's been remarkable progress in the central part of the United States around the New

Madrid area and the Charleston, South Carolina area as well as some other communities up and down the East Coast.

In closing, I would like to urge you to do something because if you save just one life, you've made up this 600K of additional expense for one year in a large city like New York.

PHIL GOLDSTEIN (*Borough Superintendent for Staten Island, Department of Buildings of the City of New York, New York, N.Y.*): Although I'm with the New York City Department of Buildings I'm speaking here as a structural engineer. Whether or not we need to update the New York City Building Code continues to be controversial, and there are some questions that engineers in New York will be asking as a result of this conference.

First, we have two documents or two national reference standards that put New York City in the zone of medium hazard for earthquakes. We also have a reference standard that has been adopted in New York City—ACI-318 of 1983— including Appendix A, which requires that at least tall buildings designed for wind must be detailed for earthquakes. The Commissioner of Buildings has asked the New York Society of Consulting Engineers for advice. Their initial response was that there was no problem, and we should not enforce Appendix A. However, Appendix A is there. There is no provision of law for it not to be enforced. So, either we will have to modify the code to take Appendix A out, or start enforcing it. This is a big question.

The suggestion that we should ask ourselves what a seismic code should provide for is particularly relevant. Should we make certain that we design for life safety and accept damage, which may not cost too much? That's another question.

The matter of Local Law 10 is also relevant. This is a New York City law which requires building owners to observe their exterior walls and submit reports every five years. The walls are inspected from the outside, yet there appears to be very little supervision of such walls during construction. Perhaps that is why we have the problems that caused Local Law 10 to be enacted.

HOWARD SIMPSON (*Simpson, Gumpertz & Heger, San Francisco, Calif.*): I was chairman of the joint committee that developed the chapter on seismic design for the Massachusetts Building Code. As René Luft has said, we did deal with the question of philosophy. We decided early on that the building code is intended to serve the health, safety, and welfare of the public, and that damage control has no part in the code as far as we were concerned except as it relates to life safety.

I would like to say a word about both idealism and the nuts and bolts of decision-making. Ideally an informed citizenry makes a decision as to whether it wants to spend money in an area like New York or Massachusetts in order to upgrade construction and to make it more earthquake-resistant. We hope that the public, in making decisions, will be informed, and we hope that there is a vehicle for making their wishes known. That is a somewhat idealistic hope, but nevertheless we try to approach that particular goal.

Now, let's examine the circumstances leading to a code in Massachusetts. First, a joint ACI/ASCE committee was formed about 1968. The committee decided that we needed a chapter on seismic design; that recommendation had the support of two important societies and was submitted to the state building code commission, who almost unanimously recommended it to the legislature. The politicians, the elected officials and the appointed officials all agreed that we needed this code and presumably they thought that the public would agree. Now, if you use the National Academy of Sciences techniques as well as engineering techniques for determining the value of human life, and you take the return periods of earthquakes in Massachusetts, you would come out with something that isn't cost-effective. But even if the event might occur once in 200 years or in 500 years and several hundred buildings collapse in Boston, what would the public say? How would they look at the engineers who did nothing? So there is more to it than just a cost-benefit figure. You have to consider the political and socioeconomic significance of a catastrophic event and what the public will expect of the professional people who are charged with protecting them against just such a disaster.

As an indication that this code was not just a casual thing that went through because nobody cared, when the first draft was accepted, the sections on reinforced masonry were deleted because the Masonry Institute put up a fight. Yet the legislature wanted to get the code through and so they put it through without those provisions. During the ensuing year, I had to fight for the incorporation of provisions for reinforced masonry. I presented the case before the Building Code Commission, and the masonry people presented theirs. It was really not much of a battle. Once again, the Building Code Commission unanimously voted to incorporate the reinforced masonry provisions, and they were adopted the next year.

So, it's quite clear that society does support some effort toward addressing the problems of loss of life and of injury in an earthquake without regard to any cost-benefit analysis. How far should you go? Ultimately it has to be society that decides. How much do people want to spend? Let us hope that in deciding they will be informed and let us hope that we do our job in making the recommendations and in transmitting critical information to the public that it needs in order to make an informed decision.

GUY NORDENSON (*Ove Arup and Partners, New York, N.Y.*): A request was made in September 1984 by the New York City Building Commissioner to the New York Association of Consulting Engineers asking whether the new ANSI provisions putting New York in Zone 2 meant that we should adopt Appendix A of ACI in New York City. This would require ductile detailing for concrete.

The initial response of the NYACE was to say that it was not necessary to adopt Appendix A. But following up on that, there was an effort to reconsider the problem in a little bit more detail. I had the privilege of chairing a committee, composed mostly of seismologists, that worked for about two years on the subject. The consensus of that committee was that indeed Zone 2 requirements were appropriate for New York City.

That went back to the structural codes committee of the New York Association of Consulting Engineers with our recommendation that we start to look into having a code for New York that recognized what our construction practices are and how we can improve them without having to radically change them. In the meantime we recommend adoption of the provisions of the upcoming UBC-88.

The Structural Code Committee of NYACE agreed to it and sent it up to the board of directors of the NYACE who unanimously agreed and passed it on to the Building Commissioner last July. So, the directors of the NYACE, which is our association of consulting engineers in New York, has unanimously recommended to the Building Commission that we adopt UBC-88 for now and that we start to look carefully at what should be our own code.

NORTON REMMER (*Consulting Engineer, Worcester, Mass.*): I'm a private consulting engineer, formerly Technical Director with the Massachusetts State Building Code Commission and presently Chairman of the Seismic Advisory Committee.

I want to make three brief comments. First, all of these subjects have been discussed for a long time. I have heard them during debate of the ATC project and during the Massachusetts State Building Code process over a period of fifteen years. And such debate is necessary because every area and every group has its own concerns, unique to the area, just as the seismic risk is.

Second, for every argument I've heard there's always a counterargument that has validity. In Massachusetts the results of the adoption of one seismic code produced two benefits: the adoption of seismic provisions has improved design procedures, and buildings are designed and built better, because designers have to think more carefully about how one building acts as a unit.

Finally, we recognize that there are many things that can be done in reconstruction and rehabilitation that are very easy to do and will accomplish a great deal without much cost. We're looking at how much further we can go. An educational process is necessary in reconstruction to tell architects and engineers about details of design that can be accomplished very inexpensively but that can greatly improve the construction of existing buildings.

SPEAKER: I just wanted to make the point that if some person or society or organization goes to the trouble of doing a cost/benefit analysis and it doesn't work out on paper and yet they still decide to implement earthquake-resistant design, then implicitly they're willing to pay that price and they value the benefits over the costs. When we discuss this we can get caught in semantics.

Also someone made the point that if you lose a drop of blood a day, it adds up to many pints over many months. But I would remind you that it's qualitatively different than losing three or four pints of blood at once. And a catastrophe in which many buildings collapse at once in an earthquake is a multiple cost.

Damage Prediction for Existing Buildings

CHRISTOPHER ROJAHN

Applied Technology Council
3 Twin Dolphin Drive
Redwood City, California 94065

INTRODUCTION

Damage to existing buildings can be expected to be severe if large earthquakes occur in or near large metropolitan areas. On the basis of a relatively recent study,[2] for example, it is estimated that property losses for a magnitude 7.5 earthquake on the Newport-Inglewood fault in the Los Angeles metropolitan area would be $62.2 billion (1980) dollars, excluding losses for communication and transportation systems, dams, military installations, and consequent losses such as unemployment, loss of taxes, shutdown of factories due to loss of supplies, and automobile damage. These large expected dollar losses are of concern to facility owners, insurers, and government officials.

Methods and data for predicting building damage have been the subject of extensive research over the past several decades. These research activities have shown that building damage is dependent upon several key factors and have resulted in the development of several approaches for predicting expected damage.

The intent of the text that follows is to provide an overview of the factors that affect building damage potential, an overview of methods and data currently available for predicting building damage, and an overview of a recent study by the Applied Technology Council (ATC) which resulted in the development of comprehensive data and methodology for predicting expected earthquake damage in California. The paper concludes with an overview of preliminary methods and data currently under development by ATC that could be used to facilitate application of California damage-evaluation data to other parts of the United States.

FACTORS AFFECTING EARTHQUAKE DAMAGE

The extent to which buildings are damaged is generally dependent upon the following factors:

- Structural characteristics
- Ground-shaking severity
- Collateral hazards

These factors, particularly structural characteristics and ground-shaking severity, may interact to enhance or reduce damage potential. A clear understanding of these factors individually as well as their combined impact will improve accuracy in the application of damage-prediction methods. Following are discussions of each of these factors.

Structural Characteristics

In general, structure characteristics important for earthquake damage-evaluation purposes include: construction material, soil-foundation material, structure foundation, structural system, configuration, structural continuity, age, proximity to other structures, and exposure to prior earthquakes. Although there are exceptions, good earthquake performers normally have the following characteristics: (1) they are constructed of materials that can undergo long-duration horizontal and vertical shaking without excessive loss of stiffness or strength; (2) they are founded on firm foundation materials (the structure does not tip or settle due to foundation failure); (3) they have regular, symmetrical plan shapes (so that torsional response is not induced in the structure); (4) they are composed of structural elements that are tied together (the structural elements do not separate during earthquake shaking); (5) they are of recent age (designed since the adoption of seismic codes); (6) they are not too close to other structures (so as to avoid pounding of two structures against one another); and (7) they have not previously been damaged by an earthquake.

Damage can generally be expected to be low for buildings that have been designed in accordance with earthquake codes or that naturally have earthquake-resistant properties. Examples of the former include recently designed and constructed buildings in urban areas of California that have adopted and enforced seismic design codes; examples of the latter include one-story wood-frame houses that are anchored to their foundations and that have light-weight roof systems.

Earthquake-resistant design to ensure good performance during earthquakes begins with the identification of appropriate earthquake shaking for the site, including proper consideration of site-soil conditions. Thereafter, the structural members must be sized and connected in such a way as to provide adequate strength and toughness to resist the earthquake demand forces. In addition, irregular plan shapes that could induce torsional earthquake response should be avoided, as should abrupt changes in stiffness with height. Finally, diligent quality assurance is required during construction to ensure that the structure is built as designed.

The first seismic design and construction regulations in the United States were imposed in California in 1933 with the Field Act (for schools) and the Riley Act (for all buildings except farm structures and small residences), which were adopted shortly after the 1933 Long Beach, California earthquake. Although codes have undoubtedly had a positive impact on design and construction quality, poor performance of buildings designed using codes has been observed in several recent U.S. earthquakes, including the 1964 Alaska earthquake, the 1971 San Fernando, California earthquake, and the 1979 Imperial Valley, California earthquake.[3] Conversely, several engineered buildings designed prior to the existence of U.S. seismic building code regulations performed very well during the 1906 San Francisco earthquake.[4] Thus, although age or the existence of seismic design codes may indicate expected performance, they are not all-encompassing criteria for distinguishing earthquake-resistant construction.[1]

Ground-Shaking Severity

In general, for a given site and distance from the earthquake source (fracture in the earth's crust along which the earthquake occurs), ground-shaking severity is directly proportional to the magnitude of the earthquake. In other words, the larger the earthquake, the more severe the ground-shaking. In California, where historic earthquakes have been shallow (most are less than 15 km deep), earthquake shaking can be expected to be more severe at shorter distances from the fault rupture zone than it is at larger distances. This point is dramatically illustrated by the strong ground-motion data from the 1979 Imperial Valley, California earthquake[5] (FIG. 1). In parts of the world where earthquakes occur at greater depth (70 to 700 km), ground-shaking may be very severe at large distances from the earthquake source zone. Such was the case for the September 1985 earthquake along the Mexico coast that severely damaged hundreds of multi-story buildings in Mexico City and the 1977 earthquake in Romania that collapsed approximately 35 multi-story buildings in Bucharest approximately 100 km south of the earthquake source region.

Earthquake shaking is normally measured in terms of acceleration and, in some cases, in velocity or displacement. The shaking is cyclic in nature, and the most important characteristics of such motions (from an engineering standpoint) are: (1) amplitude; (2) frequency content; and (3) duration. Damaging levels of ground-shaking can be expected to occur for earthquakes of about magnitude 5.5 or greater when the amplitude of ground-shaking is approximately 0.20 g (g equals the force of gravity) or greater. In order for such motions to be damaging, however, they need be of sufficient duration (several cycles of high-amplitude motion) and at a frequency near that of the structure affected. The fundamental frequency of a building is dependent on its height and stiffness (in the horizontal direction). Fundamental frequencies, for example, normally range from 1 to 10 Hz (cycles per second) for 1- to 5-story buildings, and 0.5 to 2 Hz for 5- to 10-story buildings. When the fundamental frequency of the building and the dominant frequency of earthquake-induced ground-shaking coincide (or nearly so), motions in the building can be amplified by two or more times (relative to the ground motion). This phenomenon was probably responsible for much of the damage in Mexico City where the fundamental frequencies of damaged multi-story buildings were near that of the dominant frequency of ground motion on lake-bed sites (about 0.5 Hz).

Collateral Hazards

In addition to strong ground-shaking, several other phenomena, commonly regarded as collateral hazards, have been observed to significantly affect earthquake losses.[1] These include:

- Poor ground (loose sands, sensitive clays)
- Landslide
- Fault rupture
- Inundation
- Fire

Poor ground includes soil conditions such as loose sands, sensitive clays, and some lightly cemented sands.[1] Flow failures and lateral spreading in sensitive clays and settlements of loose dry sand of a few inches to a foot have been observed to be the cause of severe earthquake structural damage. A more common cause of damage, however, is liquefaction, which is a process in which loose saturated sand liquefies during the earthquake and loses its shear strength. Liquefaction is commonly manifested during earthquakes in the form of water spouts (sand and water seething from holes in the ground), the submergence or tilting and differential settlement of structures, and the re-emergence of buried tanks. A

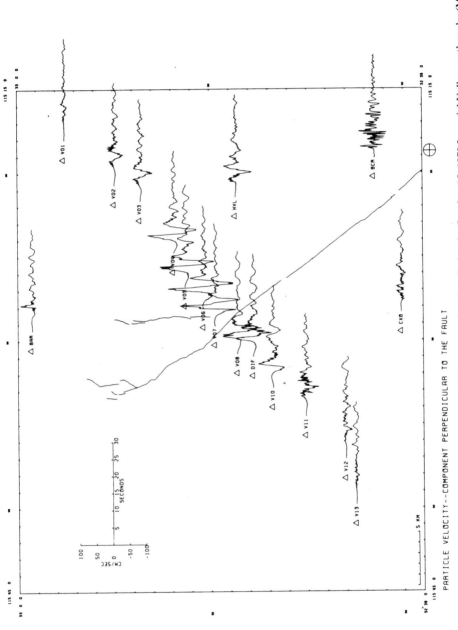

PARTICLE VELOCITY--COMPONENT PERPENDICULAR TO THE FAULT

FIGURE 1. The horizontal particle velocity recorded at USGS strong-motion sites during the October 15, 1979 Imperial Valley earthquake ($M_1 =$ 6.6) is plotted in a plan view of the source region.[5]

structure that did not sustain damage from deformation caused by ground-shaking might sustain substantial damage if it settles differentially or tilts.

A *landslide,* which is a downslope movement of a soil or rock mass due to gravity, can seriously damage a structure that is located on or at the base of the downslope movement.[1] The slope of surfaces on which landsliding has been observed to occur during earthquakes varies from quite steep to almost horizontal. In addition, the season of the year is a significant secondary factor in landslide potential because the amount of moisture present influences the load stability of most weak soils. In California, a strong earthquake during the wet winter rainy season can be expected to cause more damage from landslides than an earthquake in the dry summer.

Fault rupture can cause significant damage to structures situated immediately over simple fault breaks and to structures situated in the fault zone of more complex alluvial surficial deposits.[1] In the simpler case, the structure is literally sliced in a shearing motion with one portion of the structure moving in one direction and the other moving in the opposite direction. For the more complex case, surficial soil materials are deformed and distorted over a broad area, and structures founded in these zones are subject to potentially serious disruption.

Inundation during earthquakes can result from several causes, including tsunamis, seiches, dam/reservoir failures, and areal subsidence or tilting.[1] Tsunamis are transient sequences of long-period sea waves generated impulsively by earthquakes, coastal or submarine landslides, or volcanic phenomena. Seiches are periodic oscillations of enclosed or semi-enclosed bodies of water caused by earthquakes or landslides that disrupt the normal boundaries of lakes, bays, or other such large volumes of water. Inundation resulting from areal subsidence or tilting occurs when there is ground failure adjacent to large bodies of water. Dam/ reservoir failures can occur as a result of severe ground-shaking or collateral hazards.[1]

Fire is one of the greatest potential dangers immediately following an earthquake and, if unchecked, could lead to a major conflagration under certain circumstances. Important factors for evaluating possible conflagration include: time of day; fire-extinguishing systems in individual buildings and their probability of functioning after an earthquake; building density in the affected area; available community fire-fighting resources (manpower, equipment, and water); season of the year; wind velocity; and fire load, or number of fires. In the Los Angeles region, for example, although a number of conflagrations are to be expected after a severe earthquake, a general conflagration throughout the area is not expected because of natural separations and fire breaks created by a large number of wide streets, freeways, parks, golf courses, cemeteries, and large parking areas, plus open agricultural and undeveloped areas between many communities.[1] The situation, however, would be exacerbated should a severe earthquake occur immediately prior to or during the onset of the Santa Ana or "devil winds," which are dry, gusty wind conditions that occur most commonly during the fall and winter months, but may unpredictably take place at almost any time of the year.

METHODS AND DATA AVAILABLE TO PREDICT BUILDING DAMAGE

Earthquake damage can be expressed in a variety of ways. In general, the most revealing expressions are the ratios of dollar loss (damage factor) and buildings damaged (damage ratio). Almost any type of earthquake loss, for any facility

or component, can be rationally deduced from these expressions, which are defined as follows[1]:

$$\text{Damage factor (DF)} = \frac{\text{Dollar loss}}{\text{Replacement value}}$$

$$\text{Damage ratio (DR)} = \frac{\text{Number of buildings damaged}}{\text{Total number of buildings}}$$

The mean damage factor for a group of similar structures exposed to the same
ground-shaking intensity is defined as:

$$\text{Mean damage factor (MDF)} = \left(\frac{1}{n}\right) \sum_{i=1}^{n} \frac{(\text{Dollar loss})_i}{(\text{Replacement value})_i}$$

where n is the number of structures in the sample.

For statistical data sample sets from a given geographical study area, the
damage ratio and mean damage factor can be calculated. Plots of these parameters versus shaking intensity form what are commonly referred to as motion-
damage relationships.[6,7] An example of a mean damage-factor curve[1] is given in
FIGURE 2.

The current literature contains considerable data on the relationship between
earthquake motion and damage for various types of buildings, but relatively little
information on other structure types. Unfortunately, there is no standard method
for classifying structure types, and different methods for characterizing both motion and damage are utilized.

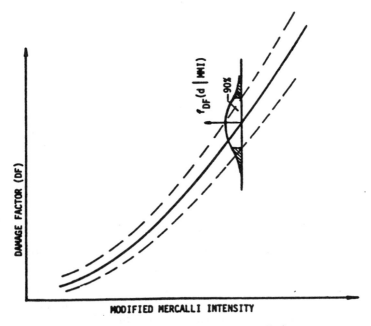

FIGURE 2. Example of DF versus MMI plot.

Observed Earthquake Effects on Buildings

Existing motion-damage relationships for buildings have been developed empirically from past earthquake investigations or from subjective judgment. Often these two methods are combined when the data available after an earthquake are incomplete.

Some of the primary sources of motion-damage relationship data for buildings are summarized in Appendix D of the ATC-13 Report.[1] These sources utilize various classification schemes, ground-motion characterizations, and damage characterizations, and are based primarily on empirical data and/or judgment.

Ground-motion characterizations utilized in these primary data sources include the Modified Mercalli Intensity,[8] Rossi-Forel Intensity,[9] and Engineering Intensity[10] scales as well as characterizations utilizing parameters derived from recorded or analytically derived strong ground-motion data (e.g., spectral acceleration, velocity, or displacement). Of these characterizations the Modified Mercalli Intensity scale is the most widely utilized.

DAMAGE PREDICTION USING ATC-13 DATA AND METHODOLOGY

The recently completed ATC-13 report[1] provides consensus opinion earthquake damage and loss estimates and companion loss estimation and inventory methodology for facilities in California. Developed for the Federal Emergency Management Agency (FEMA), the data and methodology are needed to provide input into FEMA computer simulation methodologies that estimate the economic impacts of real or hypothetical California earthquakes on the state, region, and nation. Included in the ATC-13 report are consensus opinion damage-factor estimates (expected physical damage due to ground-shaking), loss-of-function estimates, and inventory data and methods for all types of existing industrial, commercial, residential, utility and transportation facilities in California. Included also are methods to estimate deaths and injuries, methods to adjust damage-factor estimates to account for construction quality, and methods to estimate the effects of collateral hazards such as ground failure, fault rupture, and inundation. Damage-factor estimates are provided for Modified Mercalli Intensities VI through XII in the form of Damage Probability Matrices. Seven damage states are considered: 0%, 0–1%, 1–10%, 10–30%, 30–60%, 60–100%, and 100% damage. Loss-of-function estimates, which specify the time required to restore a facility to 30%, 60%, and 100% of the pre-damage usability, are provided for these same seven damage states.

Facility Classifications

Because of the comprehensive nature of the overall FEMA economic impacts investigation, it was essential that all types of industrial, commercial, residential, utility, transportation and other existing facilities in California be included in this study. These facilities have been classified in two ways: (1) by Earthquake Engineering Facility Classification, which characterizes structures in terms of their size, structural system, and type (e.g., low-rise unreinforced masonry buildings), and (2) by Social Function Classification, which characterizes facilities in terms of their economic function (e.g., commercial retail trade).

The Earthquake Engineering Facility Classification is required because earthquake-induced physical damage is dependent upon structural properties. This classification contains 78 classes of structures, including 40 building types (TABLE 1). The structure classes were selected on the basis of expected dominance in the existing inventory of California structures and on the basis of expected uniqueness in seismic performance; the structure classes were not established on the basis of inventory sampling.

TABLE 1. ATC-13 Earthquake Engineering Facility Classification: Building Types

- Wood frame, low rise
- Light metal, low rise
- Unreinforced masonry (bearing wall), low rise
- Unreinforced masonry (bearing wall), medium rise
- Unreinforced masonry (with load-bearing frame), low rise
- Unreinforced masonry (with load-bearing frame), medium rise
- Unreinforced masonry (with load-bearing frame), high rise
- Reinforced concrete shear wall (with moment-resisting frame), low rise
- Reinforced concrete shear wall (with moment-resisting frame), medium rise
- Reinforced concrete shear wall (with moment-resisting frame), high rise
- Reinforced concrete shear wall (without moment-resisting frame), low rise
- Reinforced concrete shear wall (without moment-resisting frame), medium rise
- Reinforced concrete shear wall (without moment-resisting frame), high rise
- Reinforced masonry shear wall (without moment-resisting frame), low rise
- Reinforced masonry shear wall (without moment-resisting frame), medium rise
- Reinforced masonry shear wall (without moment-resisting frame), high rise
- Reinforced masonry shear wall (with moment-resisting frame), low rise
- Reinforced masonry shear wall (with moment-resisting frame), medium rise
- Reinforced masonry shear wall (with moment-resisting frame), high rise
- Braced steel frame, low rise
- Braced steel frame, medium rise
- Braced steel frame, high rise
- Moment-resisting steel frame (perimeter frame), low rise
- Moment-resisting steel frame (perimeter frame), medium rise
- Moment-resisting steel frame (perimeter frame), high rise
- Moment-resisting steel frame (distributed frame), low rise
- Moment-resisting steel frame (distributed frame), medium rise
- Moment-resisting steel frame (distributed frame), high rise
- Moment-resisting ductile concrete frame (distributed frame), low rise
- Moment-resisting ductile concrete frame (distributed frame), medium rise
- Moment-resisting ductile concrete frame (distributed frame), high rise
- Moment-resisting nonductile concrete frame (distributed frame), low rise
- Moment-resisting nonductile concrete frame (distributed frame), medium rise
- Moment-resisting nonductile concrete frame (distributed frame), high rise
- Precast concrete (other than tilt-up), low rise
- Precast concrete (other than tilt-up), medium rise
- Precast concrete (other than tilt-up), high rise
- Long-span, low rise
- Tilt-up, low rise
- Mobile homes

NOTE: Low rise, 1–3 stories; medium rise, 4–7 stories; high rise, 8+ stories.

TABLE 2. Damage States and Damage Factor Ranges

Damage State	Damage Factor Range (%)	Central Damage Factor (%)
1 (None)	0	0
2 (Slight)	0–1	0.5
3 (Light)	1–10	5
4 (Moderate)	10–30	20
5 (Heavy)	30–60	45
6 (Major)	60–100	80
7 (Destroyed)	100	100

The Social Function Classification is required because that is the form in which structures in the existing FEMA database are listed and because this form is required as input in the economic impacts model utilized by FEMA. In addition, loss of function (or usability) is related to social function class. This classification contains 35 classes of facilities: residential (three classes), commercial (seven), industrial (eight), agricultural (one), mining (one), religious and nonprofit (one), government (two), educational (one), transportation services (four), utilities (five), communication (one), and flood control (one). These 35 facility classes were selected so as to account for all facility types listed in the four-digit Standard Industrial Classifications of the U.S. Department of Commerce.

Physical Damage Caused by Ground-Shaking

Estimates of percent physical damage caused by ground-shaking for all 78 Earthquake Engineering Facility Classifications (including 40 building types), expressed in terms of damage factor versus Modified Mercalli Intensity scale, were developed through a multiple questionnaire process involving the 13-member advisory Project Engineering Panel (PEP) and 58 other selected earthquake engineering specialists. The objective of the questionnaire was to develop damage-probability matrices (DPMs) similar in form to that suggested by Whitman, Reed, and Hong.[11] By using such DPMs, it is possible to estimate the expected dollar loss caused by ground-shaking for each facility by multiplying the damage factors for the structure and its contents by the estimated replacement values for each, respectively. The damage states and corresponding damage factor ranges defined for this project are given in TABLE 2.

After the questionnaire process, which is described in detail in the ATC-13 report, DPMs were computed for the 78 Earthquake Engineering Facility Classifications considered under this project. Examples for three classes of buildings are shown in TABLE 3. These DPMs apply to facilities having standard construction, which includes all facilities except those designated as special or nonstandard. Special construction includes (1) California elementary and secondary public school buildings, (2) post-1972 California hospitals, (3) railway bridges, and (4) any facility determined to have special earthquake damage control features. Nonstandard construction includes those structures that are more susceptible to earthquake damage than are those of standard construction. The quantitative manner in which special and nonstandard construction is treated in this project is to shift the probability of a given damage state, P_{DSI}, up or down, depending on the grade or quality of design and construction.

Weighted mean damage-factor estimates for the three building types included in TABLE 3 along with two other structure types (dams and bridges) are plotted in

TABLE 3. Damage Probability Matrices Based on Expert Opinion for Example Earthquake Engineering Facility Classifications[1]

Central Damage Factor	Modified Mercalli Intensity						
	VI	VII	VIII	IX	X	XI	XII
Low-Rise Unreinforced Masonry Bearing Wall Buildings							
0.00	a	a	a	a	a	a	a
0.50	9.1	0.6	a	a	a	a	a
5.00	90.5	55.5	10.9	0.5	a	a	a
20.00	0.4	43.4	66.0	22.4	2.0	0.1	0.1
45.00	a	0.5	22.9	65.9	35.0	10.1	3.4
80.00	a	a	0.2	11.2	62.5	83.1	50.4
100.00	a	a	a	a	0.5	6.7	46.1
Medium-Rise Moment-Resisting Steel Perimeter Frame Buildings							
0.00	22.4	1.1	a	a	a	a	a
0.50	51.3	34.0	2.5	a	a	a	a
5.00	26.3	64.9	95.4	83.1	29.5	9.2	0.2
20.00	a	a	2.1	16.9	70.5	80.7	50.6
45.00	a	a	a	a	a	10.1	49.2
80.00	a	a	a	a	a	a	a
100.00	a	a	a	a	a	a	a
High-Rise Moment-Resisting Nonductile Distributed Concrete Frame Buildings							
0.00	0.1	a	a	a	a	a	a
0.50	27.0	2.2	a	a	a	a	a
5.00	72.9	89.3	32.2	3.0	a	a	a
20.00	a	8.5	66.9	68.1	19.9	3.9	0.1
45.00	a	a	0.9	28.9	74.2	57.8	12.4
80.00	a	a	a	a	5.9	38.3	84.3
100.00	a	a	a	a	a	a	3.2

[a] Very small probability.

FIGURE 3. These data indicate the variability in expected performance, in addition to serving as a basis for estimating earthquake losses. A comparison of expected performance is also shown in TABLE 4, which ranks the 40 earthquake engineering building types considered in this project in terms of expected vulnerability when subjected to MMI VIII levels of shaking. Such data have applications in other aspects of earthquake engineering, such as the determination and ranking of building types that should be seismically evaluated in hazard abatement programs or instrumented with earthquake strong-motion instrumentation.

Losses Due to Collateral Hazards

In addition to damage caused by strong ground-shaking, collateral hazards such as ground failure, fault rupture, inundation, and fire can also cause serious damage to facilities. Initially, a literature review was conducted to ascertain existing quantitative information on the losses caused by these collateral hazards. On the basis of this information, plus judgment on the part of the project participants, methods for quantifying the impact of collateral hazards were developed. Methods are provided for estimating damage caused by:

- Poor ground/liquefaction
- Landslide
- Fault rupture
- Inundation

The estimated damage from each of these four collateral causes is defined in terms of mean damage factor, which is the same form used to describe damage due to ground-shaking.

The procedure for estimating damage caused by *liquefaction/poor ground* is based on observations of ground failure from liquefaction during the 1906 San Francisco earthquake, in which case damage on poor ground was 5 to 10 times greater than that on firm ground,[12] and regional probabilistic estimates of liquefaction potential developed recently by Legg *et al.*[13] The procedure is as follows:

$$MDF(PG) = MDF(S) \times P(GFI) \times 5$$

for surface facilities and

$$MDF(PG) = MDF(S) \times P(GFI) \times 10$$

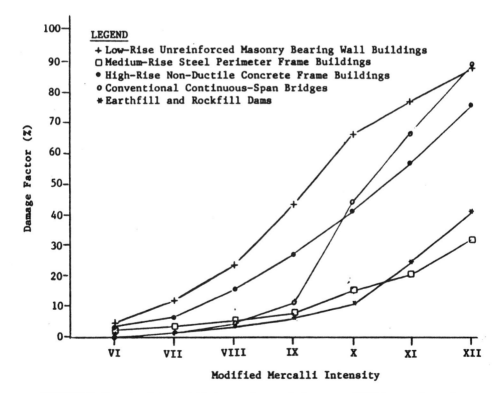

FIGURE 3. Expert opinion weighted-mean damage factor versus MMI for sample earthquake engineering facility classes. (From Rojahn *et al.*, 1986).

TABLE 4. Vulnerability Ranking of Building Types Subject to MMI VIII

Ranking	Building Type (ATC-13 Reference Number)	Expert Opinion Weighted Mean Damage Factor[a]
1	Medium-rise unreinforced masonry, bearing wall (76)	30.5
2	High-rise unreinforced masonry, with load-bearing frame (80)	25.2
3	Low-rise unreinforced masonry, bearing wall (75)	24.2
4	Medium-rise unreinforced masonry, with load-bearing frame (79)	22.3
5	Low-rise unreinforced masonry, with load-bearing frame (78)	15.7
6	High-rise moment-resisting nonductile concrete frame (89)	15.4
7	Medium-rise moment-resisting nonductile concrete frame (88)	15.1
8	High-rise reinforced masonry shear wall without frame (11)	15.0
9	Low-rise moment-resisting nonductile concrete frame (87)	14.4
10	High-rise concrete shear wall without frame (8)	13.8
11	Low-rise tilt-up (21)	12.5
12	High-rise braced steel frame (14)	11.8
13	High-rise precast concrete, other than tilt-up (83)	11.7
14	Medium-rise reinforced masonry shear wall without frame (10)	11.4
15	High-rise reinforced masonry shear wall with frame (86)	10.6
16	Medium-rise concrete shear wall without frame (7)	10.0
17	Medium-rise precast concrete, other than tilt-up (82)	9.6
18	Low-rise precast concrete, other than tilt-up (81)	9.1
19	Medium-rise braced steel frame (13)	7.8
20	Medium-rise reinforced masonry shear wall with frame (85)	7.5
21	High-rise concrete shear wall with moment-resisting frame (5)	7.5
22	High-rise moment-resisting ductile concrete frame (20)	7.5
23	Mobile homes (23)	7.2
24	Low-rise concrete shear wall without frame (6)	6.9
25	High-rise moment-resisting steel perimeter frame (17)	6.8
26	Medium-rise concrete shear wall with moment-resisting frame (4)	6.7
27	Low-rise reinforced masonry shear wall without frame (9)	6.6
28	Medium-rise moment-resisting ductile concrete frame (19)	6.2
29	Low-rise braced steel frame (12)	5.9
30	High-rise moment-resisting steel distributed frame (74)	5.6
31	Wood frame (1)	5.5
32	Low-rise reinforced masonry shear wall with frame (84)	5.3
33	Low-rise concrete shear wall with moment-resisting frame (3)	5.2
34	Medium-rise moment-resisting steel perimeter frame (16)	5.2
35	Medium-rise moment-resisting steel distributed frame (73)	5.2
36	Low-rise long span (19)	5.1
37	Low-rise moment-resisting steel perimeter frame (15)	5.0
38	Low-rise moment-resisting ductile concrete frame (18)	5.0
39	Low-rise moment-resisting steel distributed frame (72)	4.7
40	Light metal (2)	4.5

NOTE: Height definitions: Low rise, 1–3 stories; medium rise, 4–7 stories; high rise, 8+ stories.

[a] Replacement value (percentage).

TABLE 5. Ground Failure Probability Matrix for Poor Ground[1a]

Zone	Type of Deposit	VI	VII	VIII	IX	X	XI	XII
		\multicolumn{7}{c}{Probability of Ground Failure: Percent by MMI and Soil Type}						
1a	Stream channel, tidal channel	5	20	40	60	80	100	100
1b	San Francisco Bay mud and fill over bay mud	3	15	30	40	60	80	90
2a	Holocene alluvium, water table shallower than 3 m (10 ft)	2	10	20	30	40	60	80
2b	Holocene alluvium, water table deeper than 3 m (10 ft)	.5	2	5	7	12	25	40
3	Late Pleistocene alluvium	.1	.5	1	2	4	7	10

[a] Estimates are based on consensus of PEP.

for buried facilities where MDF(S) = mean damage factor caused by ground shaking; MDF(PG) = mean damage factor caused by poor ground; and P(GFI) = probability of a given ground failure intensity, taken directly, noncumulatively, from TABLE 5 for a given shaking intensity.

The procedure for estimating damage caused by *landslide,* which uses the slope failure concept proposed by Legg *et al.*,[13] is as follows:

$$MDF(LS) = \sum_{SFS=1}^{5} P_{SFI} \times CDF_{SFS}$$

where MDF(LS) = mean damage factor caused by landslide; P_{SFI} = probability of a given slope failure intensity; and CDF_{SFS} = central damage factor for a given slope failure state (TABLE 6).

The method for estimating damage caused by *fault rupture* assumes that structures astride the fault or in the drag zone can be significantly damaged as a result of fault rupture. Various mean damage factors for fault rupture, MDF(FR), are given in TABLE 7. The fault zone (TABLE 7) is assumed to be 100 meters each side of the fault (to be reduced to 5 meters if the fault trace is known) and the drag zone is assumed to be 200 meters each side of the fault (to be reduced to 100 meters if the fault trace is known). The surface fault slip displacements of 0.2, 0.6, 1, 3.5, and 10 meters (TABLE 7) are approximately representative of the slip for earthquakes of magnitudes 6, 6.5, 6.75, 7.5, and 8, respectively.

The procedure for estimating damage caused by *inundation* assumes that depth and velocity of water are the primary factors affecting damage caused by

TABLE 6. Relation between Landslide Severity and Facility Damage Factor[1a]

Central Slope Failure State	Damage Factor (Percent)
Light	0
Moderate	15
Heavy	50
Severe	80
Catastrophic	100

[a] Estimates are based on consensus of PEP.

TABLE 7. Damage Factors for Fault Rupture[1a]

Facility Type and Location	Mean Damage Factor (Fault Rupture) in Percentage for Various Fault Displacements				
	0.2 m	0.6 m	1.0 m	3.5 m	10 m
Subsurface structures					
In fault zone	50	80	100	100	100
In drag zone	20	40	60	80	100
Surface structures					
In fault zone	10	30	70	100	100
In drag zone	0	0	2	10	20

[a] Estimates are based on consensus of PEP; mean damage factor for fault rupture is additive to mean damage factor for shaking, with $MDF_{max} = 100\%$.

inundation. The damage factor for facilities exposed to high-velocity water is given in TABLE 8. The mean damage factors for inundation, MDF(I), specified in this table apply only for ground-surface structures less than 10 meters high.

The *total mean damage factor* for a facility is conservatively the sum of the mean damage factors for ground-shaking, poor ground/liquefaction, landslide, fault rupture, and inundation. The report does not provide a quantitative method for estimating damage due to fire since no method was available at the time the ATC-13 report was written that would yield a meaningful analysis for any urban area in the United States. Since then, a model by Scawthorn and Yamada for predicting the incidence and spread of fire, which is based on experience in Japan,[14] has been extended to the United States. Scenario earthquakes to which this model has been applied include events affecting greater Los Angeles and San Francisco.[15]

Applications and Limitations

The ATC-13 methodology and data can be used for estimating the expected damage to California due to earthquake shaking and collateral hazards (ground failure, fault rupture, and inundation). FEMA's application of the ATC-13 methodology and data involves a sophisticated computerized modeling approach that combines (1) a geophysical model to estimate ground-shaking for hypothetical or

TABLE 8. Damage Factor Caused by Inundation (High-Velocity Water)[1a]

Depth of Water (meters)	Mean Damage Factor (Inundation)[b] (Percent)
1	10
2	20
3	50
4	80
5	100

[a] Estimates are based on consensus of PEP; mean damage factor for inundation is additive to mean damage factor for shaking, with $MDF_{max} = 100\%$.
[b] Applies only to ground surface structures less than 10 meters high. For higher buildings, use 50% of the values in the table.

real earthquakes; (2) inventories of site-specific geophysical data (e.g., landslide, inundation, and liquefaction potential); (3) facilities inventory data and methodology; (4) ATC-13 damage and loss data and methodology; and (5) a joint supply-side/demand-side economic impacts model, which involves assessment of damage to all types of existing facilities in California as well as the economic interaction between functions housed in these facilities. Hence, application of the ATC-13 methodology and data to large regions (e.g., areas affected by moderate- and large-magnitude earthquakes) requires an extensive, sophisticated computerized modeling capability that allows manipulation and storage of large quantities of information and data. Although this requirement may be considered a disadvantage to some users, the modular design of the system provides a major advantage: models, methodologies, and data sets can be replaced at any time with improved information and without affecting other parts of the model or system.

Portions of the ATC-13 methodology and data also have direct application in other aspects of earthquake engineering and hazard assessment. For example, damage-probability matrices can be used to estimate losses to particular subsets of buildings (e.g., low-rise unreinforced masonry) or to "prioritize" building types in order of vulnerability (e.g., to identify those types that should be either inspected first in hazard abatement programs or instrumented first in earthquake strong-motion instrumentation programs).

It is essential that the reader and user of the data in the ATC-13 report be aware that the loss estimates for shaking, for collateral hazards, and for collateral losses are based on judgment, not "hard" data. It is also important to note that (1) the estimates provided are for facilities in California, where structures are designed to resist earthquakes, (2) the estimates represent average conditions, and (3) great amounts of experimental data (i.e., from actual earthquakes) are needed to verify or improve these estimates.

EXTENSION OF ATC-13 DATA TO OTHER PARTS OF THE UNITED STATES

Under the FEMA-sponsored ATC-21 project, "Development of a Handbook for Rapid Visual Screening of Seismically Hazardous Buildings," ATC is developing damage-factor modification constants that will facilitate application of the ATC-13 data to other parts of the United States. These data are needed because the ATC-21 handbook methodology, which is to be applicable nationwide, is based on data derived under the ATC-13 project.

The ATC-13 data were extended to other parts of the United States through a questionnaire process similar, though less rigorous, to that used in the ATC-13 project. Practicing design engineers on the advisory ATC-21 Project Engineering Panel, which was composed of leading practitioners from California, Washington (state), Utah, Tennessee, Massachusetts, and South Carolina, were asked to assemble a group of up to five engineers in their local region to compare the performance of specific building types in their regions to California-designed buildings. For each of the building types identified, the participants were asked to estimate the level of damage that would be sustained by a local building, given the level of damage (expressed in terms of damage factor) sustained by a California-designed building subjected to the same severity of ground-shaking. These data were averaged and used to estimate modification constants for each building type in each of the seven seismicity areas defined in the FEMA-sponsored *NEHRP-*

Recommended Provisions for the Development of Seismic Regulations for New Buildings.[16] After analysis, the first-round questionnaire results were recirculated in a second-round questionnaire to the same individuals for confirmation or revision. Modification constants derived from the second (final)-round questionnaire are provided in TABLE 9.

These modification constants (MC) can be used to change the value of the ATC-13 mean best estimates of damage factor (MB) to best estimates for each NEHRP Area (BENA) through utilization of the following equation:

$$BENA = MC*MB$$

As previously explained, users of data derived through the application of the ATC-13 data, ATC-21 modification constants, and this equation must recognize that this information is based on judgment rather than "hard" data, that any derived estimates represent average conditions, and that great amounts of experimental data (i.e., from actual earthquakes) are needed to verify or improve upon such estimates. Users of the information derived from the ATC-21 project are

TABLE 9. ATC-21 Round-2 Damage Factor Modification Constants

	NEHRP Area				
Structure Type	1,2	3	4	5	6
1. Wood frame	1.0	1.3	1.3	1.2	1.0
2. Steel moment-resisting frame	1.9	1.2	1.4	1.3	1.0
3. Steel frame with steel bracing or concrete shear walls	1.9	1.2	1.4	1.1	1.1
4. Light metal	1.1	1.1	1.3	1.3	1.2
5. Steel frame or concrete frame with unreinforced masonry infill walls	1.2	1.2	1.3	1.3	1.2
6. Concrete moment-resisting frame	2.2	1.3	1.5	1.2	1.0
7. Concrete shear wall	1.7	1.3	1.5	1.1	1.0
8. Tilt-up	2.0	1.2	1.5	1.3	1.4
9. Precast concrete frames	2.9	1.1	1.8	1.2	1.3
10. Reinforced masonry	2.9	1.1	1.3	1.1	1.0
11. Unreinforced masonry	1.1	1.2	1.0	1.0	1.0

advised to consult the ATC-21 Report *Rapid Visual Screening of Buildings for Potential Seismic Hazards: A Handbook*[17] for further explanation of the data derivation.

SUMMARY

Earthquake damage to buildings is dependent on the following factors: (1) structural characteristics, (2) ground-shaking severity, and (3) collateral hazards (poor ground, landslide, fault rupture, inundation, and fire). Methods and data from the FEMA-sponsored ATC-13 Report, *Earthquake Damage Evaluation Data for California,*[1] provide a means for predicting earthquake damage to buildings in California. These data and methodology include: consensus opinion damage-factor estimates (expected physical damage due to ground-shaking); methods to adjust damage-factor estimates to account for construction quality; and methods to estimate the effects of collateral hazards such as ground failure, fault rupture, and inundation. Under the FEMA-sponsored ATC-21 project, "Development of a Handbook for Rapid Visual Screening of Seismically Hazardous Buildings," ATC is developing damage-factor modification constants that will facilitate application of the ATC-13 data to other parts of the United States. These data are being developed through "expert-opinion" questionnaires submitted to project participants.

REFERENCES

1. APPLIED TECHNOLOGY COUNCIL. 1985. Earthquake Damage Evaluation Data for California. ATC-13 Report. Redwood City, CA.
2. STEINBRUGGE, K. V., S. T. ALGERMISSEN, H. J. LAGORIO, L. S. CLUFF & H. J. DEGENKOLB. 1981. Metropolitan San Francisco and Los Angeles earthquake loss studies: 1980 assessment. U.S. Geol. Surv. Open File Rep. 81-113.
3. STEINBRUGGE, K. V. 1982. Earthquakes, Volcanoes, and Tsunamis: An Anatomy of Hazards. Scandia America Group. New York.
4. GALLOWAY, J. D., CHAIRMAN. 1907. Report of Committee on Fire and Earthquake Damage to Buildings. Transactions ASCE **LIX**: 223.
5. U.S. GEOLOGICAL SURVEY. 1981. Earthquake hazards reduction program 1979–80. U.S. Geol. Surv. Open File Rep. 81-41. Menlo Park, CA.
6. STEINBRUGGE, K. V., F. E. MCCLURE & A. J. SNOW. 1969. Studies in Seismicity and Earthquake Damage Statistics: Appendix A. U.S. Department of Commerce, Coast and Geodetic Survey. Washington, DC.
7. SCHOLL, R. E. & J. A. BLUME. 1977. Damaging response of low-rise buildings. *In* Proceedings of the Sixth World Conference on Earthquake Engineering, New Delhi, India, Session 7D.
8. WOOD, H. O. & F. NEWMANN. 1931. Modified Mercalli Intensity scale of 1931. Seismol. Soc. Am. Bull. **21:** 277–283.
9. ROSSI, M. S. 1883. Programma dell' Osservatorio ed Archivo Centrale Geodinamico. Boll. del Vulcanismo Italiano **10:** 3–124 (Rossi-Forel scale: 67–68).
10. BLUME, J. A. 1970. An engineering intensity scale for earthquakes and other ground motion. Seismol. Soc. Am. Bull. **60.**
11. WHITMAN, R. V., J. W. REED & S. T. HONG. 1973. Earthquake damage probability matrices. *In* Proceedings of the Fifth World Conference on Earthquake Engineering. International Association for Earthquake Engineering. Rome, Italy.
12. WOOD, H. 1908. Isoseismals: Distribution of Apparent Intensity. The California Earthquake of April 18, 1906. Report of the State Investigation Commission: 220–254. Carnegie Institution of Washington.
13. LEGG, M., J. SLOSSON & R. EGUCHI. 1982. Seismic hazard for lifeline vulnerability analyses. *In* Proceedings of the Third International Conference on Micronization. Seattle, WA.
14. SCAWTHORN, C. & Y. YAMADA. 1981. Lifelines effects on post-earthquake fire risk. *In* Proceedings of the Second Speciality Conference of the Technical Council on Lifeline Earthquake Engineering, Oakland, California. American Society of Civil Engineers. New York.
15. SCAWTHORN, C. 1987. Fire Following Earthquake: Estimates of the Conflagration Risk to Insured Property in Greater Los Angeles and San Francisco. All-Industry Research Advisory Council Report. Oak Brook, IL.
16. BUILDING SEISMIC SAFETY COUNCIL. 1985. NEHRP Recommended Provisions for the Development of Seismic Regulations for New Buildings. Revised version of the ATC-3 Report, available through the Federal Emergency Management Agency, Washington, D.C.
17. APPLIED TECHNOLOGY COUNCIL. 1988. Rapid Visual Screening of Buildings for Potential Seismic Hazards: A Handbook. ATC-21 Report. Redwood City, CA.

Seismic Rehabilitation Techniques for Buildings

CHRIS D. POLAND

H. J. Degenkolb Associates, Engineers
350 Sansome Street
San Francisco, California 94104

Throughout the United States, the possibility of a damaging earthquake occurring is recognized in most areas, though few jurisdictions require specific earthquake design considerations. The vast majority of buildings in the United States, therefore, do not conform to current design standards appropriate for the earthquakes expected in their areas. Observations from damaging earthquakes over the past 100 years have demonstrated, however, that buildings do not necessarily need to meet modern design standards to be successful performers in a major seismic event. A proper seismic rehabilitation program, therefore, first begins with a detailed evaluation of the building in which the "weak links" of the structure are identified. Only those shown to be critical in past earthquakes are assessed for their susceptibility to catastrophic damage. If the level of expected damage is determined to be unacceptable, then these "weak links" need to be strengthened and/or new earthquake-resistant systems installed. A new procedure entitled ATC-14 and published in *ATC-14: A Methodology for the Seismic Evaluation of Existing Buildings* provides such a technique for identifying and assessing the "weak links" in structures on the basis of past earthquake performance. This paper discusses the ATC-14 methodology and the means of correcting the deficiencies found.

INTRODUCTION

Many existing structures were built before or during the early development of seismic design provisions. This fact becomes painfully obvious every time there is a major earthquake and the same patterns of damage are observed. Unreinforced masonry buildings without certain characteristics may collapse by a separation of their elements. A wood frame house, usually thought to be safe, can be subject to substantial damage or even collapse if it is of split-level construction or if it lacks interconnection of the various elements. Certain types of nonductile concrete frame buildings can be seriously damaged and even collapse. The 1985 Mexico City earthquake demonstrated that even large steel frame buildings can be susceptible to major damage and even collapse. Regardless of the state of present building codes, existing buildings present one of the largest risks to life safety in the event of a major earthquake anywhere in the United States.

Earthquakes have also shown that, given a specific event, serious life-threatening damage is not only concentrated in a few select types of buildings, but it is also limited to an unlucky few of those. Recall that often as many poor buildings perform well in major events as those that are substantially damaged. Speculation continues into why this is possible and local ground conditions, focusing effects, construction quality, and building size and frequency characteristics have all been suggested as possible explanations. The fact is, given a specific earthquake, experience shows that no building is automatically immune from, nor guaranteed to, experience damage because of its style of construction.

Modern building codes (hereafter referred to as the code), such as the NEHRP provisions, are written to guide the construction of new buildings. These codes gather up all of the available collective experience in the behavior of structures and present it in a way that is applicable to all forms of construction. Within the seismic provisions, building code standards are developed with life safety, damage control, and cost in mind. It is hoped that the result is a complete building system that costs slightly more to build but that has the proper strength and connection details necessary to successfully resist earthquake forces. Experience has shown that this successful performance depends on the base shear strength of the main structural elements, the strength and ductility of the connections, the building configuration, the material type, and the interconnection of the structural parts.

Unlike the traditional structural design for dead and live loads, seismic design anticipates that the buildings will be damaged after a truly major event. To design buildings to be damage-free would not only be very expensive, but would also severely limit the permissible styles of construction. New buildings are generally designed to be strong enough to resist small earthquakes without damage and major earthquakes without collapse. To accomplish this goal, the structural design based on the code involves a combination of basic lateral-force-resisting strength, with a proper structural configuration, and appropriate interconnection of the structural elements. In fact, within the code, there is a direct relationship between how a building is configured, detailed, and tied together and the amount of lateral force for which it is designed.

The current economic and political climate is making it more attractive for building owners to revitalize older structures and extend their useful life. The high cost of new construction has led many major industries, institutions, governmental agencies, and the like to renovate rather than replace their existing structures. In addition, various groups are attaching historical significance to older buildings and insisting on their preservation as a means of maintaining an important part of our heritage. Extending the useful life of buildings combined with an increased awareness of damage potential, and concern about liability is causing governmental bodies to consider and implement mandatory seismic strengthening programs. This trend, plus the observation that older non-code-complying buildings sometime perform successfully in major earthquakes, points to the need for a specific evaluation and strengthening standard for existing buildings.

Seismic rehabilitation techniques for buildings, then, need to be thought of in terms of both the evaluation of the existing lateral-force-resisting elements of the building and the addition of new elements where necessary. In this way, proper consideration and credit can be given to the available strengths of a building and these can be supplemented as necessary to achieve an acceptable result.

In spite of the current demand, there is no widely accepted procedure for the seismic evaluation of existing buildings. Many procedures have been developed and published and have found limited use. They tend toward three styles. One considers building performance on a grand scale, considering inventories of buildings in large regions on a generalized basis; this procedure should not be applied to individual buildings. Another includes the more specialized processes that are reported to be applicable to individual buildings. Most often these procedures involve numerical rating systems that score a building's general characteristics. These procedures often combine the general characteristics into a simple numerical rating, which becomes the basis of the decision-making process. These procedures are interesting but often miss the "fatal flaw" in a building and offer little guidance in the areas of the building that need seismic strengthening. Another procedure involves using the latest code provisions to evaluate the adequacy of an existing building's structural system. This procedure may be followed voluntarily or may be required by a governing agency. It may be overly conservative in some cases because it is tailored to protect life and limit damage. In other cases, it may be unconservative because the interconnection of the building's key elements are not sufficient and therefore not capable of producing the overall system performance assumed by the code.

Similarly, when the need for strengthening is identified, there is no uniform procedure for designing the strengthening elements. A proper strengthening program would consider the strengths of the existing elements, and supplement those as necessary to achieve the desired performance. It is important to realize that merely using the seismic design provisions of the code for new buildings is not appropriate. In one sense, it is overly conservative because it would lead to an entirely new lateral-force-resisting system. At the other extreme, it could lead to an inappropriate reliance on the strength of major lateral-force-resisting elements of a building that are not properly interconnected.

Proper seismic strengthening techniques begin with a detailed evaluation procedure in which a building's fatal flaws can be identified along with its inherent strengths. These need to be calibrated in a way that will permit their specific strengthening or lead to a new lateral-force-resisting system that will supplement their deficiencies.

ATC-14: EVALUATING THE SEISMIC RESISTANCE OF EXISTING BUILDINGS

The Applied Technology Council (ATC) has developed a method for evaluating specific buildings that is tailored for use by practicing structural engineers. This procedure not only leads to conclusions concerning the adequacy of the structure for a given event, but it also identifies the structure's weaknesses and therefore areas of needed rehabilitation. The technique has been structured to permit the rapid screening of a large inventory of buildings followed by detailed evaluation where necessary.

The Applied Technology Council is a nonprofit, tax-exempt corporation established in 1971 through the efforts of the Structural Engineers' Association of California. The purpose of the ATC is to assist the design practitioner in structural engineering (and related design speciality fields such as those of soils, wind and earthquake) in keeping abreast and effectively utilizing technological developments. To this end, the ATC also identifies and encourages needed research and develops consensus opinions on structural engineering issues in a nonproprietary format. The ATC thereby fulfills a unique role in funded information transfer.

ATC-14 has been developed consistent with the latest building codes, but tailored to the often nonconforming characteristics of the variety of buildings in existence. It is specifically aimed at assessing a building's life safety level of resistance, with a recommendation that all buildings be strengthened to this minimum level.

It is important to note that ATC-14 has set an evaluation standard that is less stringent than that of modern building codes. It is applicable only to existing buildings and anticipates that in the worst case a building meeting its requirements may be severely damaged and perhaps irreparable after a major earthquake. The building will have, it is hoped, provided a safe refuge during the event for its occupants. This level of performance is not acceptable for new construction because superior earthquake performance can be accomplished through proper design and at little added construction cost.

Life safety in this work has been defined broadly as avoidance of damage that would likely kill an occupant, cause injury to the point of immobility, or block any of the dedicated means of egress from the building. The ACT-14 procedure also identifies areas of potential damage, including "nonstructural" elements, but stops short of determining actual expected damage levels. It has been developed for application throughout the United States.

The project was organized around a steering committee, the ATC Project Engineering Panel, two special consultants on strong ground motion, a subcontractor, the ATC staff, and an NSF program manager.[a] The subcontractor was responsible for the development of the method under the supervision of the Project Engineering Panel. The Special Consultants developed the nation-wide strong-motion criteria to be used in the evaluation procedure.

The ATC-14 method is based on the following elements:

1. A lengthy review of the available literature, especially the published and unpublished reports of earthquake damage;
2. A State of Practice review;
3. The various seismic design provisions currently in use[1-3];
4. The tentative provisions for seismic design as recently published by the Structural Engineers Association of California[4]; and
5. Lengthy discussions between the subcontractor and the Project Engineering Panel (PEP) regarding all aspects of the project.

ATC-14 has been published as a complete document that includes the actual method, all background material, and four examples. It was written not only as a working handbook, but also as an educational tool. The persistent reader will also be rewarded with an overview of the State of Practice in this field, a discussion of ground-motion criteria, a detailed description of structural behavior in past earthquakes, and an extensive list of references and related material.

State of Practice Review

The State of Practice review involved interviews with consultants nationwide, review of the results of the Building Seismic Safety Council (BSSC) trial designs used to check ATC-3, and review of four sets of Navy Phase II evaluation reports. Each Navy evaluation was done by a different office and involved a variety of buildings at a naval base located in a seismically active area. The State

[a] The project's participants were: NSF Program Director, Dr. A. J. Eggenberger; ATC Project Director, Christopher Rojahn; ATC Technical Director, Onder Kustu; ATC Board Contact, Lawrence G. Selna; Project Engineering Panel, Daniel Shapiro, Mete A. Sozen, Boris Bresler, James I. Moore, John C. Kariotis, Raymond Chalker, James L. Noland, and G. Robert Fuller; Special Consultants, Neville C. Donovan and Eric Elsesser; and Subcontractor, H. J. Degenkolb Associates (Chris D. Poland, Loring A. Wyllie, Jr., and James O. Malley).

of Practice review showed clearly that there is a wide range of attitudes toward the evaluation of existing buildings and the expected performance. The most significant results that led to the conclusion that a new, systematic procedure was needed are as follows:

1. While all consultants interviewed treat buildings of different materials and/ or types of construction differently, most do not have a written procedure and none voluntarily follow any of the available procedures.
2. While most consultants initially perform some limited calculations to calibrate their judgment, decisions concerning the adequacy of a building and the need to do further study are based on judgment and experience from the performance of similar buildings in past earthquakes.
3. Most of the consultants interviewed use code-level forces to determine the seismic loads and allowable stresses. Very few use site-specific response spectra and none uses time history analysis.
4. Half of the consultants interviewed calculate connection capacities with respect to member capacities, while the others only check the connections for the code-level forces. None of the consultants checks the connection ductilities, although some will estimate a value on the basis of their experience.
5. The adequacy of a building in terms of providing life safety is most often judged on the basis of experience, configuration, connection details, and calculations, in that order.
6. It appears that while most consultants say that they are following what appears to be a common set of guidelines consistent with the standards of care for today, the fact is that *they are achieving drastically different results*. It appears that the phrase "based on experience" produces the variation. Each engineer has a unique collection of experiences upon which to base his/her judgment. These include his/her own design experience, awareness of technical literature, and past observations of failed conditions. Unfortunately, when an engineer is unaware that a particular type of building has performed poorly in an earthquake, and if on the surface it meets the base shear requirements of the code, he/she will most likely conclude that it is an adequate structure, when the opposite may be true.

ATC-14 is designed to overcome this deficiency. It is formatted to provide a broad base of experience regarding the actual behavior of buildings in earthquakes. It is intended as a tool that can be used to guide but not restrict an evaluating engineer so that consistent and complete thinking can be brought to bear on each seismic evaluation. It stands as a catalog of the profession's collective earthquake experience that has been incorporated with appropriate analysis and design techniques into a concise format that can guide a knowledgeable engineer in the evaluation of an existing building.

Data Collection and Screening

The method begins with data collection procedures, which are required to gather the information necessary to classify the building and perform the evaluations. The appropriate procedures for using existing documents, performing site investigations, and testing the structural materials are outlined. While the method is tailored to the evaluation of buildings on an individual basis, it is structured to permit preliminary screening of an inventory of buildings in order to identify the ones that need to be evaluated at length. Simple, small- to medium-sized, regular buildings, with a good performance record in past earthquakes are screened out. Others are identified for additional evaluation in the specific areas that they are possibly deficient.

On the basis of a brief field inspection, and a review of the available drawings, each building is classified according to an appropriate model building type. Considering the possible combinations of materials, framing types, and nonstructural elements, ninety-eight building types have been identified and grouped into fifteen model buildings. Each model building has a set of performance characteristics representative of its building type. These fifteen model building types are intended

TABLE 1. Model Building Types

I. *Wood Buildings*
　　Building 1—Wood 1—Wood-frame dwellings and light frames
　　Building 2—Wood 2—Commercial or industrial wood structures

II. *Steel Buildings*
　　Building 1—Steel moment-resisting frame buildings
　　Building 2—Braced-steel-frame buildings
　　Building 3—Light moment frame buildings with longitudinal tension only bracing
　　Building 4—Steel-frame buildings with cast-in-place concrete shear walls
　　Building 5—Steel-frame buildings with infilled walls of unreinforced masonry

III. *Cast-in-Place Reinforced Concrete Buildings*
　　Building 1—Reinforced concrete moment-resisting frame buildings
　　Building 2—Shear wall buildings
　　Building 3—Reinforced concrete frame buildings with infilled walls of unreinforced
　　　　　　　　masonry

IV. *Buildings with Precast Concrete Elements*
　　Building 1—Tilt-up buildings with precast bearing walls
　　Building 2—Buildings with precast concrete frames and concrete shear walls

V. *Reinforced Masonry Buildings*
　　Building 1—Reinforced masonry bearing wall—wood or metal deck diaphragm
　　　　　　　　buildings
　　Building 2—Reinforced masonry—precast concrete diaphragm buildings

VI. *Unreinforced Masonry*
　　Building 1—Unreinforced masonry bearing wall buildings

---

to cover the vast majority of existing buildings. Provisions are included for dealing with structures that do not fall into any specific classification given. TABLE 1 lists these fifteen model building types.

Initial Evaluation

The evaluation of a building is directed by a specific section in the book written for the related model building. It includes a building description, a summary of the performance characteristics based on past earthquake observations, paragraphs related to the expected building loads and load paths, references for examples of performance in past earthquakes, and a rapid evaluation procedure. The rapid evaluation procedure attempts to provide a quick method to assess the basic

TABLE 2. Example of "Building Description" and "Performance Characteristics" Section of ATC-14

Seismic Evaluation of Moment Resisting Cast-in-Place Concrete Buildings

Building Description: This building includes floor and roof diaphragms which are typically composed of cast-in-place concrete slabs. The slabs are generally supported by a system of beams, one-way joists, two-way waffle joists or flat slabs. Major floor framing elements may not have continuous top and bottom reinforcing. These elements may be post-tensioned. Concrete columns may have large (greater than D/2) tie spacing. The concrete frames provide the primary lateral-force-resisting system.

Performance Characteristics: Concrete moment-resisting frame buildings are typically more flexible than shear wall buildings. This low stiffness can result in large interstory drifts, which may lead to extensive nonstructural damage. If the concrete columns have a shear capacity below the moment capacity, brittle column failure can occur, possibly resulting in collapse. The following statements discuss some specific performance characteristics which these buildings may exhibit:

1. Large tie spacing in columns can lead to a lack of confinement for the concrete core and/or shear failures.
2. Insufficient column lap lengths can cause concrete to spall.
3. Location of inadequate splices for all column bars at the same section can lead to column failure.

---

strength and/or drift control provided by the structure. TABLES 2 through 5 present examples of these elements for concrete frame buildings.

The evaluation procedure consists of a collection of statements with a related concern and suggested, specific analytic technique if further study should be necessary. Each statement relates to a vulnerable area in the structural system that requires specific consideration. A separate set of statements is presented for regions of different seismicity.

The evaluation statements are written such that a positive or "true" response to a statement implies that the building is adequate in that area. If a building then passes all the related statements with true responses, it can be passed without further evaluation. It must be once again stressed that these evaluation statements are intended to flag areas of concern for the evaluating engineer and by nature must be quite conservative. They will permit buildings designed under modern seismic codes to pass without detailed evaluation.

Not all critical elements of buildings can be evaluated by simple true or false statements. The final decision regarding adequacy, need for further study, or need for strengthening still rests with the engineer, regardless of the statements. For

---

TABLE 3. Example of "Load and Load Paths" Section

Loads and Load Paths: Gravity loads are transferred from the floor slabs to floor framing elements such as one-way joists or waffle joists, or through flat slab action to large beams or columns. Concrete columns support the major floor framing elements and transfer the gravity loads to the foundation.

Lateral loads are transferred from the floor diaphragms to the moment-resisting frames. These frames may or may not be ductile, depending on their configuration and joint details. Column shear strengths that are not sufficient to develop the joint moment capacities can lead to undesirable shear failures.

Typical floor dead weights may range from 90 to 130 psf. Live loads are generally assumed to range from 40 to 100 psf, depending on the occupancy.

TABLE 4. Example of "Building Performance" Section

This section presents a short description of the performance exhibited in past earthquakes by specific buildings of this classification. In some cases, an individual building may not exactly fit into the model building classification. In these cases, the building was listed with the model building that it most closely characterizes. The listings also include the year and location of the earthquake, the reference from which this information is summarized, the approximate Modified Mercalli intensity in that area, and any reported repair costs or damage ratios. The references given are listed in accordance with Section I of Chapter 7. This list is not intended to be all-inclusive, but rather to demonstrate past performance and to provide references for more information on this building type.

1. Sheraton Universal Hotel, San Fernando 1971, Reference 7, page 307. Twenty-story structure, about 58 feet × 184 feet, with cast-in-place concrete girders and columns detailed as ductile under L.A. City Code 1966 (similar to 1970 UBC). Nineteen miles from fault, this building suffered minor damage ($2000). Peak base acceleration was 18 percent g. MM VII.
2. Bank of California, San Fernando 1971, Reference 7, page 327. Twelve-story cast-in-place ductile concrete moment frame built to 1969 L.A. Building Code. This building, which was 17 miles from fault, experienced a ground floor acceleration of 23 percent. Damaged members included columns, girder stubs, cracking in spandrels and floor slab. $40,000 repairs in $4,000,000 building. MM VII.
3. Banco Central, Managua 1972, Reference 4, page 1077. Fifteen-story, nonductile cast-in-place concrete moment frame designed for 12 percent g. Columns cracked, diaphragms tore, and partial collapse occurred. Much nonstructural damage occurred. Did not collapse.
4. Telcor Building, Managua 1972, Reference 4, page 1079. Seven-story concrete nonductile cast-in-place concrete frame, with precast beams, topping slab, and some post-tensioning in girders. Designed for earthquake. Major damage occurred, requiring the building to be torn down.

TABLE 5. Example of Rapid Evaluation Procedure

Rapid Evaluation of Reinforced Columns

Concern: Reinforced concrete frame buildings have sometimes proven to present a life safety hazard in past earthquakes because of inadequate short column shear capacity. A quick estimation of the shear stress in the concrete frame columns should be performed in all evaluations of this building type in regions of high or moderate seismicity.

Procedure: Generate the loads using the rapid evaluation procedure presented in Section 4.4.2, checking the first-floor level and all other levels where the columns could be subjected to high shear stresses. Estimate the average column shear stress, V_{AVG}, as follows:

$$V_{AVG} = \frac{n_c}{n_c - n_f} \frac{V_j}{A_c}$$

where n_c = total number of columns; n_f = total number of frames in the direction of loading; V_j = story shear at the level under consideration, determined from the loads generated by the rapid evaluation procedure; and A_c = summation of the cross sectional area of all columns in the story under consideration.

If the average column shear stress is greater than 60 psi, a more detailed evaluation of the structure should be performed. This evaluation should employ a more accurate estimation of the level and distribution of the lateral loads by using the procedures suggested in Section 4.4. Calculate the column shear capacities using the provisions of ACI 318 and compute capacity/demand (C/D) ratios.

Recommended C/D Ratio: 1.0. Many of the concerns in the following statements will address the details necessary to provide ductile column behavior.

this reason, these evaluation procedures, even the initial screening procedures, must be applied by a knowledgeable professional engineer.

Each statement carries with it a concern which explains in commentary style why the statement was written. These concerns are carefully written and intended to further assist the evaluating engineer in dealing with the issue stated. Obviously, addressing the concern takes precedent over the specific statement.

The evaluation statements are presented as two complete sets for each model building type. The first set relates to buildings in areas of low seismicity, specifically to those areas with A_a and A_v values less than or equal to .10 g. The second set relates to areas of moderate and high seismicity. The statements in the low seismicity section tend to be general in nature and require only the fundamental elements needed for seismic resistance. The statements in the high seismic section are more specific and often deal with specific characteristics of key elements and details. The engineer attempting to interpret and use the general statements in the low seismic sections can obtain additional guidance from the related statements in the high seismic section.

For statements that are "false," additional evaluation is required. This does not necessarily imply that a complete structural evaluation is necessary, or that the building is automatically deficient. In fact, the suggested procedure limits the evaluation to only the area of concern. It is offered as a suggested procedure since the responsibility for the evaluation rests with the structural engineer, who may elect to perform an alternate evaluation procedure. This is permissible as long as it addresses and leads to an opinion regarding the issue raised in the statement. Deficiencies are, therefore, identified only after an appropriate detailed evaluation has been made.

Detailed Evaluation

The evaluation procedures suggested are based on the ATC-3 and ATC 6-2 type approaches to considering the capacity of the element under review and the demand placed on that element. A recommended Capacity/Demand ratio (C/D) is listed for each statement which is based on the anticipated excess capacity available in the element and the level of overall system ductility assumed in the demand criteria. The recommended C/D ratio can then be compared to that calculated for the building being evaluated. In this way, not only can the weak links in the structural system be identified, but also their significance can be estimated. TABLE 6 presents two statements which have been included in the evaluation of concrete frame buildings.

STRENGTHENING PROCEDURES FOR EXISTING BUILDINGS

For a building to pass the ATC-14 evaluation procedure and be judged acceptable, all of the issues raised in the statements must either be "true" or the capacity/demand (C/D) ratios determined in the subsequent analysis shown to be within the prescribed limits. One of the values of this technique is that if a building does not pass the evaluation method and is judged to be unacceptable, it is clear exactly where the problem is and what its characteristics are. A strengthening technique then can be developed to correct those specific areas of weakness.

Techniques for correcting the deficiencies found during the evaluation can take on one of two major forms. Certain deficiencies can be corrected individually with repair details aimed at an acceptable C/D ratio. Other deficiencies are best corrected by adding new systems that supplement the main lateral-force-resistant system and have the direct effect of reducing the demand on the weak link. The goal is keyed to reducing the C/D ratio to an acceptable level.

An example of a strengthening technique that increases the capacity of a weak link is shown in FIGURE 1. In this condition, a high-rise, reinforced masonry residential building with highly complex shape was found to be deficient at the re-entrant corners. This turned out to be the most serious of the deficiencies noted and it was determined to be a possible source of collapse of the building. The precast corner floor system in each wing shared a common bearing wall at each of the re-entrant corners with only nominal bearing area and interconnection. The recommended solution involved adding a substantial diaphragm tie at each of those locations that would increase the capacity of the diaphragm and permit the needed transfer of tension chord forces.

TABLE 6. Example of Evaluation Statements

2.1.2 *Statement:* There are no infills of concrete or masonry placed in the concrete frames, which are not isolated from the structural elements.
Concern: Infilled walls used for partitions or walls around the stair or elevator towers, which are not adequately isolated, will alter the seismic response of the structure. Evaluation of considerations for frame structures will therefore be inappropriate.
Procedure: Evaluate the building as an infilled wall structure using the procedures of Section 4.7.3.
2.1.3 *Statement:* The lateral-force-resisting elements form a well-distributed and balanced system that is not subject to significant torsion. Significant torsion will be taken as any condition where the distance between the story center of rigidity and the story center of mass is greater than 20 percent of the width of the structure in either major plan dimension.
Concern: Plan irregularities may cause torsion or excessive lateral deflections, which may result in permanent set or even partial collapse.
Procedure: Verify the adequacy of the system by analyzing the torsional response using procedures which are appropriate for the relative rigidities of the diaphragms and the vertical elements. Compare the maximum calculated story drift with 0.005H.

An example of a strengthening technique that reduces the demand on an unacceptable element is shown in FIGURE 2. In this case, the building was a ten-story steel frame with concrete and unreinforced masonry infill walls. The steel frames had partial moment connections and were not continuous to the foundation. The tall, column-free lobby area was achieved with the use of transfer girders at the second floor. The lateral-force-resisting system of the building was provided by the concrete walls on the two property lines and the infilled unreinforced masonry wall on the two exposed, street-front elevations. The most significant problems with the building had to do with the tall first story, the discontinuous steel frame, the lack of sufficient strength and anchorage of the infilled masonry walls, and the unbalanced lateral-force-resisting system that included significant torsion. The strengthening technique selected involved adding two new full-height braced frame systems to the building that completed the lateral-force-resisting system while reducing the demand on the discontinuous steel frame

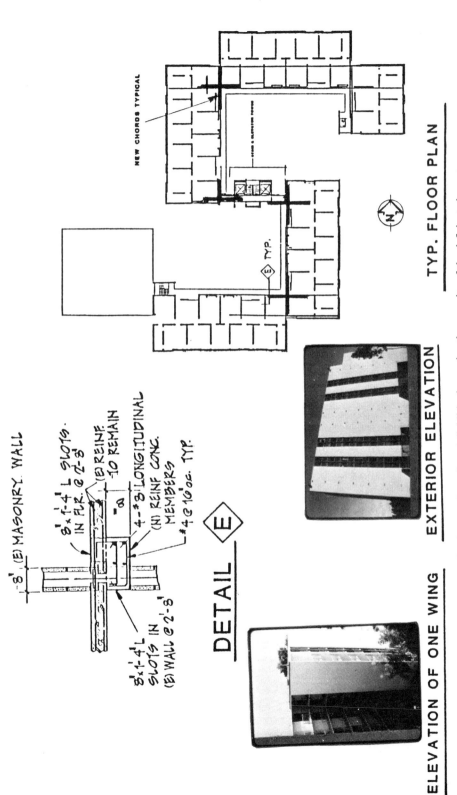

FIGURE 1. An example of correcting a ''weak link'' by increasing the capacity of the deficient element.

NEW CHORDS TYPICAL

TYP. FLOOR PLAN

N

E TYP.

(E) MASONRY WALL

8"x¼"x 4" L SLOTS.
IN FLR. @ 2'-8"

(E) REINF.
TO REMAIN

4-#8 LONGITUDINAL

(N) REINF CONC.
MEMBERS

#4 @ 16" OC. TYP.

8"

8"x¼"x 4" L
SLOTS IN
(E) WALL @ 2'-8"

DETAIL E

ELEVATION OF ONE WING

EXTERIOR ELEVATION

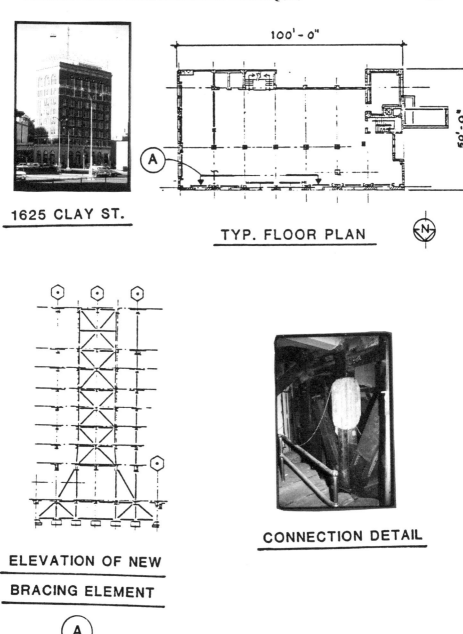

1625 CLAY ST.

TYP. FLOOR PLAN

ELEVATION OF NEW

BRACING ELEMENT

CONNECTION DETAIL

FIGURE 2. An example of correcting a "weak link" by reducing the demand on the deficient element.

system and the unreinforced masonry infill. A new anchorage system was also added to support the unreinforced masonry infill.

Extending this concept of strengthening to all possible deficiencies is beyond the scope of this paper. In concept, however, the various types of deficiencies can be organized into two general groups that relate to the styles of correction introduced above. TABLE 7 lists the common forms of deficiencies dealt with in ATC-14 and suggests a style of solution. The many other deficiencies noted for the various model buildings in ATC-14 can be dealt with in a similar manner.

The development and implementation of strengthening details should be spe-

TABLE 7. Problems and Solutions

Weak Link	Common Solution
Deficiencies Often Corrected by Increasing the Capacity of the Element or Connection	
Plan irregularities and re-entrant corners	Add diaphragm chords at the points of discontinuity without sufficient interconnection.
Diaphragm openings	Add collector elements as required to deliver the lateral force directly to the lateral force resisting element.
Connection to the foundation	Interconnect existing elements if they are judged to be sufficient, and add new foundation elements as required.
Cornices and parapets	Provide appropriate bracing to each element. The bracing must connect to an element of the structure that is capable of supporting all of the new bracing loads.
Deterioration and distress	Determine the source of deterioration and eliminate if possible. Remove and replace or restore the deteriorated elements that are a key part of the lateral-force-resisting system.
Insufficient wall to diaphragm connections	Provide new anchors that are spaced to accommodate the wall's ability to span between anchors.
Deficiencies Often Corrected by Adding Additional Elements to the Lateral-Force-Resisting System or Providing an Entirely New System (These corrective measures serve to reduce the demand on the critical weak links to an acceptable level.)	
Excessive torsion	Add new lateral-force-resisting elements to eliminate the torsion.
Vertical strength discontinuity	Extend the discontinuous lateral-force-resisting element and eliminate the discontinuity.
Vertical mass or geometric irregularity	Add to the entire lateral-force-resisting system as needed to accommodate the demands on the system.
Veneer over a means of egress	Replace with a properly anchored and reinforced veneer.
Insufficient strength of existing system	Add new elements that are complementary with the existing ones such that the combination of elements can provide the needed strength. It is permissible to use a new bracing system that is less rigid than the existing one as long as the existing system can yield and continue to carry its yield load without significant loss in strength or vertical load carrying ability. An example of this is the addition of steel braces to supplement concrete shear walls or concrete moment frames. The concept of supplementing the existing system applies to all elements of the lateral-force-resisting system including the diaphragms, walls, and foundations.

cifically tailored to each structural system. Often the style of correction is governed by the constructability of the new elements, the need to work while the building is occupied, and the need to minimize distribution to the various architectural, mechanical and electrical systems of the building. Cost is often not the primary criterion for determining the rehabilitation technique.

SUMMARY

Because of the large life safety hazard posed by existing buildings, and the lack of evaluation consistency amongst engineers, the development of a comprehensive methodology for the seismic evaluation and strengthening of existing buildings is necessary. The ATC-14 project is intended to meet this need. The method starts with the collection of information about the structure and classifies the building into one of fifteen building types. The evaluation procedure begins as a qualitative investigation to determine whether the building might possess any of the life-threatening performance characteristics that similar structures have demonstrated in previous earthquakes. If any potential life safety hazards are identified, an appropriate detailed evaluation is recommended with permissible capacity/demand ratios suggested. Even though the final decision regarding adequacy still rests with the structural engineer and is based on his experience, it is anticipated that this procedure will draw some consistency to the evaluation process and broaden the experience base upon which this ultimate and all-important decision is made.

When the need for strengthening is identified, ATC-14 provides specific information on the location of the key deficiencies. These deficiencies are normally corrected by either increasing capacity or providing supplementary lateral-force-resisting elements that serve to reduce the demand on a "weak link" to an acceptable level. Occasionally, this will require an entirely new lateral-force-resisting system. However, most often it does not.

The proper seismic strengthening of a deficient building begins with a detailed evaluation of the existing lateral-force-resisting elements, whether or not they were specifically designed for lateral forces. From an appropriate evaluation of the available system, its adequacy is determined and, if found to be insufficient, its weaknesses are specifically defined in terms of "weak links." Corrective measures then should be undertaken to strengthen and/or supplement the weak links while taking full advantage of the strength available in the existing system. In this manner, the seismic safety of a structure can be improved to an acceptable level for the minimum possible cost and least disruption.

REFERENCES

1. INTERNATIONAL CONFERENCE OF BUILDING OFFICIALS. 1985. Uniform Building Code, 1985 ed. Whittier, California.
2. AMERICAN CONCRETE INSTITUTE. Building Code Requirements for Reinforced Concrete, Including Commentary, ACI 318-83. Detroit, MI.
3. AMERICAN INSTITUTE OF STEEL CONSTRUCTION. 1978. Specification for the Design, Fabrication, and Erection of Structural Steel Buildings, with Commentary, eighth ed. Chicago, IL.
4. STRUCTURAL ENGINEERS ASSOCIATION OF CALIFORNIA. 1985. Tentative Lateral Force Requirements. Seismology Committee. Sacramento, CA.
5. APPLIED TECHNOLOGY COUNCIL. 1987. ATC-14: A Methodology for the Seismic Evaluation of Existing Buildings. Redwood City, CA.

Costs and Benefits of Seismic Rehabilitation

M. ELISABETH PATÉ-CORNELL

*Department of Industrial Engineering and
Engineering Management
Stanford University
Stanford, California 94305*

INTRODUCTION

Seismic provisions of building codes apply to new structures at the time of construction. Therefore the costs involved can be relatively low. The rehabilitation of existing structures is often more expensive, and in areas where earthquakes are not of major concern to the public, the legislators are often reluctant to impose seismic rehabilitation measures. Yet in some regions of the Eastern United States major earthquakes have occurred and decisions must be made as to what constitutes economically and politically justifiable seismic rehabilitation. Three questions must be addressed:

- What buildings should be reinforced?
- What level of reinforcement should be required?
- When should reinforcement be required.

Answers to the first and the second questions depend on the type of structure and the type of use of the buildings. For example, residential buildings of unreinforced masonry that are vulnerable and have a high level of occupancy may require more attention than commercial storage buildings. Also, the timing of the upgrading is important because the costs may be much lower at the time of another modification, such as extensive remodeling, or when the use of the building is less critical. The question is whether delaying the reinforcement is reasonable, that is, if the risk incurred until the building can be reinforced more cheaply is acceptable.

The techniques of cost-benefit analysis[1] can shed some light on this decision problem within the limits of the economic reasoning. Noneconomic considerations might include political reasons to require seismic reinforcement at a given time. There may also be some redistribution effects that do not appear in global estimates of costs and benefits. But to the extent that economic efficiency is of concern to the decision-maker, the proposed model permits him or her to check the consistency of this particular safety measure with other protective options and to compare the desirability of this expenditure with other possible options such as investment or consumption.

This paper presents a model of cost-benefit analysis for the seismic rehabilitation of existing buildings.[2] The model is illustrated by an application to the costs

392

and benefits of upgrading some particularly vulnerable buildings in the Boston area at the time of remodeling. Several key features of the method are discussed because they are critical to the validity of the cost-benefit approach altogether. First, the cost-effectiveness of seismic upgrading is measured by the implicit "cost per life saved," corresponding to the proposed reinforcement level. Secondly, the benefits are measured by the expected value of the avoided losses, which assumes that the decision-maker is "risk neutral." Thirdly, all future benefits in the computation of the cost per life saved are discounted to reflect the opportunity cost to society, that is, the benefits that would be provided by other kinds of societal investments, including those related to the improvement of public health and safety.

A COST-BENEFIT MODEL FOR THE SEISMIC REHABILITATION OF EXISTING BUILDINGS

Immediate Reinforcement

The costs of reinforcement are of two types.[3] The *direct costs* are determined by the physical characteristics of the building (such as type of structure, size, and location) and the desired level of reinforcement. The *indirect costs* represent the loss of the use of the building during the reinforcement, or the additional time of non-use if reinforcement is coupled with other type of work. For residential buildings, it is the cost of temporary housing for the regular occupants. For commercial or industrial buildings, it represents the loss of production or the loss of sales during the reinforcement period. Note, however, that this effect may amount to a transfer (as opposed to a net societal loss) if there is some unused production capacity in the country, and if production can occur elsewhere. In addition to this primary economic loss, there may be some *secondary economic loss* in the sectors of the economy that depend on the interrupted production as input to their own activity. The secondary loss occurs in the case where these inputs cannot be found elsewhere in the national economy.[4]

The benefits are the avoided future losses: human casualties and economic quantities, including property damage and loss of production. These two quantities are kept separate through the analysis. The expected value of future losses in earthquakes is computed with and without reinforcement. The benefits are measured by the difference between these two figures. Here again, the future losses in earthquakes can be divided among direct and indirect losses. In the computation of the indirect losses, the issue is how long the service or the production from a damaged facility will be interrupted following an earthquake, and, therefore, how long it will take for rehabilitation at a time of heavy demand for repair and reconstruction. In addition to these primary economic losses, there may be a secondary economic loss as described above, and its avoidance is part of the benefits of reinforcement.

An important economic aspect of a seismic reinforcement regulation is its effect on the building market. This effect depends on the economic situation and on the elasticity of the demand in the particular type of building use (for example, the housing market or the market for office buildings). Elasticity effects are analyzed in detail in the illustration presented below. Because of the sheer volume of capital involved, small variations of demand may have large repercussions on the economic life of an area.

Reinforcement at Time of Lower Costs

Reinforcement can be delayed for two economic reasons: to combine the work with other types of repair or modification, which decreases the direct cost; or to wait until the level of activity in the particular type of building is lower and, therefore, to decrease the indirect cost. An example of the first type of delay is to require seismic upgrading at the time of a major remodeling. Delays of the second type may be desirable for buildings of seasonal use (e.g., schools) or for industrial buildings for which a backup is planned in the future.

In both cases, the question is whether the delay is acceptable, that is, whether the individual risk and the societal risk incurred in the interim are generally considered acceptable.

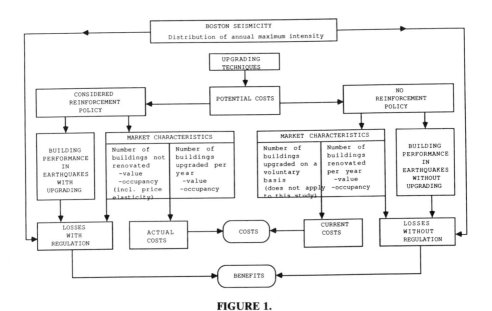

FIGURE 1.

Elements of the Cost Effectiveness Model

The following model was developed to measure the costs and the benefits of a regulation requiring seismic upgrading for a particular type of building at the time of a major remodeling with change of use (from commercial into residential) in the Boston area.[2,5] It is a cost-benefit analysis under uncertainty, over a 50-year period, of a given level of reinforcement whose effect is to reduce the proportion of lives lost and property damage under different levels of seismic intensities. The benefits are computed by reckoning difference between the losses with and without seismic upgrading. The difficulty is to define the target properly, that is, the population of buildings that would not be upgraded on a voluntary basis, but would be under a regulation although with some delay due to the extra costs involved (elasticity effect). The method is described in FIGURE 1.

The local seismicity is described by the probability distribution of the annual maximum intensity in the area. In this simplified model, a particular city is considered as one seismic zone and this probability distribution is assumed to be the same in all points of the city. In reality, different site conditions may affect the level of damage for a given peak ground acceleration.

A measure of a building's performance in earthquakes is given for each relevant intensity (measured here on the Modified Mercalli scale) by the loss ratios: damage ratio and casualty ratio. The damage ratio is the proportion of the building's market value to be spent after the earthquake to restore the building to its initial use. The casualty ratio is the proportion of victims among the occupants. These ratios are treated here as random variables and described by their probability distributions in damage probability matrices.[6] For each MMI, the probability density of the two loss ratios thus describes the building's performance and the associated uncertainties.

Without upgrading, the probability distribution of the annual losses in earthquakes can be computed knowing (1) the buildings' value and occupancy after simple remodeling, (2) the current buildings' performance in earthquakes (assumed to be unchanged), and (3) the description of the local seismicity. Note that a certain fraction of the renovated buildings may be voluntarily upgraded at the time of renovation.

With a seismic reinforcement regulation, the same computation can be performed with the following modifications. First, for different levels of reinforcement, the buildings' performance under each seismic intensity is improved and therefore the loss ratios are lower than without reinforcement. Second, the number of buildings remodeled and reinforced after the regulation is smaller than it would have been without it.

The decrease in demand for renovation after an increase in price due to mandatory seismic upgrading is measured by the demand elasticity. After consulting with architects and contractors, it was decided that a good approximation for the elasticity of the demand of the remodeling market was the elasticity of the demand for new housing.[7] This approximation neglects the difference in tax advantages to the owners when reinforcing new buildings or upgrading old ones. Also, assuming that the demand reduction can be measured by the market elasticity implies that the increase in costs due to the regulation amounts to a tax. This means in turn that potential customers do not consider the additional seismic safety as a benefit for which they would be willing to pay the corresponding price voluntarily. This is a reasonable assumption if one treats separately (or assumes to be negligible) the segment of the renovation market that would have been reinforced on a voluntary basis in the absence of seismic regulation.

An important result of the elasticity analysis is that the people who will decide not to enter the renovation market because of the added cost of seismic upgrading will live in the average (often unreinforced) residential building in the city.

Mathematical Formulation

The parameters and variables of the model are the following:

$p_1(i)$: probability that a given MMI i is the maximum observed in any given year;
N: total number of buildings of the chosen category available for renovation at the beginning of year 0;
H: average number of occupants per building after renovation;

μ:　　proportion of existing buildings being renovated annually;

B:　　average market value of each building after renovation;

B':　　average market value of each building before renovation;

μ':　　proportion of buildings renovated annually if the seismic reinforcement regulation is passed accounting for the market elasticity;

K_L:　　casualty ratio without reinforcement;

$p_{KL|I}(k_L|i)$:　　conditional probability of K_L given intensity i;

K_D:　　damage ratio without reinforcement;

$p_{KD|I}(k_D|i)$:　　conditional probability of K_D given intensity i;

K'_L and K'_D, $p_{KL'|I}$ and $p_{KD'|I}$:　　corresponding variables should the reinforcement policy be adopted;

K''_L and $p_{KL''|I}$:　　casualty ratios and their probability conditional on i in the "average building," where the people excluded from the renovation market because of the cost increase are assumed to live for the rest of the considered period;

K''_D and $p_{KD''|I}$:　　damage ratios and corresponding probability conditional on i in the considered buildings before renovation;

R:　　social rate of discount.

The mathematical model is the following:
The marginal probability distribution of the loss ratio is:

$$\begin{cases} p_{KL}(k_L) = \sum_i p_I(i)p_{KL|I}(k_L|i) \\ p_{KD}(k_D) = \sum_i p_I(i)p_{KD|I}(k_D|i) \end{cases} \tag{1}$$

The annual losses without reinforcement in the considered population of buildings are at year t:

$$\begin{cases} L(t) = K_L HN\mu t \\ D(t) = K_D BN\mu t \end{cases} \tag{2}$$

The annual losses with reinforcement regulations in the considered population of buildings are at year t:

$$\begin{cases} L'(t) = K'_L HN\mu' t + HNt(\mu - \mu')K''_L \\ D'(t) = K'_D BN\mu' t + B'Nt(\mu - \mu')K''_D \end{cases} \tag{3}$$

These losses include: (1) the casualties among the occupants of the renovated reinforced buildings; (2) the casualties among the potential inhabitants of these buildings who live in the average city home because of regulation costs; (3) the property damage in the renovated reinforced buildings; and (4) the property damage in unrenovated buildings that would have been remodeled in the absence of regulation requiring seismic upgrading.

The benefits at year t therefore are:

$$\begin{cases} \Delta L(t) = HNt[(K_L - K''_L)\mu - (K'_L - K''_L)\mu'] \\ \Delta D(t) = Nt[(BK_D - B'K''_D)\mu - (BK'_D - B'K''_D)\mu'] \end{cases} \tag{4}$$

The discounted present value of the benefits over the considered fifty year period is:

$$
\begin{cases}
\Delta L = HN[(K_L - K_L'')\mu - (K_L' - K_L'')\mu'] \displaystyle\sum_{t=0}^{50} \frac{t}{(1 + R)^t} \\[4mm]
\Delta B = N[(BK_D - B'K_D'')\mu - (BK_D' - B'K_D'')\mu'] \displaystyle\sum_{t=0}^{50} \frac{t}{(1 + R)^t}
\end{cases}
\tag{5}
$$

The benefits are linear combinations of the random variables K_L, K_L', K_L'', K_D, K_D', and K_D'', the coefficients being defined by Eq. 5.

$$
\begin{cases}
\Delta L = aK_L - bK_L' + cK_L'' \\
\Delta D = a'K_D - b'K_D' + c'K_D''
\end{cases}
\tag{6}
$$

Noting S the sum of the series $\displaystyle\sum_{t=0}^{50} \frac{t}{(1 + R)^t}$ the coefficients of Eq. 6 are:

$$
\begin{cases}
a = HN\mu S \\
b = HN\mu' S \\
c = HNS(\mu' - \mu)
\end{cases}
\qquad
\begin{cases}
a' = NB\mu S \\
b' = NB\mu' S \\
c' = NB'S(\mu' - \mu)
\end{cases}
\tag{7}
$$

The expected value of the benefits is thus:

$$
\begin{cases}
EV(\Delta L) = aEV(K_L) - bEV(K_L') + cEV(K_L'') \\
EV(\Delta D) = a'EV(K_D) - b'EV(K_D') + c'EV(K_D'')
\end{cases}
\tag{8}
$$

in which $EV(.)$ is the expected value of the corresponding random variable.

The annual costs of seismic upgrading (over and above the costs of renovation) are computed as a function of the renovated surface. Let A be the average area of the renovated buildings and C_A the cost per surface unit. The annual cost of seismic upgrading is:

$$
C(t) = NAC_A\mu
\tag{9}
$$

The total discounted cost over the considered 50-year period is thus:

$$
C = \sum_{t=0}^{50} \frac{C(t)}{(1 + R)^t}
\tag{10}
$$

If the cost C is smaller than the expected value of the avoided damage, the risk-neutral decision-maker may adopt the regulation on its financial merit alone. Otherwise, he may consider as a measure of the cost effectiveness of seismic upgrading the average net cost per life saved, which is:

$$
CPL = \frac{C - EV(\Delta D)}{EV(\Delta L)}
\tag{11}
$$

To compare the cost effectiveness of two reinforcement policies for the same type of building, the decision-maker should consider the marginal cost per life saved, which is the ratio of the increase of net cost to the increase of the number of avoided casualties. This computation is repeated for each considered level of reinforcement.

Application to Some Buildings in Boston

There has been in Boston a trend of remodeling old industrial buildings into residential and office space units. For these changes of use, current code standards do not require seismic upgrading. Yet these buildings are generally vulnerable to earthquakes and will end up with a high rate of occupancy. At the time of remodeling, seismic upgrading is less expensive than it would be otherwise. In spite of the rarity of earthquakes, the risk reduction might therefore be significant at low cost.

The case considered here is that of unreinforced masonry warehouses typical of the mill-type commercial buildings constructed in New England between 1860 and 1900. There are about 150 such buildings in Boston, each having an average floor area of 80,000 square feet. We assumed that, in absence of seismic regulation, 2 percent of these buildings would be reinforced every year, and that after reinforcement the average occupancy would be 300 persons per building. We used a price elasticity factor of 0.85, which reduces the rate of remodeling from 2 percent per year to about 1.8 percent. The price elasticity was estimated as the average of the price elasticities for renters' demand and for owners' demand, since it appeared that renting and occupation by owner were equally likely. Finally, we used a social rate of discount of 7.5 percent as an average measure of the opportunity cost incurred in real terms by society.

The seismic hazard in the area was estimated through a weighted average of the results of several studies.[8,9] The figures used were the annual probabilities of exceeding different values of MMI:

$$\begin{cases} \text{VI:} & 6.8 \times 10^{-3} \\ \text{VII:} & 1.6 \times 10^{-3} \\ \text{VIII:} & 4.2 \times 10^{-4} \end{cases}$$

The different policies of seismic upgrading that were considered and their costs per horizontal square foot in 1980 dollars are the following:

Policy 1: Provide new floor and roof diaphragm; tie new roof and floor diaphragm to framing and existing masonry walls.
    At the time of major remodeling: $1
    Otherwise: $5.5

Policy 2: Tie the building together; add a sufficient number of internal walls to get some ductility and provide foundations for those new walls.
    At the time of major remodeling: $1.5
    Otherwise: $6.5

Policy 3: Tie the building together; reinforce the existing masonry walls to get some ductility, and reinforce the parapets.
    At the time of major remodeling: $2
    Otherwise: $7.4

The performances of the buildings with and without reinforcement were characterized by a shift in the probability distribution towards lower loss ratios under the three MMIs considered here (see Paté[2] for the numerical values). The results over fifty years are shown in TABLE 1.

A reinforcement regulation based on Policy 1 can be economically justified for unreinforced masonry warehouses at the time of remodeling into residential units because the cost per life saved ($240,000) is consistent with protective measures generally adopted in the public sector. Policy 2 might appear too costly for regulation in the corresponding range of individual risks, but may be consistent with private choices and some individuals' willingness to pay for their own protection. Policy 3 offers virtually no additional protection to the buildings' occupants and, although it does decrease the amount of property damage, these savings do not justify the marginal costs.

TABLE 1. Costs and Benefits of Seismic Upgrading of Some Buildings at the Time of Major Renovation

Policy	Total Cost ($M)	EV (ΔD) ($M)	EV (ΔL) Undiscounted	EV (ΔL) Discounted	Cost per Life Saved[a]
1	3.1	1.9	37	4.9	0.24
2	4.6	2.02	41	5.5	2.56
3	6.2	2.11	42	5.7	—

[a] In million dollars (1980). Net for policy 1; marginal for policies 2 and 3.

COST EFFECTIVENESS AND VALUE OF LIFE

The rationale for the use of the cost per life saved in safety regulations is based on the concept of economic efficiency.[10,11] The first question is whether the costs of regulation matter in such a debate. If they do, one is left with a problem of allocation of limited resources for the optimization of a social welfare function. Possible answers depend on the nature of the risk involved, the level of individual risk, and the desirability of alternative use of public and private funds.

The search for an adequate balance of risks and benefits is a matter of political philosophy as well as an economic issue. Risk management in the public sector can be approached from two perspectives. It can be left to the individual preferences, private negotiation, and tort law. The alternative is regulation that reduces individual options but may protect some people who would otherwise be put at risk by the activity of others who are undeterred by the prospect of legal punishment after the fact.

There is still some resistance in the political arena to the notion that the costs of human safety might matter in regulatory decisions and that some costs are simply unacceptable. Yet the safety legislation in the United States has evolved in the last 20 years from zero-risk positions that proved untenable economically and had to be softened by court decisions, to "best available technology," which also proved impractical, to the final recognition that there are unavoidable trade-offs between costs and safety levels.

The Individual Viewpoint

From the point of view of the individual, there are basically three ways he may choose to spend his after-tax money: immediate consumption, investment in productive activities (which is equivalent to delayed consumption), and investments in health and safety that permit him to enjoy extended consumption. To the extent that there is generally a positive correlation between consumption and longevity, and, furthermore, that there may be a trade-off between risk-taking and the quality of life, there is a limit to the amount that the individual wants to pay for risk-reduction. The "value of life" that is revealed by the choice of an individual who is willing to spend N dollars to eliminate a probability of death of Δp is equal to $N/\Delta p$. In no way does this figure represent an amount paid for a sure life or death outcome. Indeed, it depends on the level of individual risk that the person faces, and also on the nature of the hazard. There is no reason, for instance, to assume that the individual is indifferent between dying from the "painless black pill" of the literature[12] and dying from a long and painful disease.

If the individual's goal is to maximize his longevity, his "value of life" is the result of the balance between the longevity effects of consumption (immediate and delayed) and the extra longevity that is acquired through investments in health and safety. If he wants to maximize another utility function involving not only longevity but also some measure of his quality of life, the "value of life" is one of the results of this optimization. One key characteristic of this figure is (1) that it varies from person to person, (2) that it depends on the probability of death Δp, and (3) that it also depends on the characteristics of the considered hazard.

To capture the dependency between the "value of life" and the level of individual risk that one proposes to eliminate, Howard[12] has constructed a taxonomy of risks that takes into account the delegation of risk-taking decisions from the individual to someone else, for example, a government agency. The range of individual risk levels is divided into four domains in which the risk is: delegable (with or without compensation), undelegable, and unacceptable regardless of the pay-offs. In the range of delegable risks, the "value of life" $N/\Delta p$ was found to be roughly constant. For example, for a white male in his thirties, this number was on the order of two million 1980 dollars.

The Public Sector Perspective

In the public sector, the problem practically is the following: (1) In what range is a particular individual risk accepted without compensation? And, (2) what is the maximum "cost per life saved" that is economically justifiable for additional safety regulations? If the goal of the regulator is to save the maximum number of lives, he must compare the cost per life saved for a considered safety investment to the risk-reduction results that can be achieved by alternative health and safety measures *in the same range of individual risk*. He may want, for example, to consider the figures that have been inferred from past regulations or are used as guidelines in some industries.

Comparison of costs per life saved across economic sectors requires some caution. For example, although the two problems of reducing an existing risk at a given cost and creating a risk for a benefit are economically symmetrical, there are major philosophical differences between these two types of situations: the individual subjected to a seismic risk mitigation regulation can always increase his protection level, whereas a person who is exposed against his will to the risk of a new industrial plant may feel that he has no risk-reduction option except to move. Seismic risk mitigation is in the first category. In that case, the political issue is a trade-off between social justice (requiring that no one be exposed to unreasonable levels of risk) and freedom of the individual to choose the balance between safety and costs that best fits his income and his preferences.

Consistency of standards across the different economic sectors is desirable for two types of reason: social equity and economic efficiency. Consistency is better achieved through consistency of the decision-making process rather than by requiring that the maximum acceptable cost per life saved be the same across the board.[13] Information about implicit cost per life saved for other regulatory measures allows the legislator to check whether the discrepancies generally observed are justified by differences in the circumstances or are simply the unwanted byproduct of the political process.

In designing a safety regulation, the administrator faces the task of "aggregating" the preferences of the individuals in order to decide what constitutes (1) an acceptable level of individual risk and (2) an acceptable cost per life saved or the so-called value of life. Different approaches have been taken in the past to the question of the value of life[14]:

- The "human capital approach" is an attempt to estimate the value of an individual to society as a function of his discounted future earnings (including or excluding consumption according to the different authors). This method obviously ignores the non-marketable contributions of people to society and has little value for public policy making.
- The willingness to pay to eliminate a risk of given magnitude is the value revealed by individuals asked to disclose their personal utility (for themselves) or their personal "social welfare function" (for others), for example, \$1 to eliminate a risk of 10^{-6}. In practice, the problem is to obtain this figure from a large number of people and to aggregate individual values into a collective one.
- The individual value of life can also be revealed by past individual actions such as the purchase of insurance or the acceptance of a risky job for increase in compensation. This value is inferred from past decisions and reveals personal utility functions for specific types of risk. The problem in using life insurance figures is that they result in awards to the family and not to the individual himself. This approach also presents the difficulty of aggregation into a social value.
- The judicial awards are indicators of what the courts have estimated in the past to be a just compensation, after the fact, for the loss of life, to the members of the victims' family. These figures vary widely according to the jury, the circumstances, and the family situation of the victim. They are difficult to interpret as guidelines for future decisions.
- The corresponding costs per life saved can also be inferred from past legislation.[15] The "value of life" can be derived as the dividing line between regulations that were passed and those that were rejected. The figures vary widely and suggest vast economic inefficiencies in the process of risk management in the public sector. Note that in many cases, the law explicitly prevents the regulatory agencies from using a "value of life" figure.

The figures generally quoted in the literature as acceptable (and accepted) costs per life saved for low-level risks (e.g., below 10^{-5}) are on the order of one to two million dollars. The opinion of the author is that higher values or lower values may be justified in some cases according to the nature of the risk involved and the degree of control of the persons exposed. It is important to compare figures and to check that special circumstances do justify potential economic differences. The key to successful risk management, however, is often an acceptable decision process involving all relevant information about risks and costs, in which the parties involved have the opportunity to enter into a fair negotiation process.

DISCOUNTING

Discounting is one of the key elements of cost-benefit analysis.[16] It is standard procedure in problems involving monetary quantities.[17] In the private sector, it reflects the equilibrium between the opportunity cost of capital and time preference in the money market. Managers tend to account for uncertainties in the return of their investments by increasing the rate of discount. When using risk analysis and decision theory, however, one accounts for uncertainty through probability, and the issue of time preference is treated separately through discounting. The appropriate rate of discount is thus a risk-free rate, also free of inflation if the outcomes are expressed in constant dollars.

When the uncertain outcomes of a policy involve not only a monetary component, but also human safety, the question is whether similar discounting techniques should apply in the estimation of the future life-saving benefits. The first issue as in the previous section is whether economic efficiency matters, that is, if the risks involved are in the range that makes them acceptable. In other cases, there may be a conflict between ethics and economics. For example, some private deals are illegal in spite of the fact that they would be Pareto-optimal if they were to occur.

If indeed economic efficiency matters, then the concepts of opportunity costs and individual time preference remain a reality.[18,19] An amount X of capital invested today in a productive economic activity will allow society to spend $X(1 + R)^t$ at a future time t. If this amount is set aside for health and safety in the future, and if in real terms the future costs of health and safety do not increase, then discounting the future benefits of risk protection simply reflect the opportunity cost of capital which, if invested, could protect more people later.

Consider the following principles:

1. A life at all times is equally valuable to society and should be equally protected under the same principles and the same laws that prevail when the decision was made.
2. Provisions should be made so that each human being can receive in the future an equal amount of life-saving technology at the time when it is needed, and it should be at least equal to that available to the current generations.

Assume first no future technological progress, which means that it will take the same amount of resources to provide future generations with the same benefits of health and safety as it does today. The principles above have the following implication. If we decide today to spend the amount $X \Delta p$ to eliminate for the present year an individual risk equal to Δp, we also want to invest today an amount of money that will yield at time t a sure return $X \Delta p$ in order to save at time t another person exposed to the same risk. If I_1 is the investment required to protect one person and R represents the return of sure capital investments in the economy, we should invest today the present value of $X \Delta p$ at time t, which is:

$$PV(I_1) = \frac{X \Delta p}{(1 + R)^t} \qquad (12)$$

If N persons are to be protected, the safety investment must be multiplied by N. Therefore, when discounting the benefits ΔL in the computation of the cost per life saved, one simply discounts the future amount to be used then for the same protection.

The rate that applies is a risk-free marginal rate of return of capital in the economy (excluding inflation for computations in constant dollars).[19] Considering the historical variations of interest rates, a risk-free pre-tax rate of return on the order of 3 to 7 percent seems generally appropriate. Slightly higher rates may be justified for short-term effects in periods of high real return.

CONCLUSION

Cost-benefit analysis can provide valuable information for seismic safety decisions because the measure of cost-effectiveness provided by the cost per life saved allows comparison with other protective investments required by regulation. It is important, however, to check first that the nature and the level of individual risks involved justify risk comparison on the basis of economic efficiency. If this is the case, the future safety benefits must be discounted to reflect the opportunity cost of capital. A risk-free, inflation-free rate of discount in the range of 3 to 7 percent is generally appropriate.

REFERENCES

1. MISHAN, E. J. 1976. Cost-Benefit Analysis. Praeger, New York.
2. PATE, M. E. 1985. Costs and benefits of seismic upgrading of some buildings in the Boston area. Earthquake Spectra 1: 721–740.
3. PATE, M. E. & H. C. SHAH. 1980. Public policy issues: Earthquake engineering. Bull. Seismol. Soc. Am. 70: 1955–1968.
4. PATE, M. E. 1980. Assessment and mitigation of earthquake effects on economic production. In Proceedings of the Seventh World Conference on Earthquake Engineering, Istanbul, Turkey. Vol. 9: 301–306.
5. MELIKOV, E. M. 1981. Economics of Seismic Safety for Existing Buildings in the Boston Area. Master's thesis, Department of Civil Engineering, M.I.T., Cambridge, MA.
6. WHITMAN, R. V., et al. 1980. Seismic resistance of existing buildings. J. Structural Div. ASCE 106 (ST7): 1573–1592.
7. POLINSKI. 1977. The demand for housing: A study in specification and grouping. Econometrica 47:
8. CORNELL, C. A. & H. A. MERZ. 1975. Seismic risk analysis of Boston. J. Structural Div. ASCE 101: 2027–2043.
9. McGUIRE, R. K. 1977. Effects of uncertainty in seismicity on estimates of seismic hazard for the East Coast of the United States. Bull. Seismol. Soc. Am. 67: 827–884.
10. MISHAN, E. J. 1971. Evaluation of life and limb: A theoretical approach. J. Pol. Econ. 79: 687–705.
11. ARTHUR, W. B. 1981. The economics of risks to life. Am. Economic Rev. 71: 54–64.

12. HOWARD, R. A. 1980. On Making Life and Death Decisions. *In* Societal Risk Assessment. R. Schwing Ed.: 89–113. Plenum. New York.
13. PATE, M. E. 1981. Consistency of safety standards across the different economic sectors. *In* Structural Safety and Reliability. Moan and Shinozuka, Ed.: 505–514. Elsevier. New York.
14. LINNEROOTH, J. 1979. The value of a human life: A review of models. Economic Enquiry **17**: 52–74.
15. GRAHAM, J. D. & J. W. VAUPEL. 1981. Value of a life: What difference does it make? Risk Anal. Vol. **1**: 89–95.
16. LIND, R. C. *et al.* 1982. Discounting for Time and Risk in Energy Policy. Johns Hopkins University Press. Baltimore, MD.
17. GRANT E. L., W. G. IRESON & R. S. LEAVENWORTH. 1982. Principles of Engineering Economy, 7th ed. Wiley. New York.
18. KEELER, E. B. & S. CRETIN. 1983. Discounting of life saving and other nonmonetary effects. Management Sci. **29**: 300–306.
19. PATE, M. E. 1984. Discounting in risk analysis: Capital vs. human safety. *In* Risk, Structural Engineering, and Human Error. Grigoriu, Ed.: 17–32. University of Waterloo Press. Waterloo, Ontario, Canada.
20. PATE, M. E. 1983. Acceptable decision processes and acceptable risks in public sector regulations. *In* IEEE Transactions on Systems, Man, and Cybernetics, Vol. SMC-13: 113–124.

General Discussion: Part VII

NORTON REMMER, CHAIRMAN (*Consulting Engineer, Worcester, Mass.*): These three papers give us an excellent overview both practically and theoretically in terms of what the engineer has to worry about, what society has to worry about, and the policy-making decisions that have to be made.

Before discussion begins I would like to tell you about a consortium organized in the New England states under the auspices of the Governor's Conference of the Seven States supported by FEMA. The purpose of the consortium is to coordinate and develop seismic policies among the seven states and to help each other in developing seismic programs and hazard-reduction programs.

I would also like to briefly mention what is happening in Massachusetts with existing buildings. The Seismic Advisory Committee for the State Building Code Commission, of which I am Chairman, is currently working on preparing for the November cycle of code changes a code for seismic retrofitting of existing buildings. Earlier Robert Whitman touched on it and showed you a matrix. The idea is to identify the capacity of an existing building on the basis of the weakest link in that building. It uses the ratio of the capacity of that member to the required capacity of that member. Depending on what is being done to the building, we maintain the ratio of the capacity of that member to its required capacity or, in some cases, we increase the ratio.

There are other factors that are also involved such as the intensity of use before and after, the number of persons exposed to risk in the building, and the type of risk with respect to the use of the building—whether it's an assembly hall or a warehouse, for example. These factors will determine the allowable ratio.

And finally, we're looking at things such as critical facilities and what measures should be taken for the rehabilitation of critical facilities. Many of the concepts are not new. Much of this was looked at in the original versions of ATC-3, and there is some very interesting and helpful material in that document. The remarkable thing is that, although the Massachusetts development went on independently of ATC-3, it parallels very closely what was developed in ATC-3 about twelve years ago.

KLAUS JACOB (*Lamont-Doherty Geological Observatory, Palisades, N.Y.*): We have heard that 1 percent damage is labeled as being slight, which makes me wonder whether there's a fundamental difference with respect to individual structures in small communities versus those in large urban settings. Let's take New York, for example. Is 1 percent of the total investment light? As a scientific definition, definitely. But, somehow we have to start thinking about how it affects the economy of a region or even of the entire nation. What kind of research takes that into account? One percent in one place is different from 1 percent somewhere else. Who's concerned with that issue? One percent in New York, which we'll be better able to assess after Charles Scawthorn presents his paper, would start at 10 billion dollars and go up from there when you consider the entire stock of buildings, probably excluding the infrastructure.

ELISABETH PATÉ-CORNELL (*Stanford University, Stanford, Calif.*): The first study that I did on the cost/benefit analysis of seismic risk mitigation was for the San Francisco Bay Area. So it was not for only one class of buildings and it was for a whole region. In this study, the losses were divided into immediate damage to buildings, primary economic losses of goods and services that are not produced because buildings and equipment are no longer there, and secondary economic effects of an interruption of economic input into other sectors of the economy elsewhere. So, what I have done is to take into account the scale effect that you're talking about through the disruption of the economic production in the nation.

Now, there may be more to it than just that, but at least these large-scale economic losses of goods and services have been taken into account.

JACOB: An additional consideration for the New York area would be the financial service industry. Is that a critical facility for the nation. Should it be treated differently or not?

PATÉ-CORNELL: Do you mean that the financial input should be treated differently? I don't believe so. To the extent that you can look at it from an input-output perspective exactly in the same way as other economic components that are inputs to industries, the same method of analysis applies. You have the same input-output problem whether you consider money or other primary inputs to an industry. And, indeed, you can analyze the effects of supply interruptions in the same way.

CHARLES SCAWTHORN (*EQE Inc., San Francisco, Calif.*): I think that Dr. Jacob is raising a different issue. Damage to a building has some amount. And then, if you take into account business interruption and other effects that ripple through, the damages increase by some amount. If you look at a region and you have a thousand buildings and a certain amount of damage, you have a point on the graph. But when all these so-called collateral or secondary effects are taken into account the damage increases exponentially. Up to a point it's still a linear problem, but at some point it becomes a catastrophe.

There is a traumatic aspect to this too. For example, we are willing to accept individual car accidents. We're willing to accept 50,000 lives lost a year in car accidents because they occur in dribs and drabs. The loss of a few drops of blood a day is no big deal even though over many months it leads to many pints lost. But if you lose four or five pints at once, it's life-threatening. Similarly our society will become traumatized when we lose two or three thousand people in one shot, which occurs overseas periodically. In the United States we haven't had a catastrophe on that order since 1906, I would guess. So, there is the aspect of trauma to all of this and in that sense it becomes nonlinear and in the realm of psychology. The only person I know of who ever looked into this question is Harold Foster at the University of British Columbia. About six or eight years ago he developed something called a disaster scale, or a trauma scale, or something like that, which took account of psychological as well as other factors, and he found that it was like an intensity scale. It was interesting, but little remarked upon because no one ever figured out its use or importance.

Yesterday Howard Simpson told us that the effects on human beings were taken into account in that they asked: what would the people of Boston think if we lost a hundred buildings or a hundred people? Intuitively, we understand that we can't accept these kinds of large blood clots in the system, to come back to the medical analogy.

PATÉ-CORNELL: In economics, the nonlinearity of what's called the utility function is a classic question. I think that for public sector-decisions we have exactly the problem talked about. Is society willing to spend more to save fifty people from a bus accident than fifty people in individual accidents? Indeed, you have a question of perception and risk attitude. Now, are you trying to minimize the actual risk to people, or are you trying to minimize the perception of the losses after the fact? That's what the issue is.

SCAWTHORN: Individuals dying in individual cars don't really threaten society, but as catastrophes start to get larger and larger there's a trauma for society as a whole—a trauma that is visceral in that it goes back to the time of cavemen. We feel that society is more and more threatened. The kind of accident that happened at Bhopal is definitely socially threatening as is the occurrence at Tangshen; in Managua it toppled a government.

PATÉ-CORNELL: What risk aversion implies for the kind of numbers that I've shown is that you are willing to accept higher and higher costs in order to avoid increments of losses. That's fine with me. The question is to know whether there are other things that we can do that will be economically more efficient in that range of costs to protect human lives and avoid economic losses.

REMMER: When you're dealing with a hazard like this you try to compare it with some other familiar hazard like fire safety. The costs and requirements for fire protection have been increasing, but there has never, until recently, been a good analysis of why we're doing it and how much "safety" we have or how much we need. The perception with fire is that although it affects a limited number of people everyone has the sense that it may affect him or her. And so most of the fire safety requirements have never been challenged. Consequently there has always been a question of how much fire safety we have and whether we have more than we need if we analyzed it properly. Earthquakes are harder to perceive for the public, as we know, because they are rare occurrences and the public has no feel for the risk; most have no personal experience with it. And so, you are dealing with a much more vague risk, and it is more difficult to support the imposition of requirements in that kind of a situation.

KAREN WEISSMAN (*Princeton University, Princeton, N.J.*): I have a question relative to the rehabilitation of structures. Since a large number of existing structures in the East were not built to withstand earthquakes, has anybody thought about the concept of installing some sort of base isolation? Would that be a financially feasible rehabilitation procedure, as opposed to strengthening the structure itself?

RESPONSE: Base isolation systems are available, and a lot of studies are being done; we've done them ourselves during rehabilitation. They entail some very large problems with construction and the economics of base isolation work for only a small portion of structures. But it looks like a very viable technique for bridges.

Basically, base isolation involves installing a very soft spot right underneath the building. There has to be a foundation down under it, and a full foundation for the building just above it.

If the building is on some kind of a pile system that you can get into—because that means that the loads of the building are coming down at discrete points and you can work with one of those each time—then there is a pretty good chance that it might work. If you don't have that kind of building, then you've got to deal with bearing walls and the isolators have to be put in regularly and it gets quite expensive.

If the building doesn't have a basement or a crawl space, and if it's not used, you have to create that kind of space. In the couple of applications we looked at, we had miners practically planning to tunnel to create a 10-foot area underneath the building to put this whole base-isolated structure under it. So, base isolation is a wonderful idea. There are some uncertainties in the analytical procedures that we're using; we are not certain how it's going to behave close to a fault. But constructability is really the key issue for using base isolation in rehabilitation.

Experience with Changes in Design Requirements in the Boston Area

RENÉ W. LUFT

Simpson Gumpertz & Heger, Inc.
221 Main Street
San Francisco, California 94105

INTRODUCTION

The Commonwealth of Massachusetts adopted its first State Building Code in August 1974. This Massachusetts State Building Code[1] (referred to hereafter as the Code) was issued by the State Building Code Commission, which was appointed by the Governor of the Commonwealth, the Honorable Francis W. Sargent. The commission, after initial deliberations and public hearings, decided to base the Code on the BOCA Basic Building Code/1970 and Accumulated Supplement 1973.[2] Three articles in the current Code (fourth edition) were written specifically for Massachusetts. These are Article 1—Administration and Enforcement, which was modified from the corresponding BOCA article to correspond with state laws; Article 7—Structural and Foundation Loads and Stresses, which was rewritten in its entirety; and Article 22—Repair, Alteration, Addition, and Change of Use of Existing Buildings, which was newly developed for the third edition of the Code. BOCA, the Uniform Building Code,[3] and the Standard Building Code[4] have no parallels to Article 22.

Article 7 was written by three advisory committees appointed by the State Building Code Commission. These committees were the Seismic Advisory Committee, the Loads Advisory Committee, and the Soils Advisory Committee. The Seismic Advisory Committee wrote the Earthquake Load section, which contains the first earthquake design provisions developed specifically for a state in the Eastern United States.

The Seismic Advisory Committee (the Committee) was initially appointed jointly by the Boston Society of Civil Engineers and the Massachusetts Section of the American Society of Civil Engineers in July of 1973. Later, when the State Building Code Commission was created, the Seismic Advisory Committee came under its umbrella. The author was Secretary and later Chairman of the Seismic Advisory Committee.

To date, the Seismic Advisory Committee has written three versions of the Earthquake Load section. The first version was adopted on 6 December 1974 as an amendment to the first edition of the Code. The second version was adopted in 1979 within the third edition of the Code. The first two versions were adopted verbatim by the State Building Code Commission, after public hearings. The third version was published in 1983, submitted to the Commonwealth, and is currently under consideration.[5]

This paper summarizes the purpose and design philosophy of the Earthquake Load section of the Massachusetts State Building Code; it describes the requirements that have had a clear impact on the design of buildings and other structures in the Commonwealth, and specifically the Boston area; and it discusses the evolution of accepted standard engineering practice of earthquake design.

DESIGN CONCEPT

The first step in developing earthquake regulations for Massachusetts was to agree on a purpose. The purpose had to follow the public policy objective of reducing the earthquake risk from buildings. The agreed-upon purpose, as stated in the Code, is "to protect life safety by limiting structural failure" in the event that the design earthquake occurs. The purpose does not include limiting damage or nonstructural failures. The purpose was defined after extensive seismic risk studies for Massachusetts indicated that the probable maximum earthquake intensities for Massachusetts were similar to those for Zone 3 in California, and that the return periods for such earthquake intensities in Massachusetts were substantially longer than those in California.[6] The return period for the design earthquake

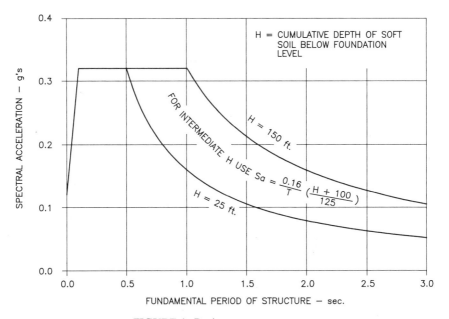

FIGURE 1. Design response spectrum.

in Massachusetts is now estimated to be in the range of 500 to 1000 years, substantially shorter than previous estimates of 5000 years. A cost/benefit study for such a return-period range indicates that the cost of providing earthquake resistance to buildings and structures exceeds the anticipated savings from reduced building damage occurring in the event of the design earthquake. Absent an economic incentive, protection of life is the only valid purpose for designing against such a long-return-period earthquake.

The design earthquake on firm ground is defined by a Newmark-Hall elastic response spectrum for 5% damping, 0.12 g peak ground acceleration, and 5.2 in/sec maximum ground velocity.[7] The firm-ground response spectrum is modified to account for cumulative depth of soft soil below the foundation level. The design response spectrum in the Code is reproduced as FIGURE 1. The ground-velocity-dependent region of the spectrum is inversely proportional to the period T. The design spectrum is similar to that published by the Structural Engineers Association of California (SEAOC) in 1985.[8]

Studies of different design concepts of earthquake reinforcement showed that the smallest incremental cost for earthquake protection of new buildings occurs when the building is detailed in a manner that will allow large cyclic deformations to occur without collapse. The specified detailing for buildings thus follows "ductile design" concepts.

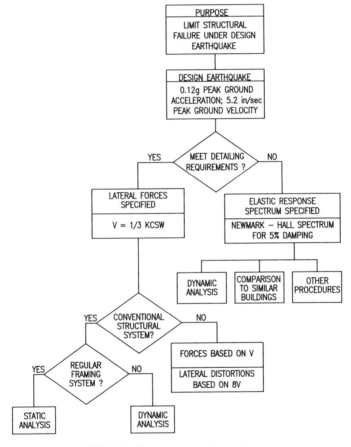

FIGURE 2. Design concept flow diagram.

To prevent collapse during an earthquake, a building must have a minimum level of lateral strength and stiffness, in addition to deformation capabilities, to ensure stability under gravity loads in the extreme deformed position, that is, under actual P-delta moments. Such lateral strength and stiffness in a building are achieved by designing to resist specified levels of lateral loads, and to limit the drift expected to occur.

The Code allows designers to use one of two approaches to achieve the purpose of limiting structural failure under the design earthquake. In the first approach, the design and detailing requirements are prescribed. The goal of these requirements is to arrive at structures with sufficient ductility to absorb the anticipated earthquake deformations. A majority of engineers follows the Code's approach of complying with the specified design and detailing requirements; these requirements have now, 13 years after the Code was first promulgated, become the accepted design practice. In the second approach, a designer must show that there is negligible risk to life safety if a building or structure experiences the design earthquake previously described. The designer can comply with this approach by performing a dynamic analysis of the building, and showing that the building can safely withstand the displacements and distortions computed in the dynamic analysis. The Code also allows the designer to compare the building being designed with a similar building that has withstood an actual earthquake equal to or stronger than the design earthquake and which building has similar foundations and subsoil conditions. Few engineers follow the alternate code approach.

This paper will concentrate on the Code's first approach because the experience with the Earthquake Load section in Boston is essentially based on it. In this first approach, a designer may select a conventional structural system and follow the design and detailing requirements given in the Code for such a system, or the designer may use "other materials or methods of construction" and show that the building can safely withstand eight times the distortion that is computed by an elastic analysis for the prescribed forces.

The conventional structural systems for which the Code includes detailing requirements are reinforced concrete, structural steel, reinforced masonry, wood frame and heavy timber construction, and prefabricated reinforced concrete. By omission, the following fall in the category of "other materials or methods of construction" for earthquake-resistance purposes: plain concrete, unreinforced masonry, cold-formed steel and cold-formed stainless steel, gypsum wallboard, and fiberboard.

FIGURE 2 shows a flow diagram, modified from one developed by the Seismic Advisory Committee, that summarizes the design concept.

LATERAL FORCES

If, and only if, a building or structure meets the design and detailing requirements of the Code, can it be designed for the code-prescribed base shear of

$$V = 1/3KCSW$$

This expression for base shear V is similar in format to that of the 1976 edition of the Uniform Building Code. In this expression, the zone factor is 1/3, the K-factor and the C-factor are similar to those of the 1973 edition of the Uniform Building Code, the S-factor is similar to the soils factor in the Recommended Provisions of the National Earthquake Hazards Reduction Program (NEHRP), and W is the weight of the building.[9]

The Code is specific in stating that the provisions of the section on minimum earthquake forces are applicable only to buildings and structures meeting the design and detailing requirements. Economy of structure results when designers utilize the low lateral earthquake forces prescribed by the Code, rather than the significantly higher forces that would follow from an elastic dynamic analysis for the design earthquake. Therefore, most buildings and structures in Massachusetts comply with the detailing requirements aimed at achieving ductile structures. When wind forces govern over earthquake forces, the higher wind forces must be used, but the earthquake detailing must still be followed.

The earthquake provisions in the Massachusetts State Building Code differ conceptually from those in the Uniform Building Code and in NEHRP. The Code, for a particular construction system, specifies only one level of design forces and corresponding detailing requirements; UBC and NEHRP allow for design options. This means, for example, that in Massachusetts all moment-resisting space frames must meet the detailing requirements specified in the Code. In both the Uniform Building Code and NEHRP, two types of moment-resisting space frames are defined: One, called a special or ductile moment-resisting space frame, is designed for a lower level of lateral forces than the ordinary moment-resisting space frame. The K-factor for a moment-resisting space frame building in Massachusetts is 0.67. Under the Code, there are no moment-resisting space frame buildings using a K-factor of 1.0. Designers from other states and those familiar with the Uniform Building Code have been confused by the absence of moment-resisting space frames designed for lateral forces using a K-factor of 1.0.

The C-factor in the expression for base shear in the Code, being based on the C-factor of the 1973 edition of the Uniform Building Code, leads to lower lateral earthquake forces than those computed using the current editions of the Uniform Building Code and NEHRP. Since 1976, the Seismic Advisory Committee has been aware that other codes specify higher lateral earthquake forces for Massachusetts than the Code; however, the Committee has consistently refused to raise the level of prescribed lateral earthquake forces because the essence of the Code's earthquake design lies in the detailing requirements for ductility.

Irregular framing systems must be analyzed by a dynamic analysis to determine the distribution of base shear. The dynamic analysis must use the shape of the elastic response spectrum, but the ordinates of the spectrum may be scaled down so that the base shear equals V, the base shear prescribed for static analysis.

If a building does not meet the design and detailing requirements of the Code, it may be designed by a dynamic analysis using the specified elastic response spectrum. The resulting base shear is many times larger than the base shear prescribed for static analysis.

DETAILING REQUIREMENTS

The detailing requirements are the essence of the earthquake design provisions of the Massachusetts State Building Code. Detailing is specified to achieve the necessary deformation capability under cyclic loads. The intent is to prevent failure of members or their connections and thus to prevent local or global structural failures.

The evolution of the Earthquake Load section between the first and third editions is indicative of the design experience that occurred subsequent to the introduction of earthquake design requirements into the Code. The first edition of the Code was based on SEAOC, third edition with commentary, published in 1973.[10] The first edition of the Code did not include detailing requirements for masonry because of strong and successful lobbying by the masonry-producing organizations. By the time the third edition of the Code was issued, the concept of earthquake design was well established in the Boston area. Reinforced masonry requirements were thus incorporated into the Code. During preparation of the third edition of the Code, the 1976 edition of the Uniform Building Code and Appendix A of ACI 318-77[11] were both available. The Seismic Advisory Committee studied the reinforced-concrete-detailing requirements in these documents and decided that a ductile moment-resisting concrete frame was too severe a requirement for Massachusetts, but that reinforced concrete frames designed under the body of ACI 318-77 would not provide the desired ductility. The Seismic Advisory Committee, therefore, developed a set of detailing requirements intended to lead to an intermediate moment frame. At that time, the concept of intermediate moment frames in reinforced concrete had not yet been developed by ACI and was also not included in ATC 3-06.[12]

The steel moment-resisting frame detailing requirements are in essence those for a ductile frame. The Seismic Advisory Committee reviewed the literature and consulted with researchers performing work on steel moment frames, but could find no information that would allow for the definition of an intermediate steel moment frame. The Committee, therefore, retained in the third edition the steel-detailing requirements from the first edition, which specified that all moment frames must be ductile steel moment frames. The Tentative Lateral Force Requirements published by SEAOC in 1985 include requirements for ordinary moment frames that effectively make them intermediate moment frames. The Seismic Advisory Committee is considering these requirements for a future edition of the Code.

Reinforced Concrete

The design and detailing requirements for reinforced concrete were developed by the Seismic Advisory Committee on the basis of the best information available at that time. The design and detailing requirements arrived at specify the use of ACI 318 (except Appendix A) and a series of requirements modified from Appendix A of ACI 318-77. The design intent was to have positive and negative bending strength throughout all frame members to absorb cyclic loading, and to have sufficient shear resistance and concrete confinement to prevent concrete shear failures or buckling of longitudinal reinforcement. Some requirements in the Code are listed below:

> Minimum ultimate compressive strength of concrete is specified as 3,000 psi, and reinforcing steel has to comply with ASTM A 615, Grade 40 or Grade 60.

> Flexural members of moment-resisting space frames must be provided with top and bottom reinforcement throughout their full length. Positive moment strength at a column connection must be at least 25% of the required negative moment strength at that column connection. The requirement for top and bottom bars came directly from Appendix A of ACI 318-77, but the requirement for positive moment strength at column connections was reduced from 50% in Appendix A of ACI 318-77 to 25%. The later Appendix A of ACI 318-83 includes a new section for frames in regions of moderate seismic risk. These requirements state that the positive moment strength at column connections shall be not less than 33% of the required negative moment strength.

For flexural members of moment-resisting space frames, web reinforcement perpendicular to the longitudinal reinforcement is required throughout the length of each member with a maximum spacing of 3/4 d unless a smaller spacing is required by ACI 318-77 (d, in inches, is distance from tension reinforcement to extreme compression fiber). Stirrup ties spaced at d/4 are required for a distance of 1½ d from the face of each support. Appendix A of ACI 318-83 contains similar but slightly more stringent requirements for frames in regions of moderate seismic risk. In Appendix A, web reinforcement spacing is 1/2 d instead of 3/4 d, and stirrups (not stirrup ties) are required over a distance of 2 d instead of 1½ d.

For columns of moment-resisting space frames, special transverse reinforcement is required within a specified distance from the face of the top and bottom joints. The required amount of transverse reinforcement is that given in Appendix A of ACI 318-77 for ductile moment frames. The maximum spacing of ties in tied columns is 4 inches. The corresponding requirements for frames in regions of moderate seismic risk in Appendix A of ACI 318-83 are less demanding.

Reinforced concrete shear walls must be designed for the strength formulas given in SEAOC (1973), except that shear stresses in shear walls of buildings without a moment-resisting space frame must be computed on the basis of twice the nominal earthquake forces. Shear walls in buildings with a dual bracing system require special vertical boundary elements. The requirements for these vertical boundary elements are those of SEAOC (1973).

Reinforced concrete braced frame members subject to axial stresses must have special transverse reinforcement as required for columns for the full length of the member. These requirements are similar to those of SEAOC (1973).

Taken in their totality, the detailing requirements for concrete in the Code are reasonably stringent for a region of moderate risk. Recent documents such as the SEAOC Tentative Lateral Force Requirements (October 1985) and NEHRP (1985) base the detailing requirements on Appendix A of ACI 318-83.

Structural Steel

The detailing requirements for moment-resisting space frames are those included in the SEAOC (1973) document for steel ductile moment-resisting space frames. The requirements for nondestructive welding tests are more detailed and specific in the Code than in SEAOC (1973).

The lack of an option in detailing requirements means that in Massachusetts all steel moment-resisting space frames must be detailed to be ductile as defined by SEAOC (1973). As a result of these detailing requirements, moment-resisting space frames in Massachusetts commonly employ continuity plates and less frequently doubler plates at beam-to-column intersections; beam and column sections tend to be compact because of the local buckling and slenderness ratio requirements; and beam-to-column connections utilize full penetration welds for the flanges and shear plates for the web.

Reinforced Masonry

The masonry design requirements first appeared in the third edition of the Massachusetts State Building Code. The requirements were developed by the Seismic Advisory Committee on the basis of the best information then available, and can be considered as intermediate ductility requirements for masonry. Important design and detailing requirements are given below:

> Bearing walls, shear walls, exterior walls, chimneys, and parapets constructed of masonry must comply with reinforced masonry requirements and have steel reinforcement in two directions. In masonry bearing and shear walls, the principal reinforcement in one direction must be spaced a maximum of 4 feet on center and in the other direction the spacing must be a maximum of 6 feet on center. The spacing of reinforcement, allowing a maximum grid of 4 feet by 6 feet is more liberal than the corresponding requirement in SEAOC (1973).

> Nonstructural masonry walls which enclose stairwells or elevator shafts must be designed as partially reinforced masonry with a spacing of reinforcement not exceeding 6 feet on center. In SEAOC (1973) all elements of masonry must be of reinforced masonry, whereas in the Code, certain walls must be reinforced masonry, other walls must be partially reinforced masonry, and nonstructural walls, except those enclosing stairwells or elevator shafts or exterior walls, may be unreinforced masonry.

> Tied columns must have ties of a size and spacing equal to those required for concrete columns.

> Masonry walls must be anchored to floors and roofs providing lateral support for the wall. Anchors must provide positive direct connections for the design forces or for 200 pounds/linear foot of wall. Anchors in hollow masonry units or cavity walls must be embedded in a reinforced grouted structural element of the wall.

The reinforced masonry requirements in the Code initially faced strong resistance from the masonry industry. But today, after the third edition of the Code has been in effect for more than eight years, reinforced masonry has become a fact of life in Massachusetts and the accepted standard engineering practice.

Timber

The detailing requirements allow the use of wood, timber, and diaphragms of lumber and plywood. The design of all wood and plywood members must conform to accepted engineering practice. Specific restrictions on wood and timber detailing are given below:

> Axial and shear forces in wood members must be transferred by positive connections and adequate anchorage. Toe-nailing or nails subject to withdrawal are not permitted for earthquake connections.

> Sheathing material may be used as tension ties, provided the tension force does not produce cross-grain bending or cross-grain tension in peripheral members or other framing to which the sheathing connects.

The timber-detailing requirements are based on SEAOC, fourth edition, which was first published in 1974. In contrast to other currently available specifications such as NEHRP, the Code does not recognize sheathing used for bracing other than lumber and plywood diaphragms; that is, fiberboard, particle board, and gypsum wallboard are not currently considered as earthquake-resisting materials within wood-framed buildings.

FOUNDATION DESIGN

The Massachusetts State Building Code contains design requirements for soils and foundations subject to earthquake loads, and it contains a screening test for liquefaction-susceptible soils.

The Code defines Class A and Class B soils. The class of soil must be considered for establishing the earthquake base shear, for the transmission of the base shear between the structure and the soil or rock, and to determine the need for interconnecting pile, pier, and caisson caps by ties. The Seismic Advisory Committee developed most of the foundation requirements. Following is a description of some foundation requirements for earthquake loads:

> Provisions must be made to transmit the earthquake base shear, acting in any direction, between the structure and the surrounding soil or rock. The Code specifies five methods for transmitting the base shear. The Code limits lateral soil pressures to one-third of the passive pressure, and prohibits the reliance on bottom friction for buildings supported on piles, piers, or caissons and underlain by Class B soils. These provisions are important in a region where the engineering profession is unfamiliar with earthquake design requirements, since they point out the need to transmit the computed earthquake loads to the foundations. The Seismic Advisory Committee in 1983 proposed modifications to this code section.

> Pile, pier, and caisson caps must be interconnected by ties when the caps overlie a Class B soil. This provision is similar to one in SEAOC (1974) except that the requirement in the Code applies only for foundations overlying Class B soils.

> Retaining walls must be designed to resist an earthquake force per unit length of $0.045\gamma_t H^2$ in addition to all other soil pressures (γ_t is soil unit weight; H is wall height). The earthquake force from the retaining wall backfill is distributed as an inverse triangle over the height of the wall. Provisions similar to this one are not found in other national earthquake codes.

> Liquefaction potential of saturated, clean, fine to medium sands must be evaluated on the basis of a screening test. The screening test is based on the standard penetration resistance in blows per foot. The Code contains a figure that gives the liquefaction potential of clean, fine to medium sands considering the depth below ground surface at which the standard penetration resistance is measured, and the depth to ground water. This figure was derived from empirical data and drawn for the maximum ground acceleration of 0.12 g corresponding to the design earthquake. Provisions similar to this one are not found in other national earthquake codes.

DESIGN EXPERIENCE

More than 13 years have elapsed since the first Earthquake Load section was incorporated into the Massachusetts State Building Code. During the time the Code has been in effect, developers, contractors, architects and engineers, and building officials have all to a smaller or larger extent become familiar with the notion that buildings in Massachusetts are designed to resist earthquakes.

The provisions, as previously described, are sufficiently specific to allow all parties involved in the development, design, and construction of buildings to incorporate earthquake-resistance and ductile detailing into buildings. Design firms, from the small one- or two-person office to the large engineering corporation, design buildings following the Earthquake Load section of the Code. There is no question about the fact that engineers perform earthquake calculations for the base shear specified in the Code, and that building officials check whether earthquake design features have been incorporated. The questions that exist relate to the interpretation of the detailing requirements and in their incorporation into the design. The more direct and clear a detailing requirement, the higher the compliance by designers and the enforcement by building officials. Below are given some observations on the use of detailing requirements.

Concrete

Physical requirements for reinforced concrete materials are followed, as are the straightforward design requirements, such as providing two top and two bottom reinforcement bars throughout the length of flexural members. Concrete confinement and stirrup requirements, on the other hand, especially at the joints of moment frames, are confusing to detailers, resulting in a spread of interpretations of the design requirements. Many engineers believe that the detailing is too severe for low- and mid-rise buildings, and this belief affects compliance.

Steel

Since only one type of moment-resisting space frame in structural steel is specified by the Code, engineers are familiar with the requirements and generally follow them. The actual connections between beams and columns in moment frames, consisting of welds, bolts, seat angles, web angles, and others, are usually designed to develop in the beam the full plastic capacity of the beam. Disagreement exists between designers on the treatment of the joint or panel zone. At one extreme, designers will provide properly sized continuity plates and doubler plates at the column-to-beam panel zone, while at the opposite extreme, designers will disregard the design of the panel zone. In Boston, where many tall moment frame buildings have been built over the last ten years, it is common to see continuity plates and infrequent to see panel zone doubler plates in steel moment frames.

The Code excludes the use of cold-formed steel to resist earthquake forces. Nevertheless, light industrial and storage buildings are erected in Massachusetts using these materials. These buildings rely on diagonal roof and wall bracing or light gauge metal roof and wall panels to carry the base shear. The American Iron and Steel Institute has for a long time advocated the use of cold-formed steel structures in the state, arguing that such construction is accepted by other national codes.

A recent article[13] describing an expansion of the Hawthorne Country Club in North Dartmouth, Massachusetts, describes how the steel frame carries the lateral wind load moments. The article does not mention earthquake loads, and the typical connection detail shown in FIGURE 1 of the article does not appear to comply with the Code detailing requirements. The extent of Code compliance by designers of low-rise buildings outside of Boston is questionable. While many designs meet the letter of the Code, some steel frames include constructions of poor earthquake-resistance, such as weak-column/strong-beam, or eccentric beams relative to the columns. Many engineers feel that detailing of moment frames is too severe; the design community has some considerable sentiment toward lessening detailing requirements for low- to mid-rise moment frames.

Masonry

When the Code was first adopted, the masonry provisions requiring reinforced masonry were the most controversial. Today, about eight years after the reinforced masonry provisions were adopted, the parties to the building process recognize this requirement and provide reinforcement in two directions in those walls specified by the Code. During the past five years, the Seismic Advisory Committee has received no official communication from the masonry industry regarding difficulties in implementing the Code's provisions.

Timber

The timber design requirements do not impose major restraints, but designers often misinterpret them. Diaphragms are specified in the Code to be of lumber and plywood. Other American Plywood Association performance-rated panels such as structural particle board, wafer board, or oriented strand board, while not approved, are also used in building walls and floors.

One- and two-family dwellings are exempt from seismic requirements and typically not designed to resist earthquake loads. The technology of wood-frame construction, therefore, does not incorporate many elements found in California construction. Plywood floor and roof diaphragms and shear walls are commonly used, but they are not nailed at close enough intervals and do not include elements such as blocking (except when required as sheathing), anchorage of shear walls for moment resistance (sill anchor bolts, however, are used), shear wall chord members, and the corresponding tension ties across floors, or collectors/drag strut members. The absence of this technology follows through to town houses and two- or three-story wood-frame condominiums, which are not exempt from seismic requirements. Perhaps of greater concern is that designers of light wood-frame construction also design heavier timber structures in which they fail to incorporate the seismic detailing necessary to tie the building together. In roof trusses, for example, missing may be diaphragm continuity at the ridge, continuous chords at the edges, or hold-down anchors.

By comparison with the concrete-detailing requirements, the timber requirements are more general and less well understood. The specificity of a statement in the concrete section, such as "provide top and bottom reinforcement in flexural members," is in contrast to the generality of a statement in the timber section, such as "transfer shear forces in wood members by positive connections." In a region where earthquake design is new and not well understood, specific requirements would be more effective (for example, "provide diaphragms with chord members, provide horizontal diaphragms with collectors at openings in shear walls, or provide anchorage to shear wall chord members").

CONCLUSIONS

The Earthquake Load provisions of the Massachusetts State Building Code have had a clear impact on the design and construction of buildings in Massachusetts. Earthquake load analysis and earthquake-detailing requirements are followed by most engineers. Earthquake design is today an accepted engineering practice. This practice has led over the last 13 years to important changes in the way buildings are analyzed, designed, and constructed. The Earthquake Load section of the Code, therefore, is achieving one of the public policy objectives, namely, to reduce the earthquake risk from building stock by introducing earthquake-resistance into new buildings as they are built. As the fraction of post-1975 buildings to total building stock increases over time, the average earthquake risk in Massachusetts will decrease. For a region with long-return-period earthquakes, a slow decrease in earthquake risk over time is sound public policy. Further, considering the economic boom in Massachusetts over the last several years, and the consequent inflation of real estate prices, the incremental cost for earthquake protection of buildings has had no appreciable negative effect on the construction industry and the state economy.

In summary, in Massachusetts the public policy of slowly reducing the earthquake risk from buildings is being implemented at a cost that has been no detriment to the growth of the economy. The building community from developers to building officials recognizes the Code's requirement for earthquake-resistant design and detailing, and follows this requirement to an extent consistent with public policy.

REFERENCES

1. THE COMMONWEALTH OF MASSACHUSETTS, STATE BUILDING CODE COMMISSION. 1974, 1975 and 1979. State Building Code, first, second, and third eds. Sections 716–720. Boston, MA.
2. BUILDING OFFICIALS & CODE ADMINISTRATORS INTERNATIONAL, INC. 1970. The BOCA Basic Building Code/1970, fifth ed. Chicago, IL.
3. INTERNATIONAL CONFERENCE OF BUILDING OFFICIALS. 1973 and 1976. Uniform Building Code, 1973 and 1976 Eds. Whittier, CA.
4. SOUTHERN BUILDING CODE CONGRESS INTERNATIONAL, INC. 1985. Standard Building Code. Birmingham, AL.
5. SEISMIC DESIGN ADVISORY COMMITTEE, R. W. Luft, Chairman. 1983. Proposed Changes To Earthquake Design Sections, Massachusetts State Building Code. J. Boston Soc. Civil Engrs./A.S.C.E. 69(2): 209–234.
6. LUFT, R. W. & H. SIMPSON. 1979. Massachusetts Earthquake Design Requirements. J. Boston Soc. Civil Engrs./A.S.C.E. 66(2): 67–87.
7. NEWMARK, N. M. & W. J. HALL. 1969. Seismic design criteria for nuclear reactors. In Proceedings of the Fourth World Conference on Earthquake Engineering 2: 37–50. Santiago, Chile.
8. STRUCTURAL ENGINEERS ASSOCIATION OF CALIFORNIA, SEISMOLOGY COMMITTEE. 1985. Tentative Lateral Force Requirements. Sacramento, CA.

9. BUILDING SEISMIC SAFETY COUNCIL. 1985. NEHRP-Recommended Provisions for the Development of Seismic Regulations for New Buildings, Vol. 2. Washington, DC.

10. STRUCTURAL ENGINEERS ASSOCIATION OF CALIFORNIA, SEISMOLOGY COMMITTEE. 1973. Recommended Lateral Force Requirements and Commentary, third ed. with Commentary. Sacramento, CA.

11. AMERICAN CONCRETE INSTITUTE. 1977 and 1983. Building Code Requirements for Reinforced Concrete, ACI 318-77 and ACI 318-83. Detroit, MI.

12. APPLIED TECHNOLOGY COUNCIL. 1978. Tentative Provisions for the Development of Seismic Regulations for Buildings, ATC 3-06. National Bureau of Standards (U.S.), Spec. Publ. 510. U.S. Government Printing Office. Washington, DC.

13. LAW, F. M. 1987. It was steel for the Hawthorne Country Club. Modern Steel Construction **27**(6): 18–22.

Seismic Building Code Efforts in the Central United States

WARNER HOWE

Gardner & Howe, P.C.
Consulting Structural Engineers
Memphis, Tennessee 38119

This article reviews our experience with code revisions in the Memphis area and the elements being considered in reaching those conclusions.

BACKGROUND

During about 150 years of recorded history, Memphis has not had a seriously damaging earthquake nor has anyone been injured or killed by an earthquake there, although a magnitude 6 earthquake in 1843 did extensive chimney damage in the Memphis area, in northern Mississippi, and in eastern Arkansas.[1-5] Memphis and the surrounding areas have never had or enforced building code provisions for earthquake protection. Memphis first had locally drafted building codes, and then in 1965 adopted the BOCA Basic Building Code, a national model code, but changed in 1979 to the Southern Standard Building Code (SBC). Both of these national codes had seismic provisions, but when adopted locally these seismic provisions were systematically excluded. Today Memphis and Shelby County have a jointly administered building code (SBC 1982) with no mandated seismic requirements for building construction.

Against that background it must be remembered that for a period of more than a year, in 1811–1812, a series of more than 1000 earthquakes occurred near New Madrid, Missouri, which included three earthquakes considered to be the largest in magnitude and area of damage ever recorded in the continental United States, including the 1906 San Francisco Earthquake. Damage during the 1811–1812 earthquakes was reported over a wide area of the Eastern United States that included Washington, D.C. and Boston, Massachusetts, where church bells rang more than 1000 miles away. The mighty Mississippi River flowed backwards for a short period. The Tiptonville area was uplifted and 18,000-acre Reelfoot Lake was formed.[6] Since that event, a number of moderate earthquakes have been recorded on other portions of the New Madrid fault system causing local damage but no loss of life.

In recent years microtremors are being recorded almost weekly on the New Madrid fault by arrays of very sensitive seismographs operated by Otto Nuttli[1-5] of St. Louis University and Arch Johnston of Memphis State University.[7] These recordings have given seismologists and geologists a good indication of the location and extent of the fault zone. Furthermore, this record indicates an active fault system capable of producing major quakes but with a much less frequency of occurrence than in the more active seismic zones on the West Coast continental plate boundary. Therefore, the probability of damaging earthquakes occurring during the normal life expectancy of building structures in the Central United States is considerably less than in the more active zones, but there is still a possibility of a large earthquake occurring at any time. That possibility prudently must not be completely ignored.[8-11]

This dichotomy has placed Memphis officials in a dilemma as to how to address this seismic hazard. At one extreme are those who feel that the seismic hazard can be ignored; at the other, are those who feel that at least the more important facilities should be able to resist a maximum credible shake. Unfortunately, in the past, national model building code earthquake design standards have been developed for conditions in the more seismically active regions of California, where large earthquakes have a high probability of occurring during the lifetime of buildings. These design standards are not totally appropriate for the less seismically active regions because of their different earthquake risk. Although, in the past ten years research has produced valuable knowledge regarding the causes, source areas, probability of occurrences, and potential ground-motion intensities in the Central United States, not all of these considerations have yet been incorporated into the national model building codes. The existing dilemma must be resolved and the latest state-of-the-art technology integrated into the local building code.

Memphis did not exist as a city at the time of the 1811–1812 earthquakes; only a few settlers and Indians may have lived on the Chickasaw Bluffs overlooking the Mississippi River. Memphis began to grow and prosper by the mid 1800s because of cotton farming and river transportation including excellent steamboat landing facilities at the foot of the bluff. The alignment of the Mississippi River and the formation of the bluffs, as well as the cause of earthquakes in this region, are related to the geologic development of the Mississippi Embayment, which extends from the southern tip of Illinois southward to the Gulf of Mexico and from the Tennessee River on the east to the Ozark Mountains on the west. This area in geologic time had been a shallow sea embayment with sedimentary sand and silt deposits approximately 2000 feet thick at Memphis. The embayment has been downwarping through geologic time and in turn rides upon and in line with the New Madrid rift in the basement igneous rock.

The Mississippi River and its flood plane is superimposed on the embayment deposits and has meandered from the Chickasaw Bluffs on the east all the way across the embayment to the Ozarks on the west. Aeolian deposits, about 20 feet thick at Memphis, cap and form the bluffs of West Tennessee which are the eastern boundary of the flood plane. The New Madrid Rift (earthquake source zone) extends from about Marked Tree, Arkansas (35 miles northwest of Memphis) northeastward through New Madrid, Missouri, and into southern Illinois.

The 1906 San Francisco earthquake had a fault break of more than 100 miles in length and probably 30 to 40 miles deep with a maximum displacement on the fault of more than 20 feet in the epicentral area. The 1811–1812 New Madrid Earthquakes undoubtedly had similar fault break areas but no surficial evidence existed because of the deep embayment and river alluvial deposits. It can be assumed, however, that the maximum displacement took place in the central portion of the rift zone or in the Bootheel of Missouri, as supported by the reported earthquake damage surrounding New Madrid, Missouri.

Seismic ground motions can originate anywhere along the fault zone but the larger waves from a major earthquake probably will originate in the central portion or the Bootheel area of Missouri about 100 miles north of Memphis. This means that the near-field shorter period waves may be smaller than the more distant, longer period ground wave from the epicentral area. Furthermore, these more distant (seismic) waves will tend to have longer more harmonic periods, be of longer durations, and thus, have more devastating effect on taller structures (similar to the Mexico City 1985 earthquake, but probably less pronounced).

MEMPHIS BUILDING CODE ADVISORY BOARD

The Memphis Building Code Advisory Board (BCAB) was first instituted after the adoption of the BOCA Basic Building Code in 1965. The Advisory Board had been made aware of the seismic hazard and disposed to do something about it, as soon as national building code seismic standards became available which appropriately address the unique seismic problems in Memphis.

REGIONAL EARTHQUAKE RISK STUDY

Their first assist came in 1974 with the publication of *Regional Earthquake Risk Study* made by M & H Engineering and Memphis State University for the Mississippi, Arkansas, Tennessee Council of Governments (MATCOG)[12,13] funded by the Federal Department of Housing and Urban Development (HUD). This comprehensive report covered a study of the regional and local geology, earthquake ground motions, earthquake scales, seismic history and seismicity of the MATCOG area, topographic effects, secondary hazards, population and construction forecasts, damage parameters, life loss and injury projections, and public policy options. Unfortunately this report had limited local attention from public policy makers.

ATC-3-06 TENTATIVE PROVISIONS

A major potential assist came in 1978 with the Applied Technology Council's (ATC-3-06) Project "Tentative Provisions."[14] The ATC is a nonprofit research organization sponsored by the Structural Engineers Association of California (SEAOC). The ATC-3-06 project was funded by the National Science Foundation (NSF) and supported by the National Bureau of Standards (NBS). This project assembled 85 recognized experts in their fields to research the then current state-of-the-art seismic technology and prepare a document that would be applicable in all seismic regions of the United States.

A new type of risk maps, based upon the statistical probability of earthquake occurrence in each seismic area, was generated by the U.S. Geological Survey.[15] Using the approach taken by building codes in general for wind, floods and other natural hazards, a "design earthquake" with a 90% probability of not being exceeded in a 50-year period was selected to give equal probabilities of seismic ground motion throughout the country. The concept of contour maps of ground motion attenuated with distances from discrete source zones was also introduced. Two modified maps were developed, one for short-period (A_a) near-field ground motions affecting low-rise structures and the other for the longer period motions (A_v) affecting taller structures and potential far-field effects.

The ATC noted that "because of the many new concepts and procedures included in these Tentative Provisions, they should not be considered for code adoption until their workability, practicability, enforceability, and impact on costs

[are] evaluated by providing and comparing building designs for the various design categories included in the document."[14]

NBS ASSESSMENT & IMPLEMENTATION PLAN

In November 1978, NBS issued a "Plan for the Assessment and Implementation of Seismic Design Provisions for Buildings." This plan included four tasks:

1. A thorough review by all interested building industry organizations of the ATC-3-06 "Tentative Provisions";
2. The conduct of trial designs to establish the technical validity of the new provisions and to predict their economic impact;
3. The establishment of a mechanism to encourage consideration and adoption of the new provisions by organizations promulgating national standards and model codes; and
4. Educational, technical, and administrative assistance to facilitate implementation and enforcement.

FEDERAL EMERGENCY MANAGEMENT AGENCY (FEMA)

During this same period, other events significant for this effort were taking place. In October 1977, Congress passed the Earthquake Hazards Reduction Act (P.L. 95-124) and the National Earthquake Hazards Reduction Program (NEHRP) was released in June 1978. The concept of a separate federal agency to coordinate all emergency management functions resulted in the creation of the Federal Emergency Management Agency (FEMA); FEMA became the implementing agency with NSF retaining its research-support role.

BUILDING SEISMIC SAFETY COUNCIL (BSSC)

The emergence of FEMA as the agency responsible for implementation of P.L. 95-124 and the President's program on earthquake hazard mitigation also required establishment of a mechanism for obtaining a broad public and private consensus on housing and building design and construction regulatory provisions and the means to be used in their promulgation. A series of meetings was held between representatives of the original participants in the ATC-3-06 project, FEMA, the American Society of Civil Engineers, and the National Institute of Building Sciences (NIBS). From these meetings, the concept of the Building Seismic Safety Council (BSSC) was born. In the spring of 1979, the BSSC was established under the auspices of NIBS. A Charter and Organizational Rules and Procedures were thoroughly debated and agreed upon.

Building regulations established to protect the health, safety and welfare of the public involve highly complex technical issues which have an impact on the socioeconomic interests of many segments of the total building community. In order to serve best the public interest, the development of building regulations must involve a consensus process which assures consideration of the latest technological research available as well as its potential impact on all segments of the building community. A truly consensual process will also minimize the possibility of undue pressure from highly vocal special interest groups. Fortunately, such a process for consensus exists in the BSSC Program for developing provisions for mitigating seismic hazards.

NIBS-BSSC

In 1972, the Congress made provisions in the National Housing Act for the creation of the National Institute of Building Sciences (NIBS). The Institute was to have representation and input from both the public and private sectors of the building community, but with predominant control in the private sector; initially the Institute was to be funded by the federal government, but afterwards to be supported jointly by both sectors. Recognizing the need for such a truly consensual organization to address the special problems being raised by the National Earthquake Hazards Reduction Program, NIBS fostered the creation of the Building Seismic Safety Council (BSSC) patterned after the Institute's membership arm, its Consultative Council. This has been a truly unique prototypical process and an important development for the building industry and the building regulatory system in this country.

The National Model Building Code organizations have and will continue to depend on such national consensus efforts to keep their codes up to date with state-of-the-art technology. This process is working well for the building industry and should be supported by the total building community. This is why the local seismic study committee has recommended the adoption of the most recent edition of the Standard Building Code, which includes this latest consensus technology.

The BSSC provided the mechanism envisioned in Task 3 of the 1978 NBS plan (i.e., a forum for standards and model code bodies; professional societies; building industry and trade organizations; elements of federal, state, and local governments; the research community; and consumer groups to participate equally in the development of a consensus on improved building seismic safety provisions and the means for their promulgation through the existing public and private building regulatory systems throughout the nation). A joint BSSC-NBS committee then was formed to conduct the review called for in Task 1 of the NBS plan. The review effort concluded in 1980 and resulted in 198 recommendations for changes in the ATC-3-06 "Tentative Provisions." Funds for both the BSSC and NBS in support of this activity were provided by FEMA.

BSSC TRIAL DESIGNS

As the review effort drew to a close, the BSSC and NBS created a joint committee to develop criteria by which the trial designs called for in Task 2 of the 1978 NBS plan could be evaluated and to recommend a specific trial design program plan. An Overview Committee was appointed and completed the revised report, which was released in November 1982 as a "Plan for a Trial Design Program to Assess Amended ATC-3-06 Tentative Provisions for the Development of Seismic Regulations for Buildings" (NBSIR 82-2589/ BSSC 82-1). The result of the Committee's revision effort was released in December 1982.[16]

BSSC PROGRAM ON IMPROVED SEISMIC SAFETY PROVISIONS

All of these efforts provided the basis for the BSSC Program on Improved Seismic Safety Provisions. The basic objectives of this program were to:

1. Estimate the economic impact that various sectors of the building community would experience in changing from current local practice to the upgraded Provisions;
2. Evaluate the practicality of the amended Provisions insofar as designers, builders, and regulatory officials are concerned; and
3. Establish the technical validity of the amended Provisions (i.e., determine whether seismic safety would be enhanced by their adoption).

ASCE/ANSI A58.1-1982

In 1982 the American Society of Civil Engineers (ASCE) and the American National Standards Institute (ANSI) extensively revised the ASCE/ANSI Design Load Standard (A58.1-1972) "Minimum Design Loads for Buildings and Other Structures" patterning it after the ATC-3-06 "Recommended Provisions." The previous *maximum* intensity risk map was discarded and replaced by the ATC-3, Effective Peak Velocity—related acceleration (A_v) *probability* map. However, the previous seismic Zones 1 through 4 with their related "Z" factors (3/16, 3/8, 3/4 and 1) were maintained. This negated the contour concept of the ATC-3-06 document and caused an inappropriate jump of 100% in design ground motion when crossing the line between Zones 2 and 3. For cities like Memphis, which are located on the 0.2-g contour which separates Zones 2 and 3, this presented a serious dilemma as to the appropriate interpretation of their earthquake risk when using ANSI A58.1 standard.

The Memphis Seismic Study Committee appointed by the BCAB two years ago considered using the ASCE/ANSI A58.1-82 Load Standard, but were finding that many modifications and loose interpretations were required to make it appropriate for Memphis. Furthermore, it is a load standard, not a design standard as required for code enforcement.

CONDUCT OF THE BSSC PROGRAM

During the previously mentioned BSSC Trial Design Program, 17 professional design organizations from nine cities were retained to prepare the trial designs for the following building types:

1. Low-, mid-, and high-rise residential buildings,
2. Mid- and high-rise office buildings,
3. One-story industrial buildings, and
4. Two-story commercial buildings.

Each building was designed twice: once according to the prevailing local code and once according to the amended BSSC Provisions. Basic structural designs, partial structural designs and partial nonstructural element designs (complete enough to assess the cost of the nonstructural portion for the building) were prepared, and design and construction cost estimates were developed. "For the

29 trial designs conducted in the 5 cities (Chicago, Ft. Worth, Memphis, New York, and St. Louis) whose local building codes currently have no seismic design provisions, the average projected increase in total building construction costs was 2.1 percent."[17]

Two Memphis engineering firms were retained to prepare trial designs for six prototypical buildings (Table 7 in Ref. 17). The estimated additional total cost of construction ranged from 1.8% to 4.8% for five of the six buildings. For the sixth design (M18) the estimated additional cost was 13.8%, but this was heavy concrete, flat-plate, moment-frame design. The Memphis average (excluding M18) was a 3.44% increase in total cost, a moderate price to pay for earthquake protection.

Phase II of the BSSC Program concluded with publication of:

1. A draft version of "The NEHRP Recommended Provisions," which included three parts: the draft Provisions, Commentary to the provisions, and an Appendix;
2. An Overview of Phases I and II of the BSSC program including the Overview Committee's analysis of the results and the executive summaries from the reports of the design firms participating in the program; and
3. The design firms' reports.

The Draft Provisions issued at the conclusion of Phase II reflected the initial amendments to the original ATC document as well as further refinements made by the Overview Committee during Phases I and II of the program. In July 1984 this interim set of provisions was mailed to the BSSC member organizations, for balloting. Since the goal was to allow members full expression, this first ballot provided for four responses: "yes," "yes with reservations," "no," and "abstain." The "yes with reservations" and "no" votes had to be accompanied by an explanation of the reasons for the vote. The "no" votes were to be accompanied by specific alternative suggestions, if they would change the negative vote to an affirmative.

During Phase III of the program, all comments and "yes with reservation" and "no" votes received as a result of the first ballot were compiled for review. The Draft Provisions were revised to reflect the changes deemed appropriate by the BSSC Board and were again submitted to BSSC membership for balloting in August 1985.

As a result of this second ballot, virtually the entire provisions document received consensus approval, and a special BSSC Council meeting was held in November 1985 to resolve the remaining differences. The 1985 Edition of the "NEHRP Recommended Provisions" was published and distributed by FEMA.

BSSC THREE-YEAR UPDATE PROGRAM

To keep NEHRP Provisions up to date with current technology, a BSSC Three-Year Update Program got under way in January 1987, and has involved more than 85 individuals who have served on nine technical committees. By September of 1987, these committees had prepared their recommended revisions or amendments to the 1985 NEHRP Recommended Provisions for review by the Technical Management Committee (TMC). The results of the TMC review were presented to the BSSC board, which approved them in December for balloting by the BSSC member organizations. When approved by the 61 BSSC member organizations, the 1988 Amended NEHRP Provisions will be printed and distributed by FEMA.

NEHRP PROVISIONS COMMENTARY: INTENT STATEMENT

The Commentary of the Updated NEHRP Recommended Provisions has been extensively amended to clarify the intent of these provisions. That statement has been carefully reviewed through the BSSC consensus process and reads in part as follows:

> The goal of the NEHRP Recommended Provisions is to present improved criteria for the design and construction of buildings subject to earthquake ground motions in order to minimize the hazard to life for all buildings, to increase the expected performance of higher occupancy structures as compared to ordinary structures, and to improve the capability of essential facilities and high-occupancy structures to function during and after an earthquake. The primary function of these Provisions is to provide the minimum criteria considered prudent and economically justified for the protection of life and safety in buildings subject to earthquakes at any location in the U.S. These Provisions have been extensively reviewed and balloted by the building community and, therefore, are a proper source for development of building codes in areas of seismic exposure.
>
> Some design standards go farther than these Provisions and attempt to minimize damage as well as protect building occupants. For example, Title 17 and Title 21 of California's Administrative Code have added property protection in relation to the design and construction of hospitals and public schools. These NEHRP Recommended Provisions generally consider property damage primarily as it relates to occupant safety for ordinary buildings. For high occupancy and essential facilities, the damage limitation criteria are more strict in order to better provide for the life and safety of occupants and the continued functioning of critical/essential facilities. Some structural and nonstructural damage can be expected as a result of the "design ground motions" because these Provisions allow inelastic energy dissipation by utilizing the deformability of the structural system. For ground motions in excess of the design levels, the intent is that there be a low likelihood of collapse.
>
> It must be emphasized that assurance of absolute safety and no damage even in an earthquake event with a reasonable probability of occurrence cannot be achieved economically. The objective of these Provisions is therefore to present the minimum requirements to provide reasonable and prudent life safety for building occupants. For most structures, it is expected that structural damage from even a major earthquake would be repairable; however, this would depend upon a number of factors including the type, materials, and details of construction actually used.
>
> Because of the complexity of and the great number of variables involved in seismic design (e.g., the dynamic characteristics of the structure and the variability in ground motion, intensity of the earthquake, distance to the epicenter of the seismic disturbance, and soil type), these Provisions detail only the minimum criteria in general terms. Thus, the experienced structural engineer is relied upon to exercise judgment in interpreting and adapting the basic principles to a specific project.
>
> These Provisions are applicable in all sections of the U.S. exposed to earthquake ground motions because the "design earthquake" ground motion is based on an estimated 90% probability of not being exceeded in a 50 year period. This is consistent with that provided for other natural hazards such as wind, snow, storm, and floods. However, it must be emphasized that larger earthquakes are possible and may occur during the life of a structure.

In some areas the *probable* and the *maximum intensity* earthquakes are approximately the same, but this is not true for many other earthquake-prone parts of the country. In Central and Eastern U.S., the *maximum intensity* earthquake often may be two or more times larger than the *probable* earthquake. Although the probability of the *maximum* event occurring during a structure's life is very small, it can nevertheless occur at any time and most certainly will occur sometime in the future. In order to quantify this possibility, two sets of maps are presented in the Commentary; one set giving accelerations and velocities with 90% probabilities of not being exceeded in 50 years and another set giving acceleration and velocities with probabilities of not being exceeded in 250 years. Use of these maps will allow regulatory agencies to rationally appraise the possibility that larger earthquakes will occur and to modify these provisions accordingly.

SEISMIC LOAD FACTOR: EAST VERSUS WEST

Last October, the National Center for Earthquake Engineering (NCEER) held a symposium entitled, "Seismic Hazards, Ground Motions, Soil-Liquefaction and Engineering Practice in Eastern North America." Under the subject of engineering practice and code-related aspects, it was observed:

> The mid-plate conditions in eastern North America suggest a higher load factor than those in the western plate-boundary conditions. The load factor is the ratio of seismic loads from the most severe earthquake in a given region to those loads during the more probable moderate earthquake. The former would typically apply to earthquakes occurring every 500 years at western localities and every few thousands of years in the east. For these largest events (in the west), ductility of a structure is the most common mitigating measure. For the moderate events, i.e., those with ground motions that have a 10% probability of exceedence in 50 years, a structure should be capable of responding elastically. The latter loads are low in the east and moderate in the west. Therefore, ratios for ductile to elastic load capability of structures in the east should be higher than in the west, while in practice the opposite prevails.

We are not aware that there has been enough practice in the East to reach the conclusion of the previous statement: most designs have been made to UBC (West Coast) standards. A building's earthquake resistance is measured both by its ability to resist lateral forces within the elastic range and its ability to withstand deformations in the inelastic range (i.e., its ductility). As pointed out at the NCEER Symposium, the California engineers have truncated the level of ground motion, A_v, at $0.4g$ and increased the ductility requirements. In the Eastern United States it is equally rational to truncate the ductility requirements and increase the lateral force requirements in order to decrease structural and nonstructural damage in the more moderate earthquakes. This is what the Memphis Seismic Committee chose to recommend ($I = 1.25$) for the higher-occupancy structures; assembly and high-rise occupancies.

NATIONAL MODEL CODES

In their most recent revisions, all three national Model Building Codes [Uniform Building Code (UBC), BOCA Basic Building Code (BOCA) and Standard Building Code (SBC)] have adopted essentially equivalent provisions to the ATC/ NEHRP Recommended Provisions. All three have discarded their previous *maxi-*

mum intensity risk maps in favor of the ATC/NEHRP *probability* risk map. Though the three have slightly different formulations and detailing requirements, each will give essentially equivalent end results to the ATC/NEHRP Provisions if equivalent risk map design ground motions, "A_g" or "Z", and "I", and "S" factors are used in their lateral force equations. This is because all are based upon monitored building response and recorded damage experienced in structures in California which have been subjected to damaging earthquakes. Therefore, current (1987–88) model codes basically follow the latest technology as established by the ATC/NEHRP document.

MEMPHIS BCAB SEISMIC COMMITTEE RECOMMENDATION

The Seismic Study Committee of the Memphis Building Code Advisory Board (BCAB) has reviewed and recommends, with minor exceptions, adoption of the new 1988 Standard Building Code seismic provisions, having found them to be essentially in keeping with the committee's conclusions reached after two years of research. Both local and nationally recognized experts met with the committee, which had also reviewed current authoritative studies in the several fields involved.

It was found that the SBC 1988 seismic risk map (Figure 1206.1) follows the contour concept of the ATC/NEHRP Provisions maps. Furthermore, this contouring concept better addresses the Memphis situation where the 0.2-g-contour dividing Zones 2 and 3 lies through the center of the City and County. The 1988 SBC provides a unique method (Figure 1206.4.1) for converting the A contours on the map to an equivalent "Z" factor for use in the equivalent lateral force equation.

The committee also recommended that Memphis and Shelby county be considered in Zone 2 for detailing ductility requirements but using a "Z" factor of 0.5 (A = 0.2g). The committee agreed with all other provisions of the SBC 1988 except the Importance Factor ("I" = 1.25), which the SBC limits to assembly occupancies with 300 persons. The committee recommended that building structures higher than 75 feet having a calculated period T = 0.5 seconds or greater have an "I" = 1.25 also. This latter requirement is to account for the distant large earthquake with longer predominant ground motion periods and a predicted local predominant ground-motion period ranging between 0.5 to 1.0 seconds.[18]

On February 2, 1988, the Seismic Committee recommendations (see the APPENDIX) were formally presented to the Memphis and Shelby County Building Code Advisory Board. We anticipate that within the year seismic requirements for design and construction will be incorporated in the building code of Memphis—"The City of Good Abode."

REFERENCES

1. NUTTLI, O. W. 1973. The Mississippi Valley earthquakes of 1811 and 1812: Intensities, ground motion and magnitude. Seismol. Soc. Am. Bull. **63:** 227–248.
2. NUTTLI, O. W. 1974. Magnitude-recurrence relation for central Mississippi Valley earthquakes: Bull. Seismol. Soc. Am. **64:** 1189–1207.
3. NUTTLI, O. W. 1979. The relation of sustained maximum ground acceleration and velocity to earthquake intensity and magnitude: State-of-the art for assessing earthquake hazards in the United States. U.S. Army Engineer Waterways Experiment Station Miscellaneous Paper, S-73-1: 16, November.

4. NUTTLI, O. W. 1982. Damaging earthquakes of the central Mississippi Valley. *In* Investigations of the New Madrid, Missouri, Earthquake Region. F. A. McKeown & L. C. Pakiser, Eds. U.S. Geol. Surv. Prof. Paper 1236: 15–20.
5. NUTTLI, O. W. 1983. Empirical magnitude and spectral scaling relations for mid-plate and plate-margin earthquakes. Techtonophysics **93**: 207–223.
6. FULLER, M. L. 1912. The New Madrid earthquake. U.S. Geol. Surv. Bull. **494.**
7. JOHNSTON, A. C. 1981. On the use of the frequency-magnitude relation in earthquake risk assessment. *In* Proceedings of The Conference on Earthquakes and Earthquake Engineering: Eastern United States, Vol. **I**: 161–182.
8. HAYS, W. H. 1983. Proceedings of Conference XVIII, A Workshop on Continuing Actions to Reduce Losses from Earthquakes in the Mississippi Valley Area, 24–26 May, 1982, St. Louis, Missouri. U.S. Geol. Surv. and the Federal Emergency Management Agency, Open-File Rep. 83-157, Reston, Virginia.
9. HOPPER, M. G. *et al.* 1983. Estimation of Earthquake Effects Associated With a Great Earthquake in the New Madrid Seismic Zone, FEMA, Central United States Earthquake Preparedness Project CUSEPP Number 82-3. U.S. Geological Surv. Open-File Rep. 83-179.
10. HOPPER, M. G., S. T. ALGERMISSEN & E. DOBROVOLNY. 1983. Estimation of earthquake effects associated with a great earthquake in the New Madrid seismic zone. U.S. Geol. Surv. Open-File Rep. 83-179.
11. 1984. Source characteristics and strong ground motion of the New Madrid earthquakes. *In* Proceedings of the Symposium on the New Madrid seismic zone. P. L. Gori and W. W. Hays, Eds. U.S. Geol. Surv. Open-File Rep. 84-770: 330–352.
12. M & H ENGINEERING AND MEMPHIS STATE UNIVERSITY. 1974. Regional earthquake risk study: Mississippi-Arkansas-Tennessee Council of Governments/Memphis Delta Development District (MATCOG/MDDD).
13. MANN, O. C. & W. HOWE. 1973. Regional earthquake risk study: M & H Engineering, Memphis, Tennessee, Progress Report No. 2.
14. APPLIED TECHNOLOGY COUNCIL. 1978. Tentative Provisions for the Development of Seismic Regulations for Buildings. Report ATC 3-06. San Francisco, CA.
15. ALGERMISSEN, S. T. & D. M. PERKINS. 1976. A Probabilistic Estimate of Maximum Acceleration in Rock in the Contiguous United States. U.S. Geol. Surv. Open-File Rep. 76-416.
16. Amendments to ATC-3-06 Tentative Provisions for the Development of Seismic Regulations for Buildings for Use in Trial Designs (NBSIR 822626/BSSC 82-2).
17. WEBER, S. F. 1985. Cost Impact of the NEHRP Recommends Provisions on the Design and Construction of Buildings. Earthquake Hazards Reduction Series 14, FEMA-84/ June 1985.
18. SHARMA, S. & W. D. KOVACS. 1980. The microzonation of the Memphis, Tennessee area: U.S. Geol. Surv. Open-File Rep. 80-914.

ADDITIONAL REFERENCES NOT CITED IN TEXT

ALGERMISSEN, S. T., D. M. PERKINS, P. C. THENHAUS, S. L. HANSEN & B. B. BENDER. 1982. Probabilistic estimates of acceleration and velocity in the contiguous United States. U.S. Geol. Surv. Open-File Rep. 82-1033, scale 1 : 750,000.
WOODWARD-LUNDGREN AND ASSOCIATES. 1975. Geologic, Seismologic, and Earthquake Engineering Evaluation of the Veterans Administration Hospital, Memphis, Tennessee, report to Veterans Administration, Office of Construction, VA Project No. V1001 P-266.
ZOBACK, M. D., R. M. HAMILTON, A. J. CRONE, D. P. RUSS, F. A. McKEOWN & S. R. BROCKMAN. 1980. Recurrent intraplate tectonism in the New Madrid seismic zone. Science **209**: 971–976.
NATIONAL BUREAU OF STANDARDS AND BUILDING SEISMIC SAFETY COUNCIL. 1982. Amendments to ATC 3-06, Tentative Provisions for the Development of Seismic Regulations for Buildings, for Use in Trial Designs. U.S. Department of Commerce. Washington, DC.

BUILDING SEISMIC SAFETY COUNCIL. 1984. BSSC Program on Improved Seismic Safety Provisions. Building Seismic Safety Council. Washington, DC.

NEHRP Recommended Provisions for the Development of Seismic Regulations for New Buildings, 1985 Edition, FEMA.

APPENDIX

BCAB SPECIAL SEISMIC STUDY COMMITTEE
FINAL REPORT
1-12-88

At a specially called meeting of the BCAB Seismic Study Committee on January 12, 1988, at the Tennessee Earthquake Information Center, MSU, the Committee adopted the following recommendations. All actions were voted affirmative unanimously by those present (list of attendees attached)[a] with the exception of the first item below in which the Developers' representative felt obligated to support the stated position of his group, i.e., Z = 0.375. All recommendations of the Developers' group as expressed in letter dated January 5, 1988, to the Committee Chairman by Jerry Hanover, were discussed and given serious consideration as were several correspondences from the Engineering group before finalizing the following recommendations:

1. Adopt the SBC-88 Seismic provisions (Section 1206 Earthquake loads) in its entirety with the following modifications:
2. In Section 1206.4.1 Z Factor. Add:
   The factor "Z" for Memphis and Shelby County shall be 0.5 ($A_v = 0.20$g) and Seismic Zone 2.
3. In Section 1206.1 General
   Exceptions:
   (Change item 1 as follows)
   1. One- and two-story wood frame, type VI Construction in Occupancy use groups. B-Business, M-Mercantile and R-Residential are exempt from the regulations of this section.
   Table 1206.1: OCCUPANCY IMPORTANCE FACTOR I
   (Change 2nd Group, I = 1.25 delete and change to the following:)

---

[a] **Seismic Provisions Committee:** *Fire Department:* Chief Crossnine, Fire Prevention Bureau, Memphis Fire Department; *Engineers:* Mr. Mike Banker, Allen & Hoshall; *Heavy Industry:* Mr. William E. Capelle, Project Engineer Designer, E.I. DuPont; *Educational:* Mr. Billy Evans, Assistant Superintendent, Department Plant Manager; *Hospitals:* Mr. Maurice Elliott, Executive Vice President, Methodist Hospital; *Engineers-Chairman:* Mr. Warner Howe, P.E., Gardner & Howe, P.C.; *General Contractors:* Mr. Frank Inman, Inman Construction Corp.; *Earthquake Center:* Dr. Arch Johnston, Director, Tennessee Earthquake Information Center, Memphis State University; *Mortgage Bankers Association:* Mr. J. Davant Latham, Senior Vice President, Pope Mortgage; *Architects:* Mr. David McGehee, President, McGehee/Nicholson/Burke, Architects; *Memphis Insurors:* Mr. Bobby Meeks, Board Chairman, Consultants' Service Insurance Agency; *Memphis Light, Gas & Water:* Mr. Henry Winter, Memphis Light, Gas & Water; *Engineers:* Mr. John W. Smith, Chairman, Civil Engineering Department, Memphis State University; and *Building Owners, Manager, Developers:* Mr. Bill Wehrum, President, Mid-America Construction Company.

— All buildings and structures over 75 feet in height (Section 506.1.1) having a Fundamental Period, T ≥ 0.5 seconds.
— Educational Facilities with classrooms and/or lecture halls. (Add to 3rd Group, I − 1.5:)
— Hazardous facilities housing or containing quantities of toxic or explosive substances to be dangerous to the safety of the general public if released.

4. Add Section 1206.15:
   Special Structural Inspection: All structures which are required by the Code to be designed with a "ZI" factor greater than 0.5 shall be required to have special structural inspection during the construction of the structural portions of the work. A structural inspection plan shall be prepared before the issuance of a building permit. This inspection shall be performed under the direction of a qualified inspector who should be the engineer of record, a licensed architect or engineer, a testing and/or inspection agency under the direction of a licensed engineer, or a qualified inspector approved by the Building Official. The special structural inspector shall be responsible to the Building Official for all inspections and sign all required reports.

5. Transition Period Before Mandatory Compliance:
   The Seismic Study agrees with the Developers' group recommendations that a three year transitional (grace) period be allowed after enactment of these seismic provisions to allow completion of the development and design of projects in progress under current Code provisions.

6. Non-Structural Elements: (Section 1206.10)
   The Committee agrees with the developers that the SBC 88 Seismic provisions are much too comprehensive on the subject and at least some of these requirements should be reduced or deleted. However, this is a very complex issue and interrelates to other sections of the Code in such a manner that it will be difficult to appropriately sort them out without detailed intensive further study. The Committee recommends that the SBC 88 proposed section 1207.10 be left temporarily as is until a more detailed study can be made, possibly during the BCAB review process on the Code enforcement or during the grace period.

7. Essential Facilities—Private Hospitals:
   Previously the local Hospital Association has made recommendations that their type of facilities should not be required by the Building Code to have a design criteria which would not only provide life safety but be operable for post-earthquake recovery (I = 1.5). They maintained this additional cost should not be added to the already high cost of hospitalization which the patient must pay. They take no issue with the desirability of having functioning hospital facilities for disaster recovery but this should be a result of public policy which is paid for out of the public's treasury. Seismic committee voted to recommend that Essential Facilities with I = 1.5 in SBC 88 not be changed.

8. Hazardous Materials—Maintenance and Storage:
   Numerous discussions previously had been held in the Committee with our Heavy Industry representative, William Capelle, E.I. DuPont, and he was in the process of drafting appropriate recommended regulations which he neither was able to finalize nor coordinate into the SBC 88 document. These considerations should be given further study as the Code is deficient in the area of handling toxic and explosive materials which if loosed upon the community would be seriously life threatening.

These recommendations are in keeping with the current state-of-the-art recognized nationally by the National Building Seismic Safety Council sponsored by the National Institute of Building Sciences, and the three national model code groups (UBC, BOCA and SSBCI), and by the nationally recognized authorities in the fields of seismology, geology, engineering, architecture, building owners and managers, the insurance industry and others in the entire building community nationally. Therefore, we are confident that this position will stand up in a court of law as being representative of good and acceptable professional practice in Memphis and Shelby County if ever litigation should result in the future. We also believe that it will provide reasonable and prudent life safety in buildings designed under its provisions when an earthquake occurs on the New Madrid Fault System. It should be pointed out, however, that considerable damage, both nonstructural and structural, can be expected even in a moderate earthquake but total collapse and life loss will have a very low probability even in a major quake. The primary objective is safety to life; damage control is considered only as it relates to life safety.

WARNER HOWE
Chairman

Estimation of Earthquake Losses for a Large Eastern Urban Center: Scenario Events for New York City[a]

CHARLES SCAWTHORN AND STEPHEN K. HARRIS

Design and Analysis Division
EQE Engineering, Inc.
San Francisco, California 94105

INTRODUCTION

While it has long been recognized that there exists the possibility of a moderate or even large earthquake in many areas of the eastern United States, the potential damage that might ensue from such an event has generally not been considered in many areas. An area of especial interest is the New York City area (NYC) because of its large size and global importance, and its general lack of seismic design.

The New York metropolitan area has generally been considered a low seismicity area. The Uniform Building Code (UBC), for example, places New York in Zone 1, resulting in seismic loading only about 19% of that required for the highly seismic portions of California (Zone 4). Earthquakes are not unknown in this area, however, and MMI VII has been observed in historical times (e.g., the 18 December 1737 event, which reportedly caused chimneys to fall in NYC[1]). In growing recognition of the potential for a damaging earthquake in the region around New York, the recent National Earthquake Hazard Reduction Program's *Recommended Provisions for the Development of Seismic Regulations for New Buildings*[2] has placed this area in Map Area 3, increasing the lateral load requirements relative to California from 19% to 25%.

This paper presents results of ongoing work to estimate the potential losses due to a moderate but credible earthquake in or near the center of the NYC metropolitan region, for the purpose of placing in perspective the potential for destruction involved. This information is of use in planning and development of building codes.

There are several fundamental steps involved in seismic damage estimation, including (i) the estimation of the structures or building inventory at risk, (ii) the determination of the seismic hazard, and (iii) the estimation of the vulnerability of these structures to various levels of seismic shaking. We shall discuss our approach to each of these, since they directly influence our results.

[a] This work was performed under Grant NCEER 87-4004 from the National Center for Earthquake Engineering Research at the State University of New York at Buffalo.

BUILDING INVENTORY

Building inventory information is presented in TABLES 1 and 2. Building inventory for NYC has been estimated by Jones *et al.*,[3] and TABLE 2 presents findings of that investigation, wherein numbers of buildings for the example of Manhattan are crosstabulated by height (number of stories) and total floor area. Information is not presently available regarding detailed structural characteristics of this inventory, so this has been estimated by the authors on the basis of experience. TABLE 1 shows a typical estimated breakdown for Manhattan. Using this breakdown and the data in TABLE 2, we can estimate numbers of buildings by height and size for each gross structural category, presented in TABLES 2A (wood), 2B (reinforced concrete), 2C (steel), 2D (reinforced masonry) and 2E (unreinforced masonry) for Manhattan. Similar techniques were employed for the other boroughs.

TABLE 3 presents summary values for the five boroughs, by height range and basic material of construction. Values were arrived at using a uniform unit value of $100 per square foot of floor area. The total building inventory of the five boroughs is seen to be approximately $404 billion (1988 dollars).

Precise distribution of this inventory is possible using building department records, but it is a mammoth task, and is not critical to the purpose of this paper. Approximate distribution of inventory within each borough is appropriate, given the available data. Maps of land use in NYC prepared by the Sanborn Map Company were reviewed in the course of allocation of building inventory.

SEISMIC HAZARD

Seismic hazard in the northeast United States is a subject involving considerable uncertainty.[4] Major events in the New York City area include the 18 December 1737 and the 10 August 1884 earthquakes. The latter is probably the largest and best documented event. It was a strong shock, centered off Rockaway Beach about 17 miles southeast of New York's City Hall, and felt over 70,000 square miles, from Vermont to Maryland. In New York City the effects were strong but varied, frightening many but causing very little or no damage. In Manhattan, newspaper reports[5] indicated general alarm in many portions of lower Manhattan. Crockery and bottles rattled but generally did not fall. Human behavior, an important indicator of seismic intensity, was generally low-key, although in some areas people were extremely frightened. In Park Row (lower Manhattan) persons dining (the shock was at 2:07 PM) looked up, but did not leave the building. At Brooklyn's City Hall (Brooklyn was not yet a borough) "there was little notice taken of the shock by people moving in the street, and comparatively few noticed it." While the shock made front-page headlines on the first day, by the next the story was on the second page. On the basis of these descriptions, we estimate the general seismic intensity pattern in Manhattan and northwest Brooklyn as Modified Mercalli Intensity[6] (MMI) IV and approaching MMI V toward the southeast.

Current best estimates of the magnitude (based on felt area, termed M_{FA}) of the 1737 and the 1884 events are $M_{FA} = 4.5$ for the 1737 event (epicenter 41N, 73.75W) and $M_{FA} = 4.9$ for the 1884 event (epicenter 40.51N, 73.83W).[7]

In order to determine seismic intensities, the epicenter of the 1884 event has been taken as a possible location of an M6.0 event. Magnitude 6.0 has been taken as a large but possible event in the vicinity of New York City. This location is 17

TABLE 1. Building Structure Dominant Material (Fraction, by Height)[a]

Height (stories)	Total					Residential					Nonresidential				
	WOOD	RC	STL	RM	URM	WOOD	RC	STL	RM	URM	WOOD	RC	STL	RM	URM
0	0.00	0.50			0.50	0.1				0.9		0.5			0.5
1	0.10	0.39			0.51	0.1				0.9		0.4			0.5
2		0.25	0.07	0.15	0.53		0.1			0.9		0.3			0.4
3		0.15	0.02	0.12	0.70		0.1		0.1	0.8	0.1	0.3	0.1	0.2	0.4
4		0.15	0.05	0.12	0.68		0.1		0.1	0.8		0.1	0.1	0.2	0.3
5		0.10	0.06	0.12	0.72		0.1		0.1	0.8		0.3	0.2	0.2	0.3
6		0.22	0.24	0.20	0.34		0.2	0.2	0.2	0.4		0.4	0.4	0.2	0.1
7		0.31	0.31	0.34	0.05		0.2	0.2	0.5	0.1		0.4	0.4	0.2	
8		0.35	0.40	0.25			0.2	0.4	0.4			0.5	0.4	0.2	
9		0.35	0.50	0.15			0.3	0.5	0.2			0.4	0.5	0.1	
10		0.44	0.50	0.06			0.3	0.5	0.2			0.5	0.5		
11–15		0.40	0.55	0.05			0.4	0.5	0.1			0.4	0.6		
16–20		0.20	0.80				0.2	0.8				0.2	0.8		
21–40		0.09	0.91					1				0.1	0.9		
41–60			1.00					1					1		
60+			1.00										1		

NOTE: RC = reinforced concrete; STL = steel; RM = reinforced masonry; and URM = unreinforced masonry.

[a] Estimated, for Manhattan.

TABLE 2. Manhattan Building Inventory (as of 1972): Total Number of Residential and Nonresidential Buildings—All Materials[a]

Height (stories)	Floor Area (1000 sq ft)												Sum
	0–1	2	4	6	10	15	25	50	100	500	1000	1000+	
0	329	261	986	147	361	133	38	13	16	17	6	4	2311
1	525	507	616	332	270	106	62	32	17	15		2	2484
2	178	534	560	360	341	172	208	111	40	24	2	1	2531
3	33	951	5216	549	386	193	176	122	57	38	1	2	7724
4	5	86	3986	3166	1358	303	252	174	79	57	2	2	9470
5		5	349	2675	6227	4092	1342	562	139	66	2	1	15,460
6		1	22	57	440	1630	1850	1308	517	117	6	1	5949
7			2	4	47	191	241	313	122	52	3	3	957
8			1	1	4	37	55	166	94	51	3	2	415
9					1	4	46	200	141	48	4	2	446
10						3	27	97	124	79	4		336
11–15						3	42	354	844	860	56	17	2176
16–20						1	1	40	222	725	53	12	1054
21–40									1	4	10	2	17
41–60									3	60	30	42	135
60+										5		6	11
Sum	1070	2345	11,738	7291	9435	6868	4340	3492	2416	2218	182	99	51,494

[a] From Jones et al.[3]

TABLE 2A. Manhattan Building Inventory (as of 1972): Number of Residential and Nonresidential Buildings—Wood[a]

Height (stories)	Floor Area (1000 sq. ft.)												Sum	% WOOD
	0–1	2	4	6	10	15	25	50	100	500	1000	1000+		
0	53	51	62	33	27	11	6	3	2	2			250	0.00
1														0.10
2														
3														
4														
5														
6														
7														
8														
9														
10														
11–15														
16–20														
21–40														
41–60														
60+														
Sum	53	51	62	33	27	11	6	3	2	2			250	

[a] Based on data from Jones et al.[3]

TABLE 2B. Manhattan Building Inventory (as of 1972): Number of Residential and Nonresidential Buildings—Reinforced Concrete[a]

Height (stories)	Floor Area (1000 sq. ft.)												Sum	% RC
	0–1	2	4	6	10	15	25	50	100	500	1000	1000+		
0	164	130	492	73	180	66	19	6	8	8	3	2	1151	0.50
1	203	196	238	128	104	41	24	12	7	6		1	960	0.39
2	44	132	139	89	85	43	52	28	10	6			628	0.25
3	5	141	773	81	57	29	26	18	8	6			1144	0.15
4	1	13	583	463	199	44	37	25	12	8			1385	0.15
5		1	35	268	623	409	134	56	14	7			1547	0.10
6			5	13	97	361	409	289	114	26	1		1315	0.22
7			1	1	14	59	74	96	37	16	1		299	0.31
8					1	13	19	58	33	18	1	1	144	0.35
9							16	69	49	17	1	1	154	0.35
10						1	12	42	54	35	2	1	147	0.44
11–15						1	17	142	338	344	22	7	871	0.40
16–20								8	44	145	11	2	210	0.20
21–40											1		1	0.09
41–60														
60+														
Sum	417	613	2266	1116	1360	1068	839	849	728	642	43	15	9956	

[a] Based on data from Jones et al.[3]

TABLE 2C. Manhattan Building Inventory (as of 1972): Number of Residential and Nonresidential Buildings—Steel[a]

Height (stories)	Floor Area (1000 sq. ft.)												Sum	% Steel
	0–1	2	4	6	10	15	25	50	100	500	1000	1000+		
0														
1														
2	13	40	41	27	25	13	15	8	3	2			187	0.07
3	1	23	126	13	9	5	4	3	1	1			186	0.02
4		4	184	146	63	14	12	8	4	3			438	0.05
5			22	169	392	258	85	35	9	4			974	0.06
6			5	14	107	395	449	317	125	28	1		1441	0.24
7			1	1	14	59	74	96	37	16	1		299	0.31
8					2	15	22	66	38	20	1		165	0.40
9					1	2	23	100	71	24	2	1	224	0.50
10						2	14	49	62	40	2	1	170	0.50
11–15						1	23	194	462	470	31	9	1191	0.55
16–20							1	32	178	580	42	10	844	0.80
21–40									1	4	9	2	16	0.91
41–60									3	60	30	42	135	1.00
60+										5		6	11	1.00
Sum	14	67	379	370	613	766	722	908	994	1257	119	72	6281	

a Based on data from Jones et al.[3]

TABLE 2D. Manhattan Building Inventory (as of 1972): Number of Residential and Nonresidential Buildings—Reinforced Masonry[a]

Height (stories)	Floor Area (1000 sq. ft.)												Sum	% Reinforced Masonry
	0–1	2	4	6	10	15	25	50	100	500	1000	1000+		
0														
1														
2	26	79	83	53	50	25	31	16	6	4			373	0.15
3	4	118	647	68	48	24	22	15	7	5			958	0.12
4	1	11	491	390	167	37	31	21	10	7			1166	0.12
5		1	40	310	721	474	155	65	16	8			1790	0.12
6			4	11	88	326	370	262	103	23	1		1188	0.20
7			1	1	16	65	82	106	41	18	1		331	0.34
8					1	9	14	41	23	13	1	1	103	0.25
9						1	7	31	22	7	1		69	0.15
10							2	6	8	5			21	0.06
11–15							2	19	45	46	3	1	116	0.05
16–20														
21–40														
41–60														
60+														
Sum	31	209	1266	833	1091	961	716	582	281	136	7	2	6115	

[a] Based on data from Jones et al.[3]

TABLE 2E. Manhattan Building Inventory (as of 1972): Number of Residential and Nonresidential Buildings—Unreinforced Masonry[a]

Height (stories)	Floor Area (1000 sq. ft.)												Sum	% Unreinforced Masonry
	0–1	2	4	6	10	15	25	50	100	500	1000	1000+		
0	165	131	494	74	181	67	19	7	8	9	3	2	1160	0.50
1	269	260	316	170	139	54	32	16	9	8	1	1	1274	0.51
2	94	283	297	191	181	91	110	59	21	13	1	1	1342	0.53
3	23	669	3669	386	272	136	124	86	40	27	1	1	5434	0.70
4	3	59	2728	2167	930	207	172	119	54	39	1	1	6480	0.68
5		4	252	1929	4491	2951	968	405	100	48	1	1	11,150	0.72
6			7	19	148	548	622	440	174	39	2		1999	0.34
7					2	9	11	15	6	2			45	0.05
8														
9														
10														
11–15														
16–20														
21–40														
41–60														
60+														
Sum	554	1406	7763	4936	6344	4063	2058	1147	412	185	9	7	28,884	

[a] Based on data from Jones *et al.*[3]

TABLE 3. Building Value ($ Billions) Summary: Five Boroughs of the City of New York

Height	Structural Material					
	Steel	RC	URM	RM	Wood	Total
<5 stories	.6	7.2	88.1	36.0	67.2	199.0
5–10 stories	22.5	22.8	49.3	21.9	0	116.5
10–20 stories	48.9	24.8	0	2.8	0	76.5
>20 stories	11.8	.1	0	0	0	11.9
All heights	83.8	54.6	137.4	60.6	67.2	403.9

NOTE: RC = reinforced concrete; URM = unreinforced masonry; RM = reinforced masonry.

miles southeast of New York City Hall. Further, for purposes of examining the consequences, an M6.0 event has been placed at hypothetical epicenters of 5 and 11 miles southeast of City Hall (that is, closer to Manhattan, and under parts of Brooklyn).

Attenuation of ground shaking away from the epicenter has been modeled using Campbell's Central United States (CUS) relation.[8] Additionally, a factor was developed for this study to account for effects of shaking duration and its effect on damage. This factor was simply $F_d = \log(Td/To)$, where Td is the estimated duration of an earthquake of magnitude M, and To is Td determined for an earthquake of M6.5. Typical F_d (and corresponding magnitudes) are:

Magnitude	F_d
5	−.75
6	−.23
7	+.21
8	+.60

Soil conditions are an important influence in ground-shaking intensity. Soils were mapped for the NYC area by Jacob et al.[9] in terms of soil profiles S1, S2 and S3, as defined in the UBC. These are shown in FIGURE 1. MMI was increased over those values determined from Campbell's relation by a soil factor, F_s.

Soil Profile	F_s
S1	0
S2	+0.5
S3	+1.0

Final MMI then was determined on the basis of:

$$\text{MMI}_F = \text{MMI}_C + F_s + F_d \qquad (1)$$

where MMI_F = final MMI, MMI_C = MMI as determined from Campbell, F_s = soil factor, and F_d = earthquake duration factor.

FIGURE 2 presents the MMI distribution estimated using this procedure for the 1884 event, which is seen to agree reasonably well in the NYC area with MMI estimated as discussed above from newspaper reports. FIGURE 3 presents the MMI distribution estimated using this procedure for a M6.0 event at the epicenter of the 1884 event.

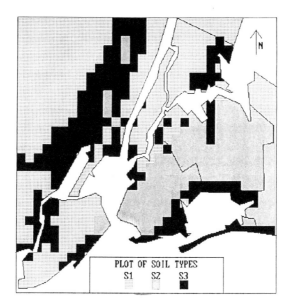

FIGURE 1. Map of earthquake effects in New York City: Plot of soil types.

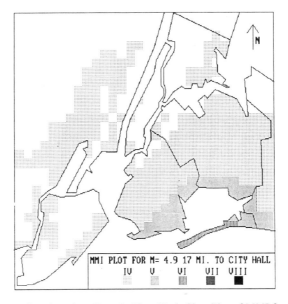

FIGURE 2. Map of earthquake effects in New York City: Plot of MMI for M = 4.9, 1884 epicenter.

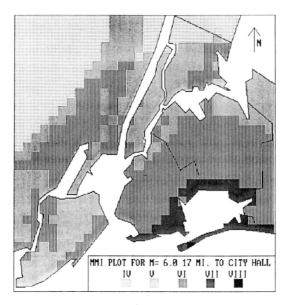

FIGURE 3. Map of earthquake effects in New York City: Plot of MMI for M = 6.0, 1884 epicenter.

SEISMIC VULNERABILITY

Very little experience exists concerning the structural vulnerability of U.S. non-California construction and the subject has been little investigated. The most extensive broad compilation to date of California construction seismic vulnerability is the recent investigation by the Applied Technology Council (ATC-13),[10] which provides mean damage estimates (and distributions thereon) for various types of construction and levels of MMI.

A recent investigation by Scawthorn into extending the ATC-13 data into regions of the U.S. beyond California involved eliciting the opinions of expert engineers familiar with local practice in various regions of the United States. Results have indicated that NYC construction seismic vulnerability may be very approximately estimated by increasing the mean damage levels of ATC-13 by about 10% on average. Actual multiplicative increases that were employed, by structural material category, are:

- Wood: 1.0 (that is, no difference between NYC and California wood building performance);
- Steel-frame buildings: 1.18 (that is, 18% increase in damage over ATC-13);
- Reinforced concrete: 1.25;
- Reinforced masonry: 1.10; and
- Unreinforced masonry: 1.0.

DAMAGE ESTIMATION

Shaking

If we employ the building inventory of TABLES 2 and the mean damage levels arrived at by adapting the ATC-13 data to NYC construction, as discussed above, then mean NYC damage levels can be approximated, and are indicated in TABLE 4.

The first section of TABLE 4, for an M6 event centered at the 1884 epicenter (i.e., 17 miles from City Hall) indicates total shaking loss of about $11.2 billion, which represents approximately 2.8% of the building value in the five boroughs. All material groups sustain damage, the highest being unreinforced masonry, as would be expected.

The mid-section of TABLE 4, for an M6 event at 11 miles from City Hall, indicates about $18.0 billion in damage, or about 4.5%, while the bottom part of TABLE 4, for an M6 quake at 5 miles from City Hall, indicates $25.9 billion in damage, or about 6.4% of total building inventory value.

These estimates of damage are for direct structural and architectural damage to building value, and do not include allowances for damage to building contents, business interruption, injuries, fire, or other agents of damage.

FIGURE 4 shows the distribution of damage for the case of the first section of TABLE 4 (i.e., an M6.0 event at the epicenter of the 1884 event, 17 miles southeast of City Hall). As would be expected, most of the damage is concentrated in the portions of the city closest to the epicenter, although there are pockets of damage at other locations, due to soil conditions and concentrations of buildings.

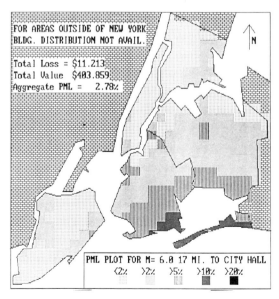

FIGURE 4. Map of earthquake effects in New York City: Plot of PML for M = 6.0, 1884 epicenter.

TABLE 4. Earthquake Damage for New York City for M = 6.0 Events 17, 11, and 5 Miles from City Hall (Billions, 1988 Dollars)[a]

Height	Steel Frame			Reinforced Concrete			Unreinforced Masonry			Reinforced Masonry			Wood Frame			All Materials		
	Loss	Value	%	Loss	Value	%	Loss	Value	%	Loss	Value	%	Loss	Value	%	Loss	Value	%
M = 6.0 Event 17 Miles from City Hall																		
<5 stories	0.001	0.574	0	0.077	7.187	1	4.371	88.050	5	0.837	35.988	2	1.389	67.200	2	6.675	198.999	3
5–9 stories	0.387	22.515	2	0.570	22.764	3	2.443	49.312	5	0.507	21.866	2	0.000	0.000	0	3.907	116.458	3
10–20 stories	0.322	48.909	1	0.277	24.785	1	0.000	0.000	0	0.032	2.774	1	0.000	0.000	0	0.631	76.467	1
>20 stories	0.000	11.816	0	0.000	0.118	0	0.000	0.000	0	0.000	0.000	0	0.000	0.000	0	0.000	11.934	0
All heights	0.711	83.814	1	0.924	54.854	2	6.814	137.362	5	1.375	60.628	2	1.389	67.200	2	11.213	403.859	3
M = 6.0 Event 11 Miles from City Hall																		
<5 stories	0.003	0.574	0	0.131	7.187	2	7.127	88.050	8	1.291	35.988	4	2.061	67.200	3	10.613	198.999	5
5–9 stories	0.598	22.515	3	0.903	22.764	4	3.997	49.312	8	0.776	21.866	4	0.000	0.000	0	6.275	116.458	5
10–20 stories	0.616	48.909	1	0.485	24.785	2	0.000	0.000	0	0.057	2.774	2	0.000	0.000	0	1.158	76.467	2
>20 stories	0.000	11.816	0	0.000	0.118	0	0.000	0.000	0	0.000	0.000	0	0.000	0.000	0	0.000	11.934	0
All heights	1.217	83.814	1	1.520	54.854	3	11.124	137.362	8	2.125	60.628	4	2.061	67.200	3	18.046	403.859	4
M = 6 Event 5 Miles from City Hall																		
<5 stories	0.012	0.574	2	0.257	7.187	4	9.095	88.050	10	1.525	35.988	4	2.401	67.200	4	13.290	198.999	7
5–9 stories	0.897	22.515	4	1.351	22.764	6	6.044	49.312	12	1.149	21.866	5	0.000	0.000	0	9.441	116.458	8
10–20 stories	1.700	48.909	3	1.346	24.785	5	0.000	0.000	0	0.144	2.774	5	0.000	0.000	0	3.191	76.467	4
>20 stories	0.000	11.816	0	0.000	0.118	0	0.000	0.000	0	0.000	0.000	0	0.000	0.000	0	0.000	11.934	0
All heights	2.610	83.814	3	2.954	54.854	5	15.139	137.362	11	2.819	60.628	5	2.401	67.200	4	25.922	403.859	6

[a] Precision implied by number of significant digits may not be warranted.

Fire

Beyond the dollar loss wrought by shaking damage, there are many other types of damage that are of interest, such as number of collapsed buildings (of interest to emergency planners), number of buildings where damage exceeds some threshold and which must be vacated (of interest to persons responsible for housing the displaced residents, or for assisting small businesses), the number of outbreaks of fire (obviously of interest to the fire department, but also of interest to the insurance industry, among others), and so on. Herein, we provide a limited examination of the post-earthquake fire potential, by examining the number of fire ignitions. Full investigation of post-earthquake fire spread and losses is beyond

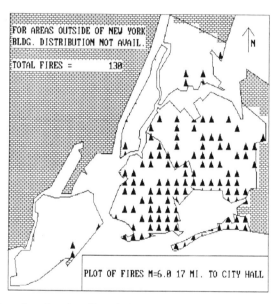

FIGURE 5. Map of earthquake effects in New York City: Plot of fires for M = 6.0, 1884 epicenter.

the scope of the present study, although the methodology exists.[11] Using techniques previously developed by Scawthorn,[12] post-earthquake fire ignitions can be estimated along lines similar to the procedures described above for structural damage. FIGURE 5 indicates a likely distribution of ignitions for the case shown in TABLE 4 (i.e., an M6.0 event at the epicenter of the 1884 event, 17 miles southeast of City Hall). About 130 ignitions would result, depending on time of day.

It is of interest to note that the New York City Fire Department has in excess of 100 fire engines in Brooklyn and Queens, and in excess of 200 engines in total,[13] so that 130 ignitions, while definitely not seen in a normal day for the fire department, might not result in large conflagrations if water supply is unimpaired.

CONCLUDING REMARKS

The above findings provide guidance to those responsible for planning and preparing for the effects of a potential earthquake located in the vicinity of New York City. While losses in the billions of dollars are to be expected, and an M6.0 earthquake would probably be a disaster unparalleled in New York's history, it would appear on the basis of this preliminary study to have effects not unlike those brought about by similar events in California (e.g., the 1 October 1987 Whittier earthquake of M5.9). An event of this size would not result in the truly mammoth losses that have been sustained under larger events, such as those of San Francisco (1906), Tokyo (1923), or Tangshan (1976). It should be noted, however, that it has not been possible to consider many important factors in the present work, such as the potential effects of liquefaction, or the very significant effects of long-period ground motion (tall building resonance). The latter effect resulted in the collapse of approximately 200 mid- and high-rise buildings in Mexico City in 1985, and should be considered in future work regarding New York City, which has perhaps the highest concentration of high-rise buildings in the world.

SUMMARY

Earthquake damage and the number of fire ignitions due to earthquakes occurring in or near New York City are examined. The five-borough building inventory of approximately $404 billion is categorized in terms of story height and structural material. Seismic intensity (MMI) is estimated using Central United States (CUS) attenuation relations, accounting for soil conditions and earthquake magnitude. Three earthquakes of magnitude 6.0 are postulated. Epicentral positions considered are (i) 17 miles southeast of City Hall, at the location of the 1884 M4.9 event offshore Rockaway Beach, (ii) 11 miles southeast of City Hall, and (iii) 5 miles southeast of City Hall, in Brooklyn. Building vulnerability is modeled using modified ATC-13 relations. For an M6.0 event at the 1884 location, approximately 2.8% (or $11 billion) damage is sustained. Further, about 130 post-earthquake ignitions occur. Although several of these ignitions may grow into large fires, first-line fire department capability should in general be able to cope with the fire situation, if water supply is not impaired.

ACKNOWLEDGMENTS

The support of the many individuals at NCEER is gratefully acknowledged. We are especially appreciative of the help of Prof. R. Ketter (Director of the Center) and Profs. M. Shinozuka, and K. Jacob. Prof. Jacob, Dr. L. Seeber, Mr. J. Armbruster, and Ms. M. Tuttle provided valuable soil data and seismological information regarding the 1884 earthquake, and this significantly contributed to this study. Chief L. Harris and Captain T. Coleman of the New York City Fire Department kindly provided information on the capabilities and disaster planning of the NYFD.

REFERENCES

1. COFFMAN, J. L. & C. A. V. HAKE. 1982. Earthquake History of the United States. Publ. 41-1 (rev. ed). U.S. Dept. of Commerce. Boulder, CO.
2. BUILDING SEISMIC SAFETY COUNCIL. 1985. NEHRP-Recommended Provisions for the Development of Seismic Regulations for New Buildings, Part 1, Provisions. Washington, DC.
3. JONES, B., D. M. MANSON, J. E. MULFORD & M. A. CHAIN. 1976. The Estimation of Building Stocks and Their Characteristics in Urban Areas: An Investigation of Empirical Irregularities. Program in Urban and Regional Studies, Cornell University. Ithaca, NY.
4. BERNREUTER, D. L. et al. 1984. Seismic Hazard Characterization of the Eastern United States: Methodology and Interim Results for Ten Sites. NUREG/CR-3756. Lawrence Livermore National Laboratory. Livermore, CA.
5. The New York Times, August 11 and 12, 1884; The Herald Tribune. August 11 and 12, 1884.
6. WOOD, H. O. & F. NEUMANN. 1931. Modified Mercalli Intensity Scale of 1931. Bull. Seismol. Soc. Am. 277–283.
7. SEEBER, L. Personal communication.
8. CAMPBELL, K. W. 1981. A ground motion model for the central United States based on near-source acceleration data. In Earthquakes and Earthquake Engineering-Eastern United States. Vol. 1. Ann Arbor Science Publishers. Ann Arbor, MI.
9. JACOB, K. et al. 1987. Preliminary Map of Soil Types in New York City. Lamont-Doherty Geological Observatory. Palisades, NY.
10. ROJAHN, C., Principal Investigator. ATC-13 (1985) Earthquake Damage Evaluation Data for California. Applied Technology Council. Redwood City, CA.
11. SCAWTHORN, C. & Y. YAMADA. 1981. Lifeline effects on postearthquake fire risk. Second Specialty Conference on Lifeline Earthquake Engineering. 187–198. American Society of Civil Engineers. Oakland, CA.
12. SCAWTHORN, C. 1987. Fire Following Earthquake, Estimates of the Conflagration Risk to Insured Property in Greater Los Angeles and San Francisco. All-Industry Research Advisory Council (AIRAC). Oak Brook, IL.
13. Fire Department of the City of New York. 1986. Annual Report and Statistics.

General Discussion: Part VIII

ROBERT WHITMAN, CHAIRMAN (*Massachusetts Institute of Technology, Cambridge, Mass.*): Various references made throughout this conference to the experience in Massachusetts reminded me of an incident that occurred during one of the debates among Democratic candidates for president during the week before the Iowa caucuses. Our governor had been talking about the economic miracle in Massachusetts, and Representative Gephardt turned to him and said, "Mike, the real miracle would be if you'd stop talking about your miracle." In like manner I hope we haven't overburdened you with comments about the Massachusetts building code "miracle."

MICHEL BRUNEAU (*Morrison Hershfield Ltd., Toronto, Canada*): For the 475-year reference period, what magnitude earthquakes and respective peak ground accelerations are expected for New York and Boston?

KLAUS JACOB (*Lamont-Doherty Geological Observatory, Palisades, N.Y.*): I can't speak for Boston, so let me concentrate on the New York City area. The best confirmation of historical and modern data on intensity was compiled by Dr. Sykes at Lamont-Doherty and seems to indicate the following: If you take a radius of about 50 miles around Central Park, then the distribution of earthquakes in the last 200 years, projected into the future, gives one magnitude-5 event every 300 years within that zone. You can move that event around within the area and, depending on where it is, it can have quite a different impact. Whether or not we should allow that any larger earthquakes occur with the same extrapolation that we can make within the range of observations to the large earthquakes, as Charles Scawthorn did, is an open question. If a b value, as the seismologists call it, of 1 would apply, that would directly translate into the same area of magnitude 6 every 3000 years. But that's getting into a back-of-the-envelope type of speculation, and it's outside the range of observations. . . .

ROBERT KETTER (*NCEER, SUNY at Buffalo, N.Y.*): It's nice to talk about one every 3000 years, but, speaking as an engineer, I find it much more enlightening to hear that there is a 10 percent probability that an earthquake will occur in so many years. You reduce your 3000 to 300 very quickly.

JACOB: Thanks for reminding me of that. This is the recurrence period. Now the codes always talk about something like 10 percent probability of exceedance in a certain time period. A 10 percent exceedance in 50 years is thought to represent a 475-year recurrence period, and I gave you a recurrence period of 300 years. So we slightly exceed what is presently manifested in the ATC-3 or ATC-14 algorithms and probability maps. We are not fully on the conservative side, but we just about hit it right on the nail.

WHITMAN: To return to my understanding of the question: if you take the probabilistic hazard maps that have been discussed at this meeting—and obviously there's a number of versions—the level of ground shaking that would be exceeded with 10 percent probability in a 50-year time period would be somewhere around a tenth of a *g* effective peak ground acceleration. This estimate represents a composite of various possible earthquakes such as earthquakes of larger magnitude occurring at a distance or ones of small magnitudes occurring close by.

CHARLES SCAWTHORN (*EQE, Inc., San Francisco, Calif.*): The fundamental point I was trying to make was that damage estimates, along with the recognition of the fact that earthquakes are a real hazard on the West Coast, have led to mitigation programs. We do not have a consensus that earthquakes are a real hazard in the eastern United States. More damage estimates are needed.

KETTER: It is my understanding, Dr. Scawthorn, that you talked about 7 billion dollars on the basis of a particular assumed scenario. This is almost identical to the insurance costs of a major hurricane moving the entire length of the eastern seaboard.

SCAWTHORN: I should also have said that those numbers are for structural damage only. They do not reflect business interruption or anything else.

WHITMAN: We've obviously been dealing with a vexing problem. We've been talking about events that are not likely to happen within our lifetime, but if they should happen, the potential impact is enormous both in terms of death and injury and in terms of potential economic consequences.

The seismic risk in this region cannot be classified as a pressing problem because of the relative unlikelihood of the event. This characterization implies that a long-term view and a long-term type of solution are appropriate, at least with regard to the life safety aspects of the problem. With respect to new construction, there are straightforward strategies that have been suggested and that make sense to implement. With regard to the existing construction problem, I believe that there are also feasible strategies, although they are not yet as obvious or as well agreed upon.

Steps taken primarily to protect life safety will help to reduce the economic impacts of damage, but certainly those who are concerned about economic consequences to their business, and to their particular building, and to their area must think of the need for special measures above and beyond the relatively straightforward requirements in codes.

Closing Remarks

KLAUS H. JACOB

Lamont-Doherty Geological Observatory
Palisades, New York 10964

CARL J. TURKSTRA

Department of Civil and Environmental Engineering
Polytechnic University
Brooklyn, New York 11201

One of the primary purposes of this conference was to assess the level of seismic hazard in the eastern United States. Is there significant risk? What do we know about it? Can we quantify it in some way?

In general, we believe a response to these questions has been provided. Within generally accepted standards of practice in seismology and engineering, there is a significant seismic hazard in this area. If some participants continue to believe that earthquakes will remain in California, it may be because they misunderstand why we have emphasized the uncertainties in the assessments. Whenever we deal with future phenomena, our knowledge is inevitably limited and we can only speak of probabilities. Also we must recognize that even if we had perfect and complete data, much of the uncertainty would remain: the Earth and earthquakes both are inherently variable, and uncertainty is an intrinsic property. Our task therefore is to make certain that we understand this uncertainty and account for it.

The numbers on hazard maps continue to be debatable. The risk is real. Its manifestations are variable. This variability cannot and must not be used, however, as an excuse for not implementing sound earthquake engineering measures at societally acceptable cost.

Our second primary purpose was to address the question of appropriate engineering responses to these hazards. We have reviewed the basis for some current design requirements and practices. We have examined possible effects of new earthquake design requirements on structural engineering practice for new buildings and we have discussed policy questions related to structural rehabilitation. We have heard from those who faced similar questions in other cities at other times and we have heard from many of the groups that would be affected by any changes in public policy.

Although there seems to be a consensus that design requirements should change, the path to implementation will be slow and complex. On the engineering side, however, there are many small things that can be done to improve the safety of structures. Minor changes in detailing practices with a few more reinforcing bars strategically placed and tied together or ductile connections for curtain walls need not involve significant economic penalties.

We hope that the leaders in the engineering design profession will begin to consider the implications of potential earthquakes when they design new structures, even though official changes in legal requirements may take some time to develop. At the very least, an increased sense of caution should develop within the engineering community in the East, just as it has over the years with related concepts such as progressive collapse and redundancy.

Major questions remain. The question of seismic rehabilitation seems overwhelmingly complex, given the serious housing problems already experienced in many eastern urban centers. At this conference, we have not addressed the questions of disaster preparedness, or disaster recovery. We have not sufficiently discussed societal and economic impacts. These are topics for a future conference.

We believe this conference has been a success and that these proceedings record a comprehensive and stimulating review of the state of the art of seismology in the eastern United States and related current concerns of the engineering profession.

On behalf of the Organizing Committee, we would like to thank the participants for their carefully prepared presentations, our session chairmen, and all those who participated in the general discussion and panel sessions.

Index of Contributors

(Numbers in italics refer to comments made in discussions.)